Biopolymers in Drug and Gene Delivery Systems

Biopolymers in Drug and Gene Delivery Systems

Editor

Yury A. Skorik

MDPI • Basel • Beijing • Wuhan • Barcelona • Belgrade • Novi Sad • Cluj • Manchester

Editor
Yury A. Skorik
Laboratory of Natural Polymers
Institute of Macromolecular
Compounds of the Russian
Academy of Sciences
St. Petersburg
Russia

Editorial Office
MDPI
St. Alban-Anlage 66
4052 Basel, Switzerland

This is a reprint of articles from the Special Issue published online in the open access journal *International Journal of Molecular Sciences* (ISSN 1422-0067) (available at: www.mdpi.com/journal/ijms/special_issues/Biopolymers_Drug_Gene_Delivery_Systems).

For citation purposes, cite each article independently as indicated on the article page online and as indicated below:

Lastname, A.A.; Lastname, B.B. Article Title. *Journal Name* **Year**, *Volume Number*, Page Range.

ISBN 978-3-0365-8711-0 (Hbk)
ISBN 978-3-0365-8710-3 (PDF)
doi.org/10.3390/books978-3-0365-8710-3

Contents

About the Editor vii

Yury A. Skorik
Biopolymers in Drug and Gene Delivery Systems
Reprinted from: *Int. J. Mol. Sci.* **2023**, *24*, 12763, doi:10.3390/ijms241612763 1

Irina V. Tyshkunova, Daria N. Poshina and Yury A. Skorik
Cellulose Cryogels as Promising Materials for Biomedical Applications
Reprinted from: *Int. J. Mol. Sci.* **2022**, *23*, 2037, doi:10.3390/ijms23042037 5

Jiayue Liu, Bingren Tian, Yumei Liu and Jian-Bo Wan
Cyclodextrin-Containing Hydrogels: A Review of Preparation Method, Drug Delivery, and Degradation Behavior
Reprinted from: *Int. J. Mol. Sci.* **2021**, *22*, 13516, doi:10.3390/ijms222413516 22

Francesca Briggs, Daryn Browne and Prashanth Asuri
Role of Polymer Concentration and Crosslinking Density on Release Rates of Small Molecule Drugs
Reprinted from: *Int. J. Mol. Sci.* **2022**, *23*, 4118, doi:10.3390/ijms23084118 43

Natallia V. Dubashynskaya, Sergei V. Raik, Yaroslav A. Dubrovskii, Elena V. Demyanova, Elena S. Shcherbakova and Daria N. Poshina et al.
Hyaluronan/Diethylaminoethyl Chitosan Polyelectrolyte Complexes as Carriers for Improved Colistin Delivery
Reprinted from: *Int. J. Mol. Sci.* **2021**, *22*, 8381, doi:10.3390/ijms22168381 52

Dmitrii Iudin, Marina Vasilieva, Elena Knyazeva, Viktor Korzhikov-Vlakh, Elena Demyanova and Antonina Lavrentieva et al.
Hybrid Nanoparticles and Composite Hydrogel Systems for Delivery of Peptide Antibiotics
Reprinted from: *Int. J. Mol. Sci.* **2022**, *23*, 2771, doi:10.3390/ijms23052771 65

Li Ming Lim and Kunn Hadinoto
High-Payload Buccal Delivery System of Amorphous Curcumin–Chitosan Nanoparticle Complex in Hydroxypropyl Methylcellulose and Starch Films
Reprinted from: *Int. J. Mol. Sci.* **2021**, *22*, 9399, doi:10.3390/ijms22179399 85

Abdulaziz I. Alzarea, Nabil K. Alruwaili, Muhammad Masood Ahmad, Muhammad Usman Munir, Adeel Masood Butt and Ziyad A. Alrowaili et al.
Development and Characterization of Gentamicin-Loaded Arabinoxylan-Sodium Alginate Films as Antibacterial Wound Dressing
Reprinted from: *Int. J. Mol. Sci.* **2022**, *23*, 2899, doi:10.3390/ijms23052899 104

Felipe López-Saucedo, Jesús Eduardo López-Barriguete, Guadalupe Gabriel Flores-Rojas, Sharemy Gómez-Dorantes and Emilio Bucio
Polypropylene Graft Poly(methyl methacrylate) Graft Poly(*N*-vinylimidazole) as a Smart Material for
pH-Controlled Drug Delivery
Reprinted from: *Int. J. Mol. Sci.* **2021**, *23*, 304, doi:10.3390/ijms23010304 121

Davide Facchetti, Ute Hempel, Laurine Martocq, Alan M. Smith, Andrey Koptyug and Roman A. Surmenev et al.
Heparin Enriched-WPI Coating on Ti6Al4V Increases Hydrophilicity and Improves Proliferation and Differentiation of Human Bone Marrow Stromal Cells
Reprinted from: *Int. J. Mol. Sci.* **2021**, *23*, 139, doi:10.3390/ijms23010139 . . . . **135**

Natallia V. Dubashynskaya, Anton N. Bokatyi, Alexey S. Golovkin, Igor V. Kudryavtsev, Maria K. Serebryakova and Andrey S. Trulioff et al.
Synthesis and Characterization of Novel Succinyl Chitosan-Dexamethasone Conjugates for Potential Intravitreal Dexamethasone Delivery
Reprinted from: *Int. J. Mol. Sci.* **2021**, *22*, 10960, doi:10.3390/ijms222010960 . . . . **147**

Anna Egorova, Sofia Shtykalova, Marianna Maretina, Alexander Selutin, Natalia Shved and Dmitriy Deviatkin et al.
Polycondensed Peptide Carriers Modified with Cyclic RGD Ligand for Targeted Suicide Gene Delivery to Uterine Fibroid Cells
Reprinted from: *Int. J. Mol. Sci.* **2022**, *23*, 1164, doi:10.3390/ijms23031164 . . . . **161**

About the Editor

Yury A. Skorik

Dr. Yury A. Skorik serves as Head of the Laboratory of Natural Polymers at the Institute of Macromolecular Compounds of the Russian Academy of Sciences, St. Petersburg, Russia (principal affiliation) and also as Head of the Analytical Chemistry Department at the Almazov National Medical Research Centre, St. Petersburg, Russia (joint affiliation). He earned his M.Sc. in Chemistry from the Department of Chemistry, A.M. Gorky Urals State University, Ekaterinburg, Russia (now the Ural Federal University) in 1995 and his Ph.D. in Physical Chemistry from the same Department and University in 1998. He held postdoctoral positions at the Faculty of Sciences, University of Porto, Portugal (2001–2004) and at the Department of Chemistry, Carnegie Mellon University, Pittsburgh, PA, USA (2004–2005). Previously, he was an Assistant Professor at the Department of Chemistry, A.M. Gorky Urals State University (1998–2000), a Lecturer at the Department of Chemistry, University of Pittsburgh, PA, USA (2005–2007), and an Associate Professor at the Department of Pharmaceutical Engineering, St. Petersburg State Chemical Pharmaceutical University, St. Petersburg, Russia (2007–2016). Skorik's current research interests focus on the chemical modification and properties of polysaccharides; biopolymer drug and gene delivery systems—polymer–drug conjugates, micro- and nanoparticles, and polyplexes; and polysaccharide biomaterials for tissue engineering and regenerative medicine—films, nonwovens, sponges, and gels. To date, Dr. Skorik has co-authored more than 120 peer-reviewed publications (h-index 32 and 2600 total citations—Google Scholar).

International Journal of
Molecular Sciences

MDPI

Editorial

Biopolymers in Drug and Gene Delivery Systems

Yury A. Skorik

Institute of Macromolecular Compounds of the Russian Academy of Sciences, Bolshoi VO 31, St. Petersburg 199004, Russia; yury_skorik@mail.ru

Recent years have seen remarkable advances in the field of drug and gene delivery systems, revolutionizing the way we approach therapeutic treatments. The design and development of these systems has become increasingly important because of their potential to improve the efficacy and safety of drug and gene therapies and to overcome several challenges faced by traditional therapeutic approaches. Many drugs have limited stability, poor solubility, or face barriers in reaching their intended targets in the body. The development of delivery systems that can encapsulate these therapeutic agents and efficiently transport them to the desired location can greatly improve their efficacy. In addition, such systems can protect the agents from degradation, increase their bioavailability, and provide controlled release, allowing for prolonged therapeutic effects. In addition, targeting specific cells or tissues with therapeutic agents is a critical aspect of personalized medicine.

Drug and gene delivery systems with targeting capabilities, such as ligand–receptor interactions or surface modifications, can selectively deliver therapeutic agents to specific cells or tissues, maximizing their effects and minimizing off-target toxicity. This targeted delivery approach holds great promise for the treatment of complex diseases such as cancer, where precision and specificity are essential. In addition to targeted delivery, the design and development of drug and gene delivery systems also play a critical role in improving the stability and pharmacokinetics of therapeutic agents. Many drugs are rapidly metabolized or cleared from the body, limiting their efficacy. However, by formulating these agents in appropriate delivery systems, their stability can be improved, and their release can be controlled, ensuring a sustained and prolonged effect.

This Special Issue aims to emphasize the significance of designing and developing drug and gene delivery systems based on biopolymers and their potential applications, as well as their impact on the future of medical treatment. When we refer to biopolymers, we are referring to a range of polymers, not only natural polymers produced by living organisms, but also semi-synthetic polymers (i.e., modified natural polymers) and synthetic polymers that are biocompatible, biodegradable, and capable of creating drug delivery systems. Both natural and synthetic biopolymers have their respective advantages and disadvantages. Natural biopolymers are preferred over synthetic biopolymers due to their biocompatibility, biodegradability, and environmental safety. Synthetic biopolymers, on the other hand, have distinctive advantages in terms of stability and adaptability to various biomedical applications.

The Special Issue contains eleven articles, including nine original research papers and two reviews. The authors come from different geographical locations, including North America (USA and Mexico), Europe (UK, Germany, Sweden, and Russia), and Asia (China, Saudi Arabia, Pakistan, Malaysia, and Singapore). The papers cover various topics related to the use of biopolymers (polysaccharides such as cellulose and its derivatives, starch and its derivatives, chitosan and its derivatives, hyaluronan, sodium alginate, agarose, arabinoxylan, heparin, cyclodextrin; peptide and proteins; as well as synthetic biopolymers) for the delivery of various drugs and therapeutic nucleic acids. The types of biomaterials studied include polymer conjugates, nano- and submicroparticles, gels, films, and implants for drug administration via intravascular, intravitreal, buccal, topical, and implantation routes.

Citation: Skorik, Y.A. Biopolymers in Drug and Gene Delivery Systems. *Int. J. Mol. Sci.* **2023**, *24*, 12763. https://doi.org/10.3390/ijms241612763

Received: 6 August 2023
Accepted: 10 August 2023
Published: 14 August 2023

The reviews of the Special Issue focus on the use of cellulose [1] and cyclodextrin (CD) [2] in the development of gels for various biomedical applications. The review by Tyshkunova et al. [1] discusses the structure of cellulose and its properties as a biomaterial, the strategies for dissolving cellulose, and the factors that influence the structure and properties of the resulting cryogels. It focuses on the advantages of the freeze-drying process and highlights recent studies on the preparation and application of cellulose cryogels in wound healing, the regeneration of various tissues (e.g., damaged cartilage, bone tissue, and nerves), and controlled release drug delivery. Liu et al. in their review [2] discuss various methods for the design and preparation of CD hydrogels and summarize the potential applications of drug-loaded CD hydrogels. As a natural oligosaccharide, CD has shown remarkable application prospects in hydrogel development. CD can be incorporated into hydrogels to form chemically or physically cross-linked networks. The unique cavity structure of CD makes it an ideal vehicle for drug delivery to target tissues. This review broadens our understanding of the development trends in the application of CD-containing hydrogels, which lays the foundation for future clinical research.

Several strategies have been explored to develop stimulus-responsive hydrogels for designing smart drug delivery platforms that can release drugs at specific targeted areas and at predetermined rates. However, few studies have investigated how innate hydrogel properties can be optimized and modulated to tailor drug dosage and release rates. Briggs et al. [3] investigated the individual and combined effects of polymer concentration and crosslinking density (controlled by chemical crosslinking with N,N′-methylenebis-acrylamide and physical crosslinking with silica nanoparticles) on drug delivery rates, using polyacrylamide gel and 5-fluorouracil as a model system. The experiments showed a strong correlation between hydrogel properties and drug release rates and demonstrated the existence of a saturation point in the ability to control drug release rates using a combination of chemical and physical crosslinkers. This study provides a basis for developing tunable hydrogel platforms, including polymeric nanocarriers and nanogels, for delivering a wide range of therapeutics.

Two of the original papers in this Special Issue focus on the development of delivery systems for peptide antibiotics polymyxin B (PMXB) and E (PMXE) [4,5]. In recent years, unsuccessful treatments of multidrug-resistant bacterial infections have caused previously dismissed antibiotics to resurface. One such group of antibiotics is the polymyxins (PMXs)—cyclic peptide antibiotics that primarily target Gram-negative pathogens. Although this group of antibiotics was introduced into clinical practice more than 50 years ago, it was soon discovered to have serious side effects such as nephro- and neurotoxicity. Nevertheless, the World Health Organization reclassified PMXs as antibiotics that are critically important for treating infections with few or no alternative options. The reintroduction of PMXs as therapeutic agents in clinical practice has spurred the search for methods to reduce their side effects [6]. A nano-sized PMXE delivery system with hydrodynamic diameters (D_h) of 210–250 nm and a ζ-potential of -19 mV has been proposed by Dubashynskaya et al. [4] for intravascular injection. This delivery system is based on a polyelectrolyte complex between hyaluronate (polyanion), diethylaminoethylchitosan (polycation), and positively charged PMXE. In vitro experiments demonstrated that both encapsulated and pure PMXE had a minimum inhibitory concentration of 1 μg/mL against *Pseudomonas aeruginosa*, indicating that encapsulating PMXE in polysaccharide carriers does not reduce its antimicrobial activity. A hybrid delivery system of core–shell nanoparticles (D_h of about 100 nm) has been proposed by Iudin et al. [5]. These hybrid nanoparticles consist of an Ag core and a poly(glutamic acid) shell that can bind PMX via electrostatic interactions. The hybrid nanoparticles showed no cytotoxicity, had low macrophage uptake, and demonstrated intrinsic antimicrobial activity. Furthermore, composite materials based on agarose hydrogel were developed, comprising both the PMX-loaded hybrid nanoparticles and free PMX (PMXB or PMXE). The antibacterial activity of PMX-loaded hybrid nanoparticles and composite gels against *P. aeruginosa* was evaluated, and the results showed that the PMX hybrid delivery system had a synergistic effect compared to either the antibiotics or Ag nanoparticles.

The development of biopolymer composite films for drug delivery is the subject of three papers in the Special Issue [7–9]. Lim et al. [7] evaluated the feasibility of using three mucoadhesive polysaccharides (hydroxypropyl methylcellulose, starch, and hydroxypropyl starch) to develop curcumin-loaded buccal films delivered in the form of chitosan nanoparticles. The results indicated that hydroxypropyl starch is the most suitable mucoadhesive polysaccharide for developing curcumin-loaded buccal films due to the superior curcumin release, good payload uniformity, minimal weight/thickness variations, high folding resistance, and good long-term storage stability of the composite films. Alzarea et al. [8] fabricated and characterized films composed of arabinoxylan (from *Plantago ovata*) and sodium alginate (from brown algae) loaded with gentamicin sulfate, an aminoglycoside antibiotic, for potential use as wound dressings. These films displayed excellent antibacterial effects and desirable properties for wound dressings, such as a tensile strength similar to human skin, mild capacity for water/exudate uptake, suitable water transmission rate, and excellent cytocompatibility. López-Saucedo et al. [9] used an alternative approach to develop antibacterial films. This was achieved via the surface modification of polypropylene films using gamma-irradiation-induced grafting with methyl methacrylate and N-vinylimidazole, which provided a suitable surface capable of loading vancomycin, a glycopeptide antibiotic. The composite multilayer film surface exhibited moderate hydrophilicity and pH-responsiveness, properties desirable for controlled drug release.

Another example of surface modification is presented in the paper by Facchetti et al. [10], where the authors proposed a simple and convenient way to modify the surface of titanium alloy (Ti6Al4V) bone implants with biopolymers consisting of whey protein isolate (WPI) fibrils and heparin or tinzaparin (low molecular weight heparin) to enhance the proliferation and differentiation of bone-forming cells. The results showed that WPI fibrils are an excellent material for biomedical coatings because they are easily modifiable and resistant to heat treatment. In addition, a heparin-enriched WPI coating improved the differentiation of human bone marrow stromal cells by increasing tissue non-specific alkaline phosphatase activity.

The development of delivery systems for the intravitreal administration of glucocorticoids for the treatment of inflammatory conditions in the posterior segment of the eye is challenging. Intravitreal delivery systems have the potential to provide several advantages, such as overcoming anatomical and physiological barriers, increasing bioavailability, and prolonging and regulating drug release over several months. The conjugation of glucocorticoids with biopolymers is a viable approach for the development of intravitreal drug delivery systems, as it prevents rapid elimination and provides targeted and controlled drug release [11,12]. Dubashinskaya et al. [13] demonstrated the potential feasibility of this approach using a dexamethasone conjugate with succinyl chitosan. The developed conjugates showed sustained and prolonged (over one month) release of dexamethasone and significant anti-inflammatory effects in TNFα-induced and LPS-induced inflammation models, suppressing CD54 expression in THP-1 cells by 2- and 4-fold, respectively. These novel conjugates of succinyl chitosan and dexamethasone show promise as ophthalmic carriers for intravitreal administration.

A promising polymeric gene delivery system based on cysteine-flanked arginine-rich peptides modified with a cyclic RGD moiety was proposed by Egorova et al. [14]. The system is designed for targeted DNA delivery to uterine fibroid cells. The carrier can form small (D_h of 100–200 nm) and stable polyplexes that effectively protect DNA from nuclease degradation. The specificity of DNA delivery to $\alpha v\beta 3$ integrin-expressing cells was confirmed by cell transfection experiments, which showed a 3-fold increase in transfection efficiency as a result of the RGD modification. Primary cells obtained from myomatous nodes of uterine leiomyoma patients were used to model suicide gene therapy by transferring the HSV-TK suicide gene, resulting in a 2.3-fold decrease in proliferative activity after ganciclovir treatment of the transfected cells. Pro- and anti-apoptotic gene expression analysis confirmed that the polyplexes stimulate uterine fibroid cell death in a suicide-specific manner. Thus, this peptide carrier can be used in further efforts to develop uterine leiomyoma suicide gene therapy.

In summary, the design and development of drug and gene delivery systems is of paramount importance in the field of therapeutics. These systems have the potential to overcome various challenges associated with traditional therapeutic approaches, improve drug and gene stability, enable targeted delivery, and enhance therapeutic efficacy. With ongoing advances in nanotechnology and biomaterials research, these systems are poised to revolutionize the way we approach medical treatments, opening up new possibilities for personalized and effective therapies.

Funding: This research received no external funding.

Conflicts of Interest: The author declare no conflict of interest.

References

1. Tyshkunova, I.V.; Poshina, D.N.; Skorik, Y.A. Cellulose Cryogels as Promising Materials for Biomedical Applications. *Int. J. Mol. Sci.* **2022**, *23*, 2037. [CrossRef] [PubMed]
2. Liu, J.; Tian, B.; Liu, Y.; Wan, J.-B. Cyclodextrin-Containing Hydrogels: A Review of Preparation Method, Drug Delivery, and Degradation Behavior. *Int. J. Mol. Sci.* **2021**, *22*, 13516. [CrossRef]
3. Briggs, F.; Browne, D.; Asuri, P. Role of Polymer Concentration and Crosslinking Density on Release Rates of Small Molecule Drugs. *Int. J. Mol. Sci.* **2022**, *23*, 4118. [CrossRef]
4. Dubashynskaya, N.V.; Raik, S.V.; Dubrovskii, Y.A.; Demyanova, E.V.; Shcherbakova, E.S.; Poshina, D.N.; Shasherina, A.Y.; Anufrikov, Y.A.; Skorik, Y.A. Hyaluronan/Diethylaminoethyl Chitosan Polyelectrolyte Complexes as Carriers for Improved Colistin Delivery. *Int. J. Mol. Sci.* **2021**, *22*, 8381. [CrossRef] [PubMed]
5. Iudin, D.; Vasilieva, M.; Knyazeva, E.; Korzhikov-Vlakh, V.; Demyanova, E.; Lavrentieva, A.; Skorik, Y.; Korzhikova-Vlakh, E. Hybrid Nanoparticles and Composite Hydrogel Systems for Delivery of Peptide Antibiotics. *Int. J. Mol. Sci.* **2022**, *23*, 2771. [CrossRef] [PubMed]
6. Dubashynskaya, N.V.; Skorik, Y.A. Polymyxin Delivery Systems: Recent Advances and Challenges. *Pharmaceuticals* **2020**, *13*, 83. [CrossRef] [PubMed]
7. Lim, L.M.; Hadinoto, K. High-Payload Buccal Delivery System of Amorphous Curcumin–Chitosan Nanoparticle Complex in Hydroxypropyl Methylcellulose and Starch Films. *Int. J. Mol. Sci.* **2021**, *22*, 9399. [CrossRef] [PubMed]
8. Alzarea, A.I.; Alruwaili, N.K.; Ahmad, M.M.; Munir, M.U.; Butt, A.M.; Alrowaili, Z.A.; Shahari, M.S.B.; Almalki, Z.S.; Alqahtani, S.S.; Dolzhenko, A.V.; et al. Development and Characterization of Gentamicin-Loaded Arabinoxylan-Sodium Alginate Films as Antibacterial Wound Dressing. *Int. J. Mol. Sci.* **2022**, *23*, 2899. [CrossRef] [PubMed]
9. López-Saucedo, F.; López-Barriguete, J.E.; Flores-Rojas, G.G.; Gómez-Dorantes, S.; Bucio, E. Polypropylene Graft Poly(methyl methacrylate) Graft Poly(*N*-vinylimidazole) as a Smart Material for pH-Controlled Drug Delivery. *Int. J. Mol. Sci.* **2022**, *23*, 304. [CrossRef] [PubMed]
10. Facchetti, D.; Hempel, U.; Martocq, L.; Smith, A.M.; Koptyug, A.; Surmenev, R.A.; Surmeneva, M.A.; Douglas, T.E.L. Heparin Enriched-WPI Coating on Ti6Al4V Increases Hydrophilicity and Improves Proliferation and Differentiation of Human Bone Marrow Stromal Cells. *Int. J. Mol. Sci.* **2022**, *23*, 139. [CrossRef] [PubMed]
11. Dubashynskaya, N.V.; Bokatyi, A.N.; Skorik, Y.A. Dexamethasone Conjugates: Synthetic Approaches and Medical Prospects. *Biomedicines* **2021**, *9*, 341. [CrossRef] [PubMed]
12. Dubashynskaya, N.; Poshina, D.; Raik, S.; Urtti, A.; Skorik, Y.A. Polysaccharides in Ocular Drug Delivery. *Pharmaceutics* **2020**, *12*, 22. [CrossRef] [PubMed]
13. Dubashynskaya, N.V.; Bokatyi, A.N.; Golovkin, A.S.; Kudryavtsev, I.V.; Serebryakova, M.K.; Trulioff, A.S.; Dubrovskii, Y.A.; Skorik, Y.A. Synthesis and Characterization of Novel Succinyl Chitosan-Dexamethasone Conjugates for Potential Intravitreal Dexamethasone Delivery. *Int. J. Mol. Sci.* **2021**, *22*, 10960. [CrossRef] [PubMed]
14. Egorova, A.; Shtykalova, S.; Maretina, M.; Selutin, A.; Shved, N.; Deviatkin, D.; Selkov, S.; Baranov, V.; Kiselev, A. Polycondensed Peptide Carriers Modified with Cyclic RGD Ligand for Targeted Suicide Gene Delivery to Uterine Fibroid Cells. *Int. J. Mol. Sci.* **2022**, *23*, 1164. [CrossRef]

International Journal of *Molecular Sciences*

MDPI

Review

Cellulose Cryogels as Promising Materials for Biomedical Applications

Irina V. Tyshkunova, Daria N. Poshina and Yury A. Skorik *

Institute of Macromolecular Compounds of the Russian Academy of Sciences, Bolshoi VO 31, 199004 St. Petersburg, Russia; tisha19901991@yandex.ru (I.V.T.); poschin@yandex.ru (D.N.P.)
* Correspondence: yury_skorik@mail.ru

Abstract: The availability, biocompatibility, non-toxicity, and ease of chemical modification make cellulose a promising natural polymer for the production of biomedical materials. Cryogelation is a relatively new and straightforward technique for producing porous light and super-macroporous cellulose materials. The production stages include dissolution of cellulose in an appropriate solvent, regeneration (coagulation) from the solution, removal of the excessive solvent, and then freezing. Subsequent freeze-drying preserves the micro- and nanostructures of the material formed during the regeneration and freezing steps. Various factors can affect the structure and properties of cellulose cryogels, including the cellulose origin, the dissolution parameters, the solvent type, and the temperature and rate of freezing, as well as the inclusion of different fillers. Adjustment of these parameters can change the morphology and properties of cellulose cryogels to impart the desired characteristics. This review discusses the structure of cellulose and its properties as a biomaterial, the strategies for cellulose dissolution, and the factors affecting the structure and properties of the formed cryogels. We focus on the advantages of the freeze-drying process, highlighting recent studies on the production and application of cellulose cryogels in biomedicine and the main cryogel quality characteristics. Finally, conclusions and prospects are presented regarding the application of cellulose cryogels in wound healing, in the regeneration of various tissues (e.g., damaged cartilage, bone tissue, and nerves), and in controlled-release drug delivery.

Keywords: cellulose; cellulose cryogel; freeze-drying; tissue engineering; regenerative medicine

Citation: Tyshkunova, I.V.; Poshina, D.N.; Skorik, Y.A. Cellulose Cryogels as Promising Materials for Biomedical Applications. *Int. J. Mol. Sci.* **2022**, *23*, 2037. https://doi.org/10.3390/ijms23042037

Academic Editor: Axel T. Neffe

Received: 20 January 2022
Accepted: 11 February 2022
Published: 12 February 2022

1. Introduction

Cryogelation is one of the newly developed protocols for the production of polysaccharide materials for biomedical purposes [1]. Polysaccharide-based cryogels form a spongy super-macroporous structure during freeze-drying, making them highly promising materials for tissue engineering [2]. The production of polysaccharide cryogels has recently become a popular approach for the development of scaffolds [3], and these matrices are readily obtained by dissolving a polysaccharide (usually cellulose) in an appropriate solvent, followed by polymer regeneration from solution (solvent removal), freezing, and freeze-drying. Figure 1 shows the scheme for producing cellulose cryogels.

At the regeneration step, the polymer passes from the dissolved state to an insoluble state, and subsequent freezing leads to ice crystal formation. The removal of the ice during freeze-drying then generates pores and leaves a cryogel with a complex three-dimensional structure [4]. The high porosity and hydrophilicity, high water retention capacity, interconnectedness of the pores, and material consistency make cryogels very similar to natural soft tissues [5], while their mechanical stability allows their use in vivo [6]. Cryogels can also stimulate the in vivo production of various natural molecules, including antibodies, and they can act as in vitro bioreactors for the expansion of cell lines and as a means of cell separation. Excellent in vivo results have been obtained using cryogels as scaffolds for tissue engineering, as cryogels can promote the resumption of growth in numerous damaged tissues [3]. However, the surface properties of tissue engineering materials affect cell affinity [1], and these properties depend on a large number of different factors, including the conditions used for polysaccharide dissolution, regeneration, and freezing.

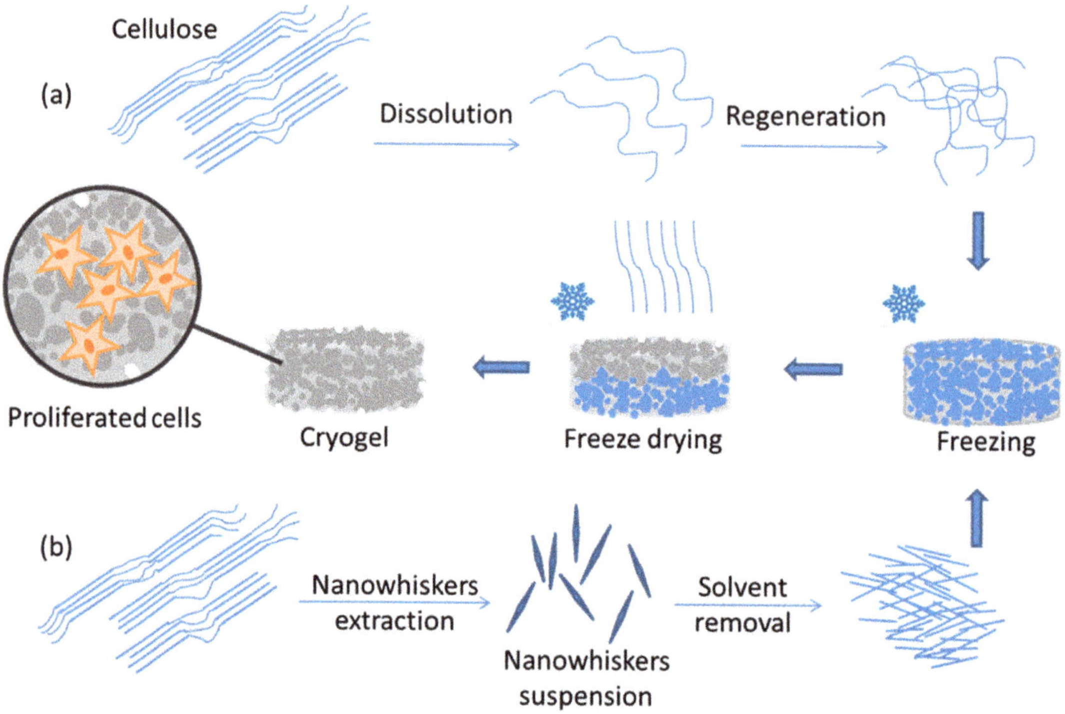

Figure 1. Ways to produce cellulose cryogel for biomedical applications: (**a**) via dissolution and regeneration; (**b**) using nanowhiskers.

Biocompatibility is one of the main requirements for cryogels used as scaffolds. The ideal scaffold should be porous, biodegradable, biocompatible, and bioresorbable and should not trigger an immune response or inflammation [7]. Consequently, scaffolds made from natural polymers have several advantages over those made from synthetic polymers, as natural polymers are bioresorbable and biocompatible, have low immunogenicity and cytotoxicity, and can stimulate intercellular interactions. By contrast, the degradation of synthetic polymers can generate harmful by-products and can have problems in terms of injection and infection [1]. Natural biopolymers, particularly cellulose, have therefore become very popular materials for the preparation of porous products used for biomedical purposes, such as wound healing, tissue engineering, and drug delivery [8,9].

Cellulose has found particular favor in biomedical sciences due to its mechanical strength, biocompatibility, and hydrophilicity, making it a promising polysaccharide for the production of biocompatible porous cryogels [10–12]. Cellulose-based materials have been proposed for a variety of biomedical applications [13,14] because, unlike other polysaccharides, cellulose is relatively bioinert and is not biodegraded in the human body. Thus, newly regenerated tissue cannot displace a cellulosic scaffold, which can be an advantage in tissue engineering. Cellulose materials have found applications in the regeneration of bone [15], neural [16], and cartilage [17] tissues, as well as in wound dressings [18]. The bioinertness of cellulose also meets the requirement that a scaffold material should not induce foreign body responses [19].

This review considers the preparation of cellulose-based cryogels using the freeze-drying technique, and presents data on the use of these cryogels for biomedical purposes. The structural features and properties of cellulose and the difficulties associated with the dissolution of cellulose are reviewed. Information on the methods for dissolving cellulose and producing cellulose cryogels is presented. The influence of various factors on the structure and properties of the produced materials is discussed, and the advantages of the freeze-drying process are analyzed. Recent studies on the production and use of cellulose cryogels for biomedical purposes are summarized, and the main quality parameters of these cryogels are presented. The current status and prospects for the use of cryogels in tissue engineering are discussed.

Previously published reviews on the biomedical application of cryogels have provided much information on various polysaccharide cryogels, but cellulose cryogels have received relatively little attention. The available reviews on cellulose cryogels contain information on production methods and the characterization of properties and morphology, without indicating possible directions for biomedical application [20,21]. Other reviews consider cellulose cryogels to be sorbents [12,22]. Reviews that focus on the biomedical applications of cryogels contain information on many polysaccharide cryogels, with little [2,3,23,24] or no mention of cellulose cryogels [1]. A review of nanocellulose sponges and their biomedical applications has been published [25]. By contrast, we present data on the use of different cellulose types (cellulose of various origins). This review is especially focused on analyzing data on cellulose cryogels obtained by freeze-drying, offering information on the use of various cellulose types for producing biomedical cryogels. The information presented starts with the structural features of cellulose and its solvents for the production of cryogels and ends with data on the biomedical applications of various cellulose cryogels.

2. Cellulose as a Source for Producing Biomedical Materials

Cellulose is a promising raw material for the production of functional biomedical materials [10]. Cellulose can be shaped in many different ways: into beads [26], fibers with a diameter from tens of nm to tens of microns [27], films (cellophane), porous foams (sponges), and aerogels [20,28,29]. The morphology and properties of these objects can be very different.

Cellulose, as a biomedical material, has certain advantages over other traditional biopolymers, including its prevalence (it can be isolated from various natural materials), availability, low toxicity, renewability, and biocompatibility, making the development of cellulose-based cryogels a promising research direction [30]. The freeze-drying of cellulose hydrogels imparts a complex heterogeneous structure to cellulose, creating useful building blocks for complex hierarchical structures [31]. Porous cellulose materials are very attractive for a variety of biomedical applications, including controlled drug release, tissue engineering scaffolds, matrices for cell growth, biosensors, and antibacterial wound dressings [12,31–35]. Each of these applications requires materials with a specific morphology, pore size distribution, specific surface area, and material density. However, the complex supramolecular structure of cellulose creates difficulties in its dissolution and processing into biomedical products.

Cellulose consists of anhydroglucose units ($C_6H_{10}O_5$) linked by β-glycosidic ($1 \rightarrow 4$) bonds and has a high crystalline content [36]. The hydroxyl groups in the cellulose macromolecule are involved in intra- and intermolecular hydrogen bonds (Figure 2b), which lead to the formation of various ordered crystal structures.

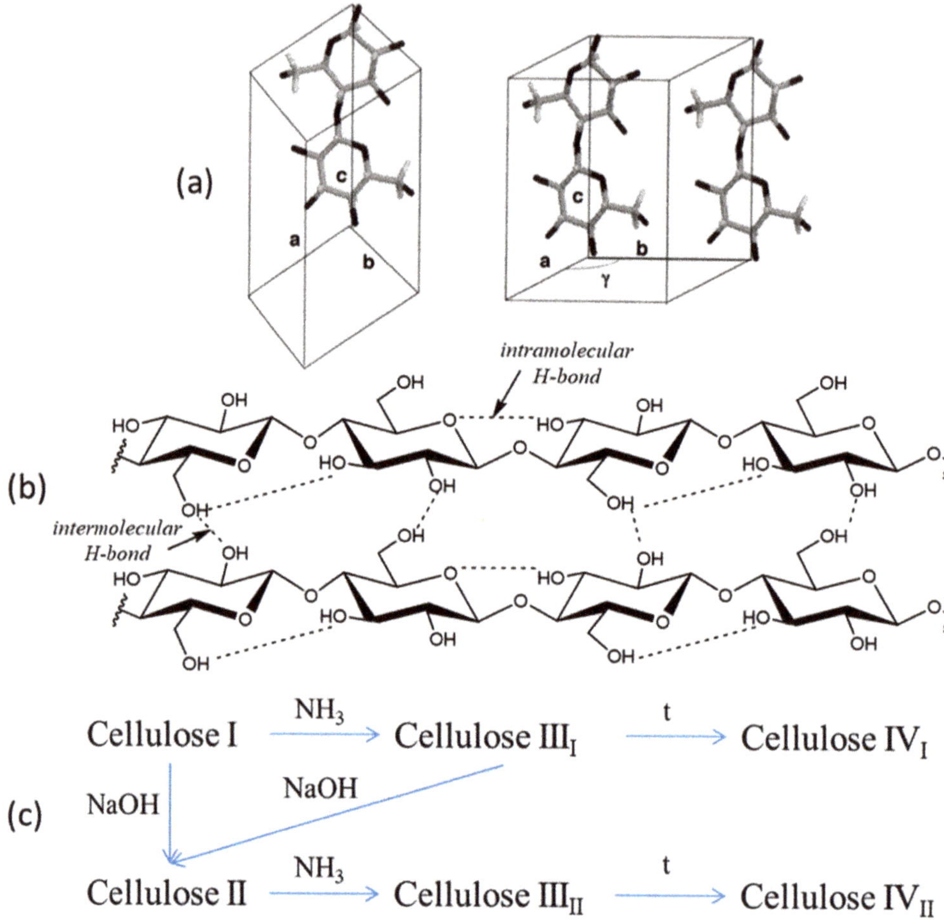

Figure 2. Complex molecular structure of cellulose: (**a**) native cellulose I unit cells, triclinic Iα and monoclinic Iβ [37] © 2022 National Academy of Sciences; (**b**) H-bond network of cellulose I; (**c**) polymorph transitions.

Crystalline allomorphs of cellulose I, II, III, and IV are distinguished according to their X-ray diffractometry and solid-state ^{13}C NMR spectra. Cellulose I, the most abundant form in nature, is a crystalline native cellulose, whereas cellulose II is obtained by mercerization (alkaline treatment) or regeneration (solubilization and subsequent recrystallization) (Figure 2c). Celluloses III_I and III_{II} can be formed from celluloses I and II, respectively, by treatment with liquid ammonia, and the reaction is reversible [38]. Cellulose IV_I and IV_{II} can be obtained by heating celluloses III_I and III_{II}, respectively [39]. Lightweight porous materials can be obtained from celluloses I or II [12], but most research has focused on cellulose I. The crystalline structure of cellulose I is a mixture of two different crystalline forms: cellulose Iα (triclinic) and Iβ (monoclinic) (Figure 2a) [40]. The relative amounts of cellulose Iα and Iβ vary depending on the cellulose source (for example, the Iβ form is dominant in higher plants, whereas the Iα form is typically found in algae and bacteria). Cellulose crystallites are usually about 5 nm wide; however, these crystallites are imperfect, and part of the cellulose structure is less ordered, termed amorphous. The traditional two-phase cellulose model describes cellulose chains containing both crystalline (ordered) and amorphous (less ordered) regions [41].

This added complexity of the supramolecular structure of cellulose creates difficulties in its dissolution and processing. Typical cellulose solvents include 7–9% aqueous NaOH [26,42,43], Cu-ethylenediamine (or Cd-ethylenediamine) complexes [44], LiCl/dimethylacetamide (DMAc) [45], *N*-methyl-morpholine-*N*-oxide monohydrate [46–49], molten salt hydrates, and ionic liquids [50–53] (Table 1).

Table 1. Characteristics of typical cellulose solvents.

Solvent	Advantages	Disadvantages	Reference
LiCl/DMAc	It does not cause any destruction of the cellulose, provided that destructive pretreatments are avoided (such as heating over 80 °C).	The difficulty of removing LiCl from the final products.	[45]
Ionic liquids	They completely dissolve the material's components.	Ionic liquids do not evaporate, have low volatility, which complicates their regeneration.	[50–53]
7–9%NaOH/water (7%NaOH/12%urea/water)	Cellulose gels can be obtained.	The thermodynamic quality of the solvent decreases with increasing temperature, as the number of cellulose–cellulose interactions increases more rapidly than the number of cellulose–solvent interactions; Na+ ions penetrate deeply into the cellulose structure, making it difficult to remove alkali.	[26,42,43]
Complexing compounds of Cu with ethylenediamine (or Cd-ethylenediamine complexes)	Commonly used to determine the molecular weight of cellulose.	The difficulty of removing from the final products.	[44]
N-methyl-morpholine-*N*-oxide monohydrate	Direct solvent of cellulose: *N*-methylmorpholine-*N*-oxide (NMMO) is a cellulose solvent used industrially for the spinning of cellulosic fibers (the Lyocell process). NMMO is known to change the highly crystalline structure of cellulose after dissolution and regeneration.	In theory, this dissolution process is merely physical, but in practice many side reactions might occur.	[46–49]
Concentrated phosphoric acid	Rapid dissolution, easily removed and regenerated.	Causes significant destruction of macromolecules.	[54]

However, most of these are toxic, have only limited ability to dissolve high molecular weight cellulose, and are difficult to remove from the final product [55]. The processing steps required for dissolution, gelation, and solvent removal for cellulose cryogel formation are very slow and can take several days [56]. However, one advantage of the insolubility of cellulose in water and typical organic liquids is that, with proper reinforcement, the structure of lightweight cellulose materials can be retained when they are immersed in most liquids [57].

Interest in porous biomedical materials continues to grow, as evidenced by the number of publications each year [2,58–60]. Cellulose is a promising raw material for the production of cryogels.

3. Advantages of Freeze-Drying and Factors Affecting the Structure and Properties of Cryogels

Freeze-drying allows the preservation of the micro- and nanostructure of the material and the generation of a large specific surface area (up to 300 m^2/g) in the dried state [14,61]. One advantage of freeze-drying is that it has no requirement for the use of flammable liquids (e.g., ethanol/acetone that are required for supercritical drying, which also allows preservation of the nanostructure of the material); another is that the structure of the resulting material corresponds to the structure of the frozen dispersion [62]. Ice crystals formed during the freezing of the dispersion change the distribution of particles, and subsequent drying creates pores where the ice crystals had formed [63]. The morphology of the materials obtained by freeze-drying can vary from random networks to lamellar solid structures. These different types of network structures can produce equally lightweight materials; therefore, they can be easily designed and produced with environmental friendliness and safety in mind.

The final properties of a cryogel, including its biocompatibility, mechanical, and thermal properties, and degradability, depend on many factors (Figure 3).

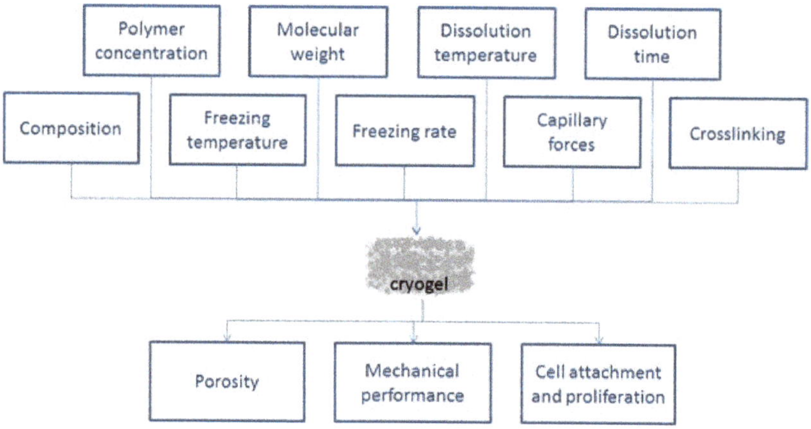

Figure 3. Influence of different factors on cellulose cryogel properties.

The chemical composition of the cryogel is probably the most important factor, since it determines the biocompatibility and degradability of the cryogel and, to some extent, affects the mechanical and thermal properties of the cryogel. The porosity and degree of crosslinking mainly affect the mechanical properties, while crosslinking itself affects the biocompatibility and degradability of the cryogel [2]. The pore size, wall thickness, and density affect the properties of cryogels [64], as thicker and higher-density walls improve their mechanical properties. The thickness and density, in turn, depend on the concentration of the polymer and the type of crosslinking in the cryogel.

The production processes used to form the cryogels also affect their structure. For example, an increase in the freezing rate or a decrease in the cryogelation temperature decreases the cryogel pore size because the solvent freezes at a higher rate, allowing for the growth of only a small number of ice crystals [65,66]. Further, a temperature gradient occurs during cryogelation, which leads to a non-uniform pore size distribution [67]. Initially, the external part of the sample is exposed to a low temperature, which leads to an increase in the freezing rate and a smaller pore size than that subsequently formed in the internal cryogel material. However, this heterogeneous pore size distribution is not an obstacle to the use of cryogels in tissue engineering, since many tissues of the human body also have heterogeneous morphology [68]. Cryostructuring, including directional freezing of cryogels, has been used to achieve varying degrees of porosity (45–75%, pore size 70–85 nm) or to equalize the porosity or anisotropy within cryogels [60].

The mechanical properties of cryogels are commonly evaluated using compression testing [3]. Reducing the pore size of cryogels has been reported to increase compressive strength [69], whereas increasing porosity increases the compressive deformation of the cryogel [70]. For one type of cryogel (injection cryogels), low compression deformation is undesirable; their ideal porosity is 91% [67].

Cryogels used in biomedical applications may require that the material eventually degrade within the body, but cryogels still need to perform their functions before this degradation occurs. Therefore, knowledge of the changes in the mechanical properties of cryogels throughout their degradation would be useful [71]. The thickness of the cryogel walls is assumed to decrease during enzymatic degradation, and in some cases, the walls are destroyed. Whether this process occurs for cryogels degraded by other mechanisms (e.g., by cleavage of disulfides [72] or chemical hydrolysis [59]) is unclear. The degradation of cryogels leads to an increase in pore size, possibly due to thinning of the pore walls and a decrease in crosslinking [73]. The mechanical properties of degraded cryogels are largely overlooked in the current literature, despite their importance for applications such as scaffold materials [71]. Due to their non-biodegradability in the human body, the main application areas of cellulose materials are bone tissue regeneration (bone implants) [15,66,67] and the production of wound dressings [18].

The following sections provide a more detailed description of some of the variables that affect the structure and properties of cryogels: the type and degree of crosslinking, the concentration and molecular weight of the polymer, the parameters of gelation and cryoconcentration, and the effects of capillary forces, temperature, and freezing rate.

3.1. Type and Degree of Crosslinking

Crosslinking can provide better mechanical performance and integrity for cellulose cryogels. The type of crosslinking affects the rigidity and degree of swelling, which in turn affects the elastic and mechanical properties and pore size of the cryogel. Methods for cryogel formation include chemical crosslinking and physical gel formation using natural or synthetic polymers [65]. Chemical crosslinking occurs during the storage of the polymer solution at a given temperature, whereas physical crosslinking occurs during the thawing step, where faster thawing results in weaker gels [6]. Physical crosslinking generates cryogels with pore sizes of less than 10 μm [74–77], whereas chemical crosslinking allows for cryogels with large pore sizes of 80–200 μm [28,68,72]. One hypothesis to explain the difference in structure formation during physical and chemical gelation of cellulose is that, during physical gelation, the chains self-associate to form a heterogeneous network with "thick" walls and pores of different sizes. By contrast, during chemical gelation, chemical bonds act as separators between the chains, thereby breaking their self-association and preventing packaging. The result is a more uniform chemical network with higher swelling and transparency when wet and lower density when dry [33].

The degree of crosslinking (i.e., the ratio of monomer to crosslinking agent) in a chemically crosslinked cryogel affects its mechanical properties. Chemical crosslinking can provide good mechanical properties; however, the compounds used as crosslinkers are often toxic, difficult to remove, and not biocompatible [78]. The effect of the amount of the crosslinking agent on the mechanical properties of cellulose cryogels is debatable, as some data show an increase in the compressive modulus with an increase in the crosslinking agent concentration [79], whereas other studies indicate an increase in the storage modulus for cellulose cryogels from 45 to 675 Pa with a decrease in the crosslinking agent concentration [80]. An increase in the crosslinking agent concentration (epichlorohydrin) also results in the formation of an inhomogeneous structure of the cellulose cryogel, whereas dense areas are observed when the pore size is 200 μm [33].

The degree of crosslinking in physically crosslinked cryogels is controlled by changing the number of freeze-thaw cycles [2]. Physical crosslinking does not use any organic solvents or toxic crosslinking agents, thereby eliminating any danger of residues in the final material and making this method very promising for biomedical applications [78]. Physical crosslinking is also easier, and this translates into cost savings. The problem with physical methods is obtaining satisfactory properties without any chemical modification while maintaining biocompatibility, biodegradability, and bioactivity [78]. However, according to some data, compared to their chemically crosslinked counterparts, physically crosslinked cryogels demonstrate greater mechanical strength [81] and crystallinity (cellulose cryogels) [33].

3.2. Concentration and Molecular Weight of the Polymer

A minimum (critical) concentration of cellulose is required to retain the integrity (shape) of the produced cryogel (i.e., to retain the integrity of the network after removing the liquid phase) [22,77,78]. This critical concentration is probably related to percolation within the precursor network [82,83], where overlap or interaction between cellulose chains results in the formation of an autonomous network [34]. At a concentration below the critical value, the network is unstable, and shrinkage increases with decreasing cellulose concentration [83]. A cellulose concentration of more than 3% in solution is required to obtain cryogels, as studies have shown that cryogels do not form at concentrations of less than 3% [28,54]. The critical concentration of the polymer also affects the mechanical properties of the produced cryogels [84].

Solutions with high polymer concentrations produce cryogels with small average pore sizes. This is due to an increase in the availability of crosslinked groups and a decrease in the availability of free water. As with conventional hydrogel formation, increasing the polymer content increases the rigidity of the cryogels. An increase in the polymer concentration also leads to a decrease in porosity and swelling of the cryogel [54,85], while decreasing the degradation rate [85].

The molecular weight of the polymer affects the structure of the cryogel. The use of polymer solutions with a lower molecular weight at the same mass concentration in a gel solution leads to the formation of larger pores compared to the use of gel solutions of polymers with a higher molecular weight [24,54,83]. Higher molecular weight polymer solutions will generate smaller pores due to the relatively smaller volume of free water that can form ice crystals in the solution. Similar observations were recorded when producing cryogels based on cellulose with various degrees of polymerization [21] compared to the concentrations of other polymers (gelatin) in a gel solution [86,87].

An increase in the degree of cellulose polymerization leads to an increase in the undissolved fraction in the solution, which reduces the content of dissolved cellulose in the matrix solution. This leads to the formation of voids in the dry matter [20]. Thus, the incomplete dissolution of cellulose with a high degree of polymerization and an increase in material heterogeneity will worsen the mechanical properties of the final cellulose composites.

3.3. Gelation and Cryoconcentration Parameters

The temperature and dissolution time (gelation) of the polymer affect the cryogel structure and properties. These parameters are typically set to values that provide the best structure and properties for each cryogel. For example, the optimal dissolution time is 24 h at room temperature for microcrystalline cellulose [54] and 16 h at room temperature for chitin [88]. An optimum temperature also exists for gelation and cryogelation for maximization of the pore size [89]. The effect of the gelation and crystallization rate of the solution on the physical properties of cryogels therefore becomes important.

To obtain a macroporous structure of cryogel by cryogelation, the gel solution must first partially crystallize to form solidified solvent crystals (pore-forming agents). This can be complicated by the action of hydrogel components that lower the freezing point of the solution (the "freezing point lowering effect") and by the effects of supercooling. To obtain a homogeneous macroporous hydrogel, the crosslinking rate of the polymer must be lower than the crystallization rate of the solvent [90]. If crosslinking occurs faster than the solvent can crystallize, a non-macroporous gel will form. Conversely, larger pores can be formed by reducing the crosslinking rate (the formation and growth of crystal pore-forming agents). The inhibitory effect of supercooling during solvent crystallization can be overcome by increasing the cooling rate. This increase leads to the formation of smaller [91] or even irregular pores [92], depending on the extent of the increase in the cooling rate.

The cryoconcentration of components in the liquid phase also affects the process of cryogel formation. For example, cryoconcentration lowers the critical concentration required for gelation, thereby allowing gel solutions with low monomer content that would not normally set at room temperature to set under cryo-conditions. Cryoconcentration can also speed up the gelation process [6]. The effect of cryoconcentration on the mechanical properties of the cryogel is of interest, given that the compaction of the polymer in the pore walls significantly increases local mechanical properties, such as elasticity.

3.4. Capillary Forces

The capillary forces between the particles of a porous material affect its density. A decrease in capillary forces decreases the density of the material, resulting in lighter materials [93]. Freeze-drying avoids capillary forces; for example, freezing at −18 °C and subsequent freeze-drying produced the lightest cellulose material (density 0.0002 g/cm^3) from a 0.1% cellulose nanofibril hydrogel [94]. The cooling rate is lower for this type of freezing (−18 °C) than when liquid nitrogen is used for freezing. This promotes the growth of ice crystals and produces a material of lower density [83,94].

3.5. Freezing Parameters

The freezing temperature affects the cryogel morphology and can result in the formation of a lamellar structure and highly porous gels with preserved micro- and nanostructure [61]. Smaller pores can be formed by lowering the temperature [95]. At lower temperatures, the solvent crystallizes more rapidly, resulting in the formation of smaller solvent crystals (pore-forming agents). However, due to the increased crystallization of the solvent, the liquid microphase becomes more concentrated, which leads to the formation of thinner and denser pore walls. A 15 °C decrease in the freezing point has been shown to cause a decrease in the pore diameter of polyacrylamide cryogels by an average of 30 μm [96]. By contrast, the pore sizes of cryogels based on polyvinyl alcohol, laminin, or gelatin crosslinked with glutaraldehyde were unaffected by the freezing point [65]. Freezing at −20 °C resulted in the formation of lamellar structures with few pores. A decrease in the pre-freezing temperature to −80 °C and −196 °C led to the appearance of more porous structures. In general, a lower pre-freezing temperature produces a more porous and less agglomerated cryogel structure [95].

Rapid cooling of the dispersion is effective for producing numerous and small ice crystals and leads to the formation of small pores (hence, a high specific surface area) [83,97]. The effects of temperature and freezing rate have been demonstrated on cellulose cryogels [98] cooled at −68 °C and −40 °C. Smaller pore sizes were obtained at the lower temperature (−68 °C) due to the higher cooling rate. Cryogels with the highest specific surface area of 201 m^2/g (i.e., the smallest pore size) were obtained at −196 °C [98]. The opposite approach (a low cooling rate) is used to increase the lightness of the cryogel [98]. Optimum freezing conditions can be determined by the initial crystallization temperature of the solvent and the freezing point for each polymer solution [99].

The structure of the cryogel will also be influenced by the type of cellulose solvent and the inclusion of various fillers or additives. Cryogel scaffolds often have more than one component and can consist of mixtures of two or more polymers or composites. Composite cryogels can be produced using both polymers and additives (nanoparticles and fibers) to obtain a material with improved physical, chemical, and biological properties. These cryogels can combine the beneficial properties of each component [59]. For example, cryogels of carrageenan/cellulose nanofibrils as carriers of antimicrobial α-aminophosphonate derivatives were produced by crosslinking with glyoxal. Cellulose nanofibrils significantly strengthened the composite material, improving its mechanical properties. Scaffolds of this material have been proposed for use as antimicrobial wound-healing materials and have been shown to be effective against *Staphylococcus aureus* infection [100].

Composite nanocellulose/gelatin cryogels with controlled porosity and network structure and good biocompatibility were obtained by chemical crosslinking of dialdehyde starch and subsequently used as carriers for the controlled release of 5-fluorouracil [101]. An increase in nanocellulose content (from 0.5 to 5 parts relative to gelatin) increased the specific surface area and porosity of the composite cryogel. The swelling coefficients first increased and then decreased with an increase in the nanocellulose content. Increasing the nanocellulose content resulted in improved drug loading and crosslinking rates.

The next section provides information on a variety of cellulose cryogels and cellulose-based composite cryogels produced using different solvents. The quality characteristics of the produced cryogels and their applications for biomedical purposes are presented.

4. Cellulose-Based Cryogels and Their Applications in Biomedicine

Cellulose cryogels, as a new generation of porous materials, are of great interest in tissue engineering, as they offer new solutions and improve existing systems and procedures [3]. In addition to their high porosity and mechanical strength, cellulose cryogels can be modified to enhance the attachment of certain other materials (e.g., extracellular matrix proteins, cultured cells, or chemicals) that can promote cell immobilization and growth [102,103] on cryogel scaffolds. Table 2 provides data on cellulose-based cryogels obtained by freeze-drying using various solvents and includes the main characteristics of the cryogels and the possible directions of their biomedical applications (Table 2).

Table 2. Cellulose cryogels for biomedical applications.

Polymer	Production	Characteristics	Application	Reference
MCC	Calcium thiocyanate tetrahydrate and water (117 °C)	Porosity 94.3% Density 84.1 kg/m^3 Surface area 23 m^2/g E 13.27 ± 1.5 a	New filter types, various biomedical applications.	[31]
MCC	8 wt% NaOH-water (cross-linking with epichlorohydrin)	Pore size up to 200 μm Density 0.04–0.121 g/cm^3	Drug release, materials with controlled morphology and porosity.	[33]
MCC/pectin	1-Allyl-3-methylimidazolium chloride	Dense network structure	Hemostatic material (had no effect on cell proliferation but offered favorable properties in liver hemostasis).	[104]
HEC	Cryogenic treatment with citric acid, freeze-drying	Interconnected pores 100–180 μm	Matrices for immobilized enzymes and cells, readily degraded in acidic conditions	[105]
HEC/polyaniline	Stirred at 40 °C in water for 20 min, sonicated		tissue engineering scaffolds, high survival and proliferation in electric field, good adhesion, spreading, and rearrangement onto materials.	[106]
CMC	Dissolved in deionized water and crosslinking with adipic acid dihydrazide and a small excess of the carbodiimide at −20 °C.	E 4.2 ± 1.4 MPa	Neural tissue engineering, cell delivery (restoration of brain tissue through delivery to the neural network).	[16]

Table 2. *Cont.*

Polymer	Production	Characteristics	Application	Reference
CMC/Col	Mixing two streams: CMC solution (2%) in deionized water with adipic acid dihydrazide, buffer solution and solution N-(3-dimethylaminopopyl)-N′-ethylcarbodiimide chloridate (EDC, in deionized water). The resulting cryogels were soaked in the collagen solution, and then soaked in the EDC solution to fix the collagen.	Porosity > 90% Uniform density	Tissue engineering, spreading and proliferation of NOR-10 fibroblasts.	[107]
CMC/Col CMC/Col/TCP	Mixing two solutions (1:2)-CMC solution (distilled water), Col solution (acetic acid). TCP was added to the final solution.	Average lamellar spaces 204 ± 95 μm (Col/CMC) and 195 ± 21 μm (Col/CMC/TCP) E 309 ± 18 kPa (Col/CMC) and 481 ± 27 kPa (Col/CMC/TCP)	Regeneration of hard tissues, non-toxic and compatible with blood.	[108]
CMC/PVA/honey	Solvent water, each layer was applied alternately with preliminary freezing of the previous.		Wound healing, showed activity against *S. aureus* compared to their counterparts without honey.	[109]
CNF (bleached softwood kraft pulp)	Mechanical defibrillation in deionized water, sonication to obtain the nanofibril aqueous gel, which then sprayed and atomized at 40 MPa, frozen in liquid nitrogen and freeze-dried.	Density 0.0018 g/cm^3 Surface area 389 m^2/g	Tissue engineering, evaluated using 3T3 NIH cells.	[110]
CNF (bleached birch Kraft Pulp)	Solvent-TEMPO, sodium bromide, NaOH. TEMPO-oxidized cellulose fibers (NaClO) were precipitated in ethanol. CNF hydrogels were obtained from the CNF films followed by solvent exchange from ethanol to tertbutanol, frozen in liquid nitrogen, and freeze-dried.	Porosity 88.0–99.7% Pore size 10–200 μm Density 0.004–0.180 g/cm^3 Surface area 158–308 m^2/g E 28–104.4 kPa	Tissue engineering, evaluated using HeLa and Jurkat cells.	[111]
CNF (cellulose powder)	CNF powder in deionized water dispersed by sonication, crosslinked with glyoxal solutions, frozen in liquid nitrogen, freeze-dried.	For CNF cryogel 35 ± 9 μm, for crosslinked cryogel 60 ± 20 μm 0.003–0.11 g/cm^3 for CNF cryogel, 0.003–0.09 g/cm^3 for crosslinked cryogel Up to 1 m^2/g 0.1 MPa for CNF cryogel, 50.8 ± 8 MPa for crosslinked cryogel	Bone tissue engineering, assayed in vitro with MG-63 cells.	[15]

Table 2. *Cont.*

Polymer	Production	Characteristics	Application	Reference
CNF/Col (wood powder of 60–80 meshes)	NCFs were sonicated, oxidized by $NaIO_4$. The dialdehyde NCFs were mixed with collagen 1:1, frozen and freeze-dried.	Porosity 90–95% Density 0.02–0.03 g/cm^3	Tissue engineering, supported fibroblast proliferation.	[18]
CNF/gelatin/chitosan (high-purity softwood cellulose)	Crosslinking in situ with genipin, frozen and freeze-dried.	Porosity 95% Pore size 75–200 μm Density 0.06–0.09 g/cm^3 E 1–3 MPa	Cartilage tissue engineering (ASC and L929 cells)	[17]
CNF/ bioactive glass	Cellulose nanofibrils (CNF) are introduced.	High porosity Pore size 96–168 μm E 24 ± 1 kPa	Bone tissue engineering (MC3T3-E1 cells and calvarial bone defect in rats in vivo)	[112]
CNF/PVA (commercial CNF)	Crosslinking with polyamide-epichlorohydrin, frozen in liquid nitrogen, freeze-dried.	Porosity 88.5–95.3%Pore size 90 and 20 μm Density 0.006–0.05 g/cm^3 Compressive strength 5–220 kPa E 0.04–8.3 kPa	Skin tissue engineering, supported fibroblast cells.	[113]
CNF)/ NIPAm (commercial bleached softwood kraft pulp)	Crosslinked and sonicated, frozen in liquid nitrogen, freeze-dried.	Density 0.01–0.14 g/m^3	Drug release.	[114]
Cellulose (wood dust from the plywood sanding)	Nanocellulose suspension from alkaline treated wood waste powders was redispersed in deionized water, frozen and freeze-dried.	Porosity 97.8–99.8% Pore diameter 3.7–8.3 nm Density 0.004–0.036 g/m^3 Surface area 419–457 m^2/g, E 7–165 kPa, Thermal performance 34–44 mW/m·K	Biomedicine, pollution filtering, thermal insulation.	[77]

MCC—microcrystalline cellulose, ECH—epichlorohydrin, HEC—hydroxyethylcellulose, CMC—carboxymethyl cellulose, ECM—extracellular matrix, EDC—*N*-(3-dimethylaminopopyl)-*N′*-ethylcarbodiimide chloridate, Col—collagen, TCP—tricalcium phosphate, TEMPO—2,2,6,6-tetramethylpiperidin-1-yl oxyl, PVA—polyvinyl alcohol, CNF—cellulose nanofibril, NIPAm—*N*-isopropylacrylamide.

The use of various solvents and cellulose dissolution techniques has produced cryogels with suitable properties, morphology, and mechanics for biomedical applications. Further, cellulose-based cryogels have shown good sorption properties; for example, keratin/cellulose cryogels have been successfully fabricated for the adsorption of oil/solvent [115]. Highly porous (more than 90%) and ultra-light (density less than 0.035 g/cm^3) cellulose/biochar cryogels have also shown high sorption capacities. The addition of 5% biochar to a cellulose cryogel yielded the highest sorption capacity, at 73 g/g of petroleum [116]. Cryogels formed from hydroxypropyl methylcellulose (HPMC) and bacterial cellulose nanocrystals (CNC) have shown good adsorption of organic pollutants [117]. Shapable cellulose nanofiber/alginate cryogels with underwater super-elasticity have been used for protein purification [118]. Highly porous (94.7–97.1%) light (density 0.016–0.028 g/cm^3) hydrophobic cellulose cryogels (unbleached long fiber of *Pinus elliottii*) have shown a high homogeneous sorption capacity (65.18 g/g) and heterogeneous sorption capacity (68.42 g/g) (solvent organosilane methyltrimethoxysilane) [119]. Thus, cellulose cryogels can be produced with different microstructures and properties, and varying the conditions of cellulose dissolution and the parameters for producing cryogels can result in cryogels with many different desirable qualities.

5. Conclusions

Due to the advantages of the freeze-drying method, interest is growing in the production of polysaccharide-based porous materials by cryogelation. The use of natural polymers for the production of cryogels, in contrast to synthetic polymers, makes it possible to create biocompatible medical materials (scaffolds) with a minimal immune response. Cellulose, due to its availability, renewability, non-toxicity, and biocompatibility, is a promising raw material for producing cryogels for biomedical applications. The production of cellulose cryogels by freeze-drying is a promising and steadily developing direction in tissue engineering. Cellulose cryogels have unique properties imparted by their interconnected super-macroporous structure and mechanical stability that make them attractive materials for a variety of applications. Much research has focused on the development of cellulose cryogels for tissue engineering. The results show that cellulose cryogels are promising tools and are applicable as scaffolds for various tissue types.

Physical and chemical parameters affect the formation of cryogels, such as the origin of the cellulose, dissolution parameters, type of solvent, temperature, freezing rate, and the inclusion of various fillers. Varying the parameters of cellulose dissolution, production technology, and freezing can change the properties of the cryogels and set the desired final characteristics of the product. Due to its complex supramolecular structure, cellulose is difficult to dissolve. Thus, an important task remains the selection of a cellulose solvent that can be easily removed from the final product prior to its use for biomedical purposes. The production of composite cryogels is promising for imparting additional properties to the cryogel (changes in morphology and mechanical properties). An important direction for research in the field of cryogels is the preservation of the properties of cryogels during their use. Cellulose cryogels have huge potential in the repair and regeneration of various tissue types, including cartilage tissue, bone tissue, and nerves, in wound healing, and in the delivery of controlled release drugs.

Author Contributions: Conceptualization, I.V.T. and Y.A.S.; writing—original draft preparation, I.V.T. and D.N.P.; writing—review and editing, Y.A.S.; funding acquisition, I.V.T. All authors have read and agreed to the published version of the manuscript.

Funding: I.V.T. was funded by the Russian Foundation for Basic Research (project 19-33-60014).

Institutional Review Board Statement: Not applicable.

Informed Consent Statement: Not applicable.

Data Availability Statement: Not applicable.

Conflicts of Interest: The authors declare no conflict of interest.

References

1. Bakhshpour, M.; Idil, N.; Perçin, I.; Denizli, A. Biomedical applications of polymeric cryogels. *Appl. Sci.* **2019**, *9*, 553. [CrossRef]
2. Jones, L.O.; Williams, L.; Boam, T.; Kalmet, M.; Oguike, C.; Hatton, F.L. Cryogels: Recent applications in 3D-bioprinting, injectable cryogels, drug delivery, and wound healing. *Beilstein J. Org. Chem.* **2021**, *17*, 2553–2569. [CrossRef] [PubMed]
3. Hixon, K.R.; Lu, T.; Sell, S.A. A comprehensive review of cryogels and their roles in tissue engineering applications. *Acta Biomaterialia* **2017**, *62*, 29–41. [CrossRef] [PubMed]
4. Arvidsson, P.; Plieva, F.M.; Lozinsky, V.I.; Galaev, I.Y.; Mattiasson, B. Direct chromatographic capture of enzyme from crude homogenate using immobilized metal affinity chromatography on a continuous supermacroporous adsorbent. *J. Chromatogr. A* **2003**, *986*, 275–290. [CrossRef]
5. Liu, P.; Chen, W.; Bai, S.; Wang, Q.; Duan, W. Facile preparation of poly (vinyl alcohol)/graphene oxide nanocomposites and their foaming behavior in supercritical carbon dioxide. *Compos. Part A Appl. Sci. Manuf.* **2018**, *107*, 675–684. [CrossRef]
6. Lozinsky, V.I.; Plieva, F.M.; Galaev, I.Y.; Mattiasson, B. The potential of polymeric cryogels in bioseparation. *Bioseparation* **2001**, *10*, 163–188. [CrossRef]
7. Kemençe, N.; Bölgen, N. Gelatin-and hydroxyapatite-based cryogels for bone tissue engineering: Synthesis, characterization, in vitro and in vivo biocompatibility. *J. Tissue Eng. Regen. Med.* **2017**, *11*, 20–33. [CrossRef]
8. Zhao, S.; Malfait, W.J.; Guerrero-Alburquerque, N.; Koebel, M.M.; Nyström, G. Biopolymer aerogels and foams: Chemistry, properties, and applications. *Angew. Chem. Int. Ed.* **2018**, *57*, 7580–7608. [CrossRef]
9. Surya, I.; Olaiya, N.; Rizal, S.; Zein, I.; Sri Aprilia, N.; Hasan, M.; Yahya, E.B.; Sadasivuni, K.; Abdul Khalil, H. Plasticizer enhancement on the miscibility and thermomechanical properties of polylactic acid-chitin-starch composites. *Polymers* **2020**, *12*, 115. [CrossRef]

10. Klemm, D.; Heublein, B.; Fink, H.P.; Bohn, A. Cellulose: Fascinating biopolymer and sustainable raw material. *Angew. Chem. Int. Ed.* **2005**, *44*, 3358–3393. [CrossRef]
11. Bajpai, S.; Swarnkar, M. New semi-ipn hydrogels based on cellulose for biomedical application. *J. Polym.* **2014**, *2014*, 376754. [CrossRef]
12. Ferreira, E.S.; Rezende, C.A.; Cranston, E.D. Fundamentals of cellulose lightweight materials: Bio-based assemblies with tailored properties. *Green Chem.* **2021**, *23*, 3542–3568. [CrossRef]
13. Jorfi, M.; Foster, E.J. Recent advances in nanocellulose for biomedical applications. *J. Appl. Polym. Sci.* **2015**, *132*, 14. [CrossRef]
14. Nemoto, J.; Saito, T.; Isogai, A. Simple freeze-drying procedure for producing nanocellulose aerogel-containing, high-performance air filters. *ACS Appl. Mater. Interfaces* **2015**, *7*, 19809–19815. [CrossRef]
15. Courtenay, J.C.; Filgueiras, J.G.; Deazevedo, E.R.; Jin, Y.; Edler, K.J.; Sharma, R.I.; Scott, J.L. Mechanically robust cationic cellulose nanofibril 3D scaffolds with tuneable biomimetic porosity for cell culture. *J. Mater. Chem. B* **2019**, *7*, 53–64. [CrossRef]
16. Béduer, A.; Braschler, T.; Peric, O.; Fantner, G.E.; Mosser, S.; Fraering, P.C.; Benchérif, S.; Mooney, D.J.; Renaud, P. A compressible scaffold for minimally invasive delivery of large intact neuronal networks. *Adv. Healthc. Mater.* **2015**, *4*, 301–312. [CrossRef]
17. Naseri, N.; Poirier, J.-M.; Girandon, L.; Fröhlich, M.; Oksman, K.; Mathew, A.P. 3-dimensional porous nanocomposite scaffolds based on cellulose nanofibers for cartilage tissue engineering: Tailoring of porosity and mechanical performance. *Rsc Adv.* **2016**, *6*, 5999–6007. [CrossRef]
18. Lu, T.; Li, Q.; Chen, W.; Yu, H. Composite aerogels based on dialdehyde nanocellulose and collagen for potential applications as wound dressing and tissue engineering scaffold. *Compos. Sci. Technol.* **2014**, *94*, 132–138. [CrossRef]
19. Hickey, R.J.; Pelling, A.E. Cellulose biomaterials for tissue engineering. *Front. Bioeng. Biotechnol.* **2019**, *7*, 45. [CrossRef]
20. Korhonen, O.; Budtova, T. All-cellulose composite aerogels and cryogels. *Compos. Part A Appl. Sci. Manuf.* **2020**, *137*, 106027. [CrossRef]
21. Buchtová, N.; Pradille, C.; Bouvard, J.-L.; Budtova, T. Mechanical properties of cellulose aerogels and cryogels. *Soft Matter* **2019**, *15*, 7901–7908. [CrossRef] [PubMed]
22. Baimenov, A.; Berillo, D.A.; Poulopoulos, S.G.; Inglezakis, V.J. A review of cryogels synthesis, characterization and applications on the removal of heavy metals from aqueous solutions. *Adv. Colloid Interface Sci.* **2020**, *276*, 102088. [CrossRef] [PubMed]
23. Eggermont, L.J.; Rogers, Z.J.; Colombani, T.; Memic, A.; Bencherif, S.A. Injectable cryogels for biomedical applications. *Trends Biotechnol.* **2020**, *38*, 418–431. [CrossRef] [PubMed]
24. Memic, A.; Colombani, T.; Eggermont, L.J.; Rezaeeyazdi, M.; Steingold, J.; Rogers, Z.J.; Navare, K.J.; Mohammed, H.S.; Bencherif, S.A. Latest advances in cryogel technology for biomedical applications. *Adv. Ther.* **2019**, *2*, 1800114. [CrossRef]
25. Ferreira, F.V.; Otoni, C.G.; Kevin, J.; Barud, H.S.; Lona, L.M.; Cranston, E.D.; Rojas, O.J. Porous nanocellulose gels and foams: Breakthrough status in the development of scaffolds for tissue engineering. *Mater. Today* **2020**, *37*, 126–141. [CrossRef]
26. Trygg, J.; Fardim, P.; Gericke, M.; Mäkilä, E.; Salonen, J. Physicochemical design of the morphology and ultrastructure of cellulose beads. *Carbohydr. Polym.* **2013**, *93*, 291–299. [CrossRef]
27. Pääkkö, M.; Vapaavuori, J.; Silvennoinen, R.; Kosonen, H.; Ankerfors, M.; Lindström, T.; Berglund, L.A.; Ikkala, O. Long and entangled native cellulose i nanofibers allow flexible aerogels and hierarchically porous templates for functionalities. *Soft Matter* **2008**, *4*, 2492–2499. [CrossRef]
28. Buchtova, N.; Budtova, T. Cellulose aero-, cryo-and xerogels: Towards understanding of morphology control. *Cellulose* **2016**, *23*, 2585–2595. [CrossRef]
29. Saylan, Y.; Denizli, A. Supermacroporous composite cryogels in biomedical applications. *Gels* **2019**, *5*, 20. [CrossRef]
30. Dong, H.; Xie, Y.; Zeng, G.; Tang, L.; Liang, J.; He, Q.; Zhao, F.; Zeng, Y.; Wu, Y. The dual effects of carboxymethyl cellulose on the colloidal stability and toxicity of nanoscale zero-valent iron. *Chemosphere* **2016**, *144*, 1682–1689. [CrossRef]
31. Ganesan, K.; Dennstedt, A.; Barowski, A.; Ratke, L. Design of aerogels, cryogels and xerogels of cellulose with hierarchical porous structures. *Mater. Des.* **2016**, *92*, 345–355. [CrossRef]
32. García-González, C.; Alnaief, M.; Smirnova, I. Polysaccharide-based aerogels—Promising biodegradable carriers for drug delivery systems. *Carbohydr. Polym.* **2011**, *86*, 1425–1438. [CrossRef]
33. Ciolacu, D.; Rudaz, C.; Vasilescu, M.; Budtova, T. Physically and chemically cross-linked cellulose cryogels: Structure, properties and application for controlled release. *Carbohydr. Polym.* **2016**, *151*, 392–400. [CrossRef] [PubMed]
34. Budtova, T. Cellulose ii aerogels: A review. *Cellulose* **2019**, *26*, 81–121. [CrossRef]
35. Abdul Khalil, H.; Adnan, A.; Yahya, E.B.; Olaiya, N.; Safrida, S.; Hossain, M.; Balakrishnan, V.; Gopakumar, D.A.; Abdullah, C.; Oyekanmi, A. A review on plant cellulose nanofibre-based aerogels for biomedical applications. *Polymers* **2020**, *12*, 1759. [CrossRef]
36. Klemm, D.; Philpp, B.; Heinze, T.; Heinze, U.; Wagenknecht, W. *Comprehensive Cellulose Chemistry: Fundamentals and Analytical Methods*; Wiley-VCH Verlag GmbH: Weinheim, Germany, 1998; Volume 1.
37. Koyama, M.; Helbert, W.; Imai, T.; Sugiyama, J.; Henrissat, B. Parallel-up structure evidences the molecular directionality during biosynthesis of bacterial cellulose. *Proc. Natl. Acad. Sci. USA* **1997**, *94*, 9091–9095. [CrossRef]
38. Hayashi, J.; Sufoka, A.; Ohkita, J.; Watanabe, S. The confirmation of existences of cellulose iiii, iiiii, ivi, and ivii by the X-ray method. *J. Polym. Sci. Polym. Lett. Ed.* **1975**, *13*, 23–27. [CrossRef]
39. Gardiner, E.S.; Sarko, A. Packing analysis of carbohydrates and polysaccharides. 16. The crystal structures of celluloses ivi and ivii. *Can. J. Chem.* **1985**, *63*, 173–180. [CrossRef]

40. Atalla, R.H.; Vanderhart, D.L. Native cellulose: A composite of two distinct crystalline forms. *Science* **1984**, *223*, 283–285. [CrossRef]
41. Nisizawa, K. Mode of action of cellulases. *J. Ferment. Technol.* **1973**, *51*, 267–304.
42. Budtova, T.; Navard, P. Cellulose in naoh–water based solvents: A review. *Cellulose* **2016**, *23*, 5–55. [CrossRef]
43. Roy, C.; Budtova, T.; Navard, P. Rheological properties and gelation of aqueous cellulose—Naoh solutions. *Biomacromolecules* **2003**, *4*, 259–264. [CrossRef] [PubMed]
44. Heinze, T.; Koschella, A. Solvents applied in the field of cellulose chemistry: A mini review. *Polímeros* **2005**, *15*, 84–90. [CrossRef]
45. Henniges, U.; Schiehser, S.; Rosenau, T.; Potthast, A. Cellulose solubility: Dissolution and analysis of "problematic"cellulose pulps in the solvent system DMAc/LiCl. *ACS Symp. Ser.* **2010**, *1033*, 165–177.
46. Rosenau, T.; Potthast, A.; Adorjan, I.; Hofinger, A.; Sixta, H.; Firgo, H.; Kosma, P. Cellulose solutions in n-methylmorpholine-n-oxide (nmmo)–degradation processes and stabilizers. *Cellulose* **2002**, *9*, 283–291. [CrossRef]
47. Rosenau, T.; Potthast, A.; Sixta, H.; Kosma, P. The chemistry of side reactions and byproduct formation in the system nmmo/cellulose (lyocell process). *Prog. Polym. Sci.* **2001**, *26*, 1763–1837. [CrossRef]
48. Rosenau, T.; French, A.D. N-methylmorpholine-n-oxide (nmmo): Hazards in practice and pitfalls in theory. *Cellulose* **2021**, *28*, 5985–5990. [CrossRef]
49. Cuissinat, C.; Navard, P. Swelling and dissolution of cellulose part 1: Free floating cotton and wood fibres in n-methylmorpholine-n-oxide–water mixtures. *Macromol. Symp.* **2006**, *244*, 1–18. [CrossRef]
50. Hermanutz, F.; Gähr, F.; Uerdingen, E.; Meister, F.; Kosan, B. New developments in dissolving and processing of cellulose in ionic liquids. *Macromol. Symp.* **2008**, *262*, 23–27. [CrossRef]
51. Wang, H.; Gurau, G.; Rogers, R.D. Ionic liquid processing of cellulose. *Chem. Soc. Rev.* **2012**, *41*, 1519–1537. [CrossRef]
52. Pircher, N.; Carbajal, L.; Schimper, C.; Bacher, M.; Rennhofer, H.; Nedelec, J.-M.; Lichtenegger, H.C.; Rosenau, T.; Liebner, F. Impact of selected solvent systems on the pore and solid structure of cellulose aerogels. *Cellulose* **2016**, *23*, 1949–1966. [CrossRef] [PubMed]
53. Zhang, J.; Wu, J.; Yu, J.; Zhang, X.; He, J.; Zhang, J. Application of ionic liquids for dissolving cellulose and fabricating cellulose-based materials: State of the art and future trends. *Mater. Chem. Front.* **2017**, *1*, 1273–1290. [CrossRef]
54. Tyshkunova, I.V.; Chukhchin, D.G.; Gofman, I.V.; Poshina, D.N.; Skorik, Y.A. Cellulose cryogels prepared by regeneration from phosphoric acid solutions. *Cellulose* **2021**, *28*, 4975–4989. [CrossRef]
55. Egal, M.; Budtova, T.; Navard, P. Structure of aqueous solutions of microcrystalline cellulose/sodium hydroxide below 0 c and the limit of cellulose dissolution. *Biomacromolecules* **2007**, *8*, 2282–2287. [CrossRef]
56. Sehaqui, H.; Zhou, Q.; Berglund, L.A. High-porosity aerogels of high specific surface area prepared from nanofibrillated cellulose (nfc). *Compos. Sci. Technol.* **2011**, *71*, 1593–1599. [CrossRef]
57. Medronho, B.; Romano, A.; Miguel, M.G.; Stigsson, L.; Lindman, B. Rationalizing cellulose (in) solubility: Reviewing basic physicochemical aspects and role of hydrophobic interactions. *Cellulose* **2012**, *19*, 581–587. [CrossRef]
58. Yahya, E.B.; Alzalouk, M.M.; Alfallous, K.A.; Abogmaza, A.F. Antibacterial cellulose-based aerogels for wound healing application: A review. *Biomed. Res. Ther.* **2020**, *7*, 4032–4040. [CrossRef]
59. Savina, I.N.; Zoughaib, M.; Yergeshov, A.A. Design and assessment of biodegradable macroporous cryogels as advanced tissue engineering and drug carrying materials. *Gels* **2021**, *7*, 79. [CrossRef]
60. Shiekh, P.A.; Andrabi, S.M.; Singh, A.; Majumder, S.; Kumar, A. Designing cryogels through cryostructuring of polymeric matrices for biomedical applications. *Eur. Polym. J.* **2021**, *144*, 110234. [CrossRef]
61. Beaumont, M.; König, J.; Opietnik, M.; Potthast, A.; Rosenau, T. Drying of a cellulose ii gel: Effect of physical modification and redispersibility in water. *Cellulose* **2017**, *24*, 1199–1209. [CrossRef]
62. Haseley, P.; Oetjen, G.-W. *Freeze-Drying*, 3rd ed.; John Wiley & Sons: Newark, NJ, USA, 2017.
63. Sehaqui, H.; Salajková, M.; Zhou, Q.; Berglund, L.A. Mechanical performance tailoring of tough ultra-high porosity foams prepared from cellulose i nanofiber suspensions. *Soft Matter* **2010**, *6*, 1824–1832. [CrossRef]
64. Plieva, F.M.; Karlsson, M.; Aguilar, M.-R.; Gomez, D.; Mikhalovsky, S.; Galaev, I.Y. Pore structure in supermacroporous polyacrylamide based cryogels. *Soft Matter* **2005**, *1*, 303–309. [CrossRef] [PubMed]
65. Henderson, T.M.; Ladewig, K.; Haylock, D.N.; McLean, K.M.; O'Connor, A.J. Cryogels for biomedical applications. *J. Mater. Chem. B* **2013**, *1*, 2682–2695. [CrossRef]
66. Okay, O. *Polymeric Cryogels: Macroporous Gels with Remarkable Properties*; Springer: Berlin/Heidelberg, Germany, 2014; Volume 263.
67. Koshy, S.T.; Ferrante, T.C.; Lewin, S.A.; Mooney, D.J. Injectable, porous, and cell-responsive gelatin cryogels. *Biomaterials* **2014**, *35*, 2477–2487. [CrossRef] [PubMed]
68. Mackova, H.; Plichta, Z.k.; Hlidkova, H.; Sedláček, O.; Konefal, R.; Sadakbayeva, Z.; Duskova-Smrckova, M.; Horak, D.; Kubinova, S. Reductively degradable poly (2-hydroxyethyl methacrylate) hydrogels with oriented porosity for tissue engineering applications. *ACS Appl. Mater. Interfaces* **2017**, *9*, 10544–10553. [CrossRef]
69. Bölgen, N.; Plieva, F.; Galaev, I.Y.; Mattiasson, B.; Pişkin, E. Cryogelation for preparation of novel biodegradable tissue-engineering scaffolds. *J. Biomater. Sci. Polym. Ed.* **2007**, *18*, 1165–1179. [CrossRef]
70. Dispinar, T.; Van Camp, W.; De Cock, L.J.; De Geest, B.G.; Du Prez, F.E. Redox-responsive degradable peg cryogels as potential cell scaffolds in tissue engineering. *Macromol. Biosci.* **2012**, *12*, 383–394. [CrossRef]

71. Zhang, H.; Zhou, L.; Zhang, W. Control of scaffold degradation in tissue engineering: A review. *Tissue Eng. Part B Rev.* **2014**, *20*, 492–502. [CrossRef]
72. Petrov, P.D.; Tsvetanov, C.B. Cryogels via uv irradiation. *Polym. Cryogels* **2014**, *263*, 199–222.
73. Meena, L.K.; Raval, P.; Kedaria, D.; Vasita, R. Study of locust bean gum reinforced cyst-chitosan and oxidized dextran based semi-ipn cryogel dressing for hemostatic application. *Bioact. Mater.* **2018**, *3*, 370–384. [CrossRef]
74. Plieva, F.M.; Karlsson, M.; Aguilar, M.R.; Gomez, D.; Mikhalovsky, S.; Galaev, I.Y.; Mattiasson, B. Pore structure of macroporous monolithic cryogels prepared from poly (vinyl alcohol). *J. Appl. Polym. Sci.* **2006**, *100*, 1057–1066. [CrossRef]
75. Vrana, N.E.; Cahill, P.A.; McGuinness, G.B. Endothelialization of pva/gelatin cryogels for vascular tissue engineering: Effect of disturbed shear stress conditions. *J. Biomed. Mater. Res. Part A* **2010**, *94*, 1080–1090. [CrossRef] [PubMed]
76. Bernhardt, A.; Despang, F.; Lode, A.; Demmler, A.; Hanke, T.; Gelinsky, M. Proliferation and osteogenic differentiation of human bone marrow stromal cells on alginate–gelatine–hydroxyapatite scaffolds with anisotropic pore structure. *J. Tissue Eng. Regen. Med.* **2009**, *3*, 54–62. [CrossRef] [PubMed]
77. Beluns, S.; Gaidukovs, S.; Platnieks, O.; Gaidukova, G.; Mierina, I.; Grase, L.; Starkova, O.; Brazdausks, P.; Thakur, V.K. From wood and hemp biomass wastes to sustainable nanocellulose foams. *Ind. Crops Prod.* **2021**, *170*, 113780. [CrossRef]
78. Zhang, H.; Zhang, F.; Wu, J. Physically crosslinked hydrogels from polysaccharides prepared by freeze–thaw technique. *React. Funct. Polym.* **2013**, *73*, 923–928. [CrossRef]
79. Qin, X.; Lu, A.; Zhang, L. Gelation behavior of cellulose in naoh/urea aqueous system via cross-linking. *Cellulose* **2013**, *20*, 1669–1677. [CrossRef]
80. Chang, C.; Zhang, L.; Zhou, J.; Zhang, L.; Kennedy, J.F. Structure and properties of hydrogels prepared from cellulose in naoh/urea aqueous solutions. *Carbohydr. Polym.* **2010**, *82*, 122–127. [CrossRef]
81. Bagri, L.; Bajpai, J.; Bajpai, A. Cryogenic designing of biocompatible blends of polyvinyl alcohol and starch with macroporous architecture. *J. Macromol. Sci.®Part A Pure Appl. Chem.* **2009**, *46*, 1060–1068. [CrossRef]
82. Heath, L.; Thielemans, W. Cellulose nanowhisker aerogels. *Green Chem.* **2010**, *12*, 1448–1453. [CrossRef]
83. Martoïa, F.; Cochereau, T.; Dumont, P.J.; Orgéas, L.; Terrien, M.; Belgacem, M. Cellulose nanofibril foams: Links between ice-templating conditions, microstructures and mechanical properties. *Mater. Des.* **2016**, *104*, 376–391. [CrossRef]
84. Teraoka, I. *Polymer Solutions: An Introduction to Physical Properties*; John Wiley & Sons: Hoboken, NJ, USA, 2002.
85. Ceylan, S.; Demir, D.; Gül, G.; Bölgen, N. Effect of polymer concentration in cryogelation of gelatin and poly (vinyl alcohol) scaffolds. *Biomater. Biomech. Bioeng.* **2019**, *4*, 1–8.
86. Tripathi, A.; Kathuria, N.; Kumar, A. Elastic and macroporous agarose–gelatin cryogels with isotropic and anisotropic porosity for tissue engineering. *J. Biomed. Mater. Res. Part A Off. J. Soc. Biomater. Jpn. Soc. Biomater. Aust. Soc. Biomater. Korean Soc. Biomater.* **2009**, *90*, 680–694. [CrossRef] [PubMed]
87. Van Vlierberghe, S.; Dubruel, P.; Lippens, E.; Cornelissen, M.; Schacht, E. Correlation between cryogenic parameters and physico-chemical properties of porous gelatin cryogels. *J. Biomater. Sci. Polym. Ed.* **2009**, *20*, 1417–1438. [CrossRef] [PubMed]
88. Tyshkunova, I.V.; Chukhchin, D.G.; Gofman, I.V.; Pavlova, E.N.; Ushakov, V.A.; Vlasova, E.N.; Poshina, D.N.; Skorik, Y.A. Chitin cryogels prepared by regeneration from phosphoric acid solutions. *Materials* **2021**, *14*, 5191. [CrossRef] [PubMed]
89. Dinu, M.V.; Ozmen, M.M.; Dragan, E.S.; Okay, O. Freezing as a path to build macroporous structures: Superfast responsive polyacrylamide hydrogels. *Polymer* **2007**, *48*, 195–204. [CrossRef]
90. Hwang, Y.; Zhang, C.; Varghese, S. Poly (ethylene glycol) cryogels as potential cell scaffolds: Effect of polymerization conditions on cryogel microstructure and properties. *J. Mater. Chem.* **2010**, *20*, 345–351. [CrossRef]
91. Gang, E.J.; Jeong, J.A.; Hong, S.H.; Hwang, S.H.; Kim, S.W.; Yang, I.H.; Ahn, C.; Han, H.; Kim, H. Skeletal myogenic differentiation of mesenchymal stem cells isolated from human umbilical cord blood. *Stem Cells* **2004**, *22*, 617–624. [CrossRef]
92. Ozmen, M.M.; Dinu, M.V.; Dragan, E.S.; Okay, O. Preparation of macroporous acrylamide-based hydrogels: Cryogelation under isothermal conditions. *J. Macromol. Sci. Part A Pure Appl. Chem.* **2007**, *44*, 1195–1202. [CrossRef]
93. Erlandsson, J.; Pettersson, T.; Ingverud, T.; Granberg, H.; Larsson, P.A.; Malkoch, M.; Wågberg, L. On the mechanism behind freezing-induced chemical crosslinking in ice-templated cellulose nanofibril aerogels. *J. Mater. Chem. A* **2018**, *6*, 19371–19380. [CrossRef]
94. Chen, W.; Yu, H.; Li, Q.; Liu, Y.; Li, J. Ultralight and highly flexible aerogels with long cellulose i nanofibers. *Soft Matter* **2011**, *7*, 10360–10368. [CrossRef]
95. Lozinsky, V.; Vainerman, E.; Ivanova, S.; Titova, E.; Shtil'man, M.; Belavtseva, E.; Rogozhin, S. Study of cryostructurization of polymer systems. Vi. The influence of the process temperature on the dynamics of formation and structure of cross-linked polyacrylamide cryogels. *Acta Polym.* **1986**, *37*, 142–146. [CrossRef]
96. Ivanov, R.V.; Lozinsky, V.I.; Noh, S.K.; Lee, Y.R.; Han, S.S.; Lyoo, W.S. Preparation and characterization of polyacrylamide cryogels produced from a high-molecular-weight precursor. Ii. The influence of the molecular weight of the polymeric precursor. *J. Appl. Polym. Sci.* **2008**, *107*, 382–390. [CrossRef]
97. Otoni, C.G.; Figueiredo, J.S.; Capeletti, L.B.; Cardoso, M.B.; Bernardes, J.S.; Loh, W. Tailoring the antimicrobial response of cationic nanocellulose-based foams through cryo-templating. *ACS Appl. Bio Mater.* **2019**, *2*, 1975–1986. [CrossRef] [PubMed]
98. Li, G.; Nandgaonkar, A.G.; Habibi, Y.; Krause, W.E.; Wei, Q.; Lucia, L.A. An environmentally benign approach to achieving vectorial alignment and high microporosity in bacterial cellulose/chitosan scaffolds. *RSC Adv.* **2017**, *7*, 13678–13688. [CrossRef]

99. He, X.; Yao, K.; Shen, S.; Yun, J. Chemical engineering science: Freezing characteristics of acrylamide-based aqueous solution used for the preparation of supermacroporous cryogels via cryo-copolymerization. *Cell. Polym.* **2007**, *26*, 145–147.
100. Elsherbiny, D.A.; Abdelgawad, A.M.; El-Naggar, M.E.; El-Sherbiny, R.A.; El-Rafie, M.H.; El-Sayed, I.E.-T. Synthesis, antimicrobial activity, and sustainable release of novel α-aminophosphonate derivatives loaded carrageenan cryogel. *Int. J. Biol. Macromol.* **2020**, *163*, 96–107. [CrossRef]
101. Li, J.; Wang, Y.; Zhang, L.; Xu, Z.; Dai, H.; Wu, W. Nanocellulose/gelatin composite cryogels for controlled drug release. *ACS Sustain. Chem. Eng.* **2019**, *7*, 6381–6389. [CrossRef]
102. Lozinsky, V.; Galaev, I.Y.; Plieva, F.M.; Savina, I.N.; Jungvid, H.; Mattiasson, B. Polymeric cryogels as promising materials of biotechnological interest. *Trends Biotechnol.* **2003**, *21*, 445–451. [CrossRef]
103. Mikhalovsky, S.; Savina, I.; Dainiak, M.; Ivanov, A.; Galaev, I. *5.03-Biomaterials/Cryogels*, 2nd ed.; Moo-Young, M., Ed.; Academic Press: Burlington, MA, USA, 2011; pp. 11–22.
104. Chen, W.; Yuan, S.; Shen, J.; Chen, Y.; Xiao, Y. A composite hydrogel based on pectin/cellulose via chemical cross-linking for hemorrhage. *Front. Bioeng. Biotechnol.* **2020**, *8*, 627351. [CrossRef]
105. Bozova, N.; Petrov, P.D. Highly elastic super-macroporous cryogels fabricated by thermally induced crosslinking of 2-hydroxyethylcellulose with citric acid in solid state. *Molecules* **2021**, *26*, 6370.
106. Petrov, P.; Mokreva, P.; Kostov, I.; Uzunova, V.; Tzoneva, R. Novel electrically conducting 2-hydroxyethylcellulose/polyaniline nanocomposite cryogels: Synthesis and application in tissue engineering. *Carbohydr. Polym.* **2016**, *140*, 349–355. [CrossRef] [PubMed]
107. Serex, L.; Braschler, T.; Filippova, A.; Rochat, A.; Béduer, A.; Bertsch, A.; Renaud, P. Pore size manipulation in 3D printed cryogels enables selective cell seeding. *Adv. Mater. Technol.* **2018**, *3*, 1700340. [CrossRef]
108. Odabas, S. Collagen–carboxymethyl cellulose–Tricalcium phosphate multi-lamellar cryogels for tissue engineering applications: Production and characterization. *J. Bioact. Compat. Polym.* **2016**, *31*, 411–422. [CrossRef]
109. Santos, G.d.S.d.; Santos, N.R.R.d.; Pereira, I.C.S.; Andrade, A.J.d.; Lima, E.M.B.; Minguita, A.P.; Rosado, L.H.G.; Moreira, A.P.D.; Middea, A.; Prudencio, E.R. Layered cryogels laden with brazilian honey intended for wound care. *Polímeros* **2020**, *30*, e2020031. [CrossRef]
110. Cai, H.; Sharma, S.; Liu, W.; Mu, W.; Liu, W.; Zhang, X.; Deng, Y. Aerogel microspheres from natural cellulose nanofibrils and their application as cell culture scaffold. *Biomacromolecules* **2014**, *15*, 2540–2547. [CrossRef]
111. Liu, J.; Cheng, F.; Grénman, H.; Spoljaric, S.; Seppälä, J.; Eriksson, J.E.; Willför, S.; Xu, C. Development of nanocellulose scaffolds with tunable structures to support 3D cell culture. *Carbohydr. Polym.* **2016**, *148*, 259–271. [CrossRef]
112. Ferreira, F.V.; Souza, L.; Martins, T.M.M.; Lopes, J.H.; Mattos, B.D.; Mariano, M.; Pinheiro, I.F.; Valverde, T.M.; Livi, S.; Camilli, J.A.; et al. Nanocellulose/bioactive glass cryogels as scaffolds for bone regeneration. *Nanoscale* **2019**, *11*, 19842–19849. [CrossRef]
113. Ghafari, R.; Jonoobi, M.; Amirabad, L.M.; Oksman, K.; Taheri, A.R. Fabrication and characterization of novel bilayer scaffold from nanocellulose based aerogel for skin tissue engineering applications. *Int. J. Biol. Macromol.* **2019**, *136*, 796–803. [CrossRef]
114. Zhang, F.; Wu, W.; Zhang, X.; Meng, X.; Tong, G.; Deng, Y. Temperature-sensitive poly-nipam modified cellulose nanofibril cryogel microspheres for controlled drug release. *Cellulose* **2016**, *23*, 415–425. [CrossRef]
115. Guiza, K.; Ben Arfi, R.; Mougin, K.; Vaulot, C.; Michelin, L.; Josien, L.; Schrodj, G.; Ghorbal, A. Development of novel and ecological keratin/cellulose-based composites for absorption of oils and organic solvents. *Environ. Sci. Pollut. Res.* **2021**, *28*, 46655–46668. [CrossRef]
116. Kunz Lazzari, L.d.; Perondi, D.; Zattera, A.J.; Campomanes Santana, R.M. Cellulose/biochar cryogels: A study of adsorption kinetics and isotherms. *Langmuir* **2021**, *37*, 3180–3188. [CrossRef] [PubMed]
117. Toledo, P.V.; Martins, B.F.; Pirich, C.L.; Sierakowski, M.R.; Neto, E.T.; Petri, D.F. Cellulose based cryogels as adsorbents for organic pollutants. *Macromol. Symp.* **2019**, *383*, 1800013. [CrossRef]
118. Tian, Y.; Zhang, X.; Feng, X.; Zhang, J.; Zhong, T. Shapeable and underwater super-elastic cellulose nanofiber/alginate cryogels by freezing-induced oxa-michael reaction for efficient protein purification. *Carbohydr. Polym.* **2021**, *272*, 118498. [CrossRef] [PubMed]
119. Lazzari, L.K.; Zampieri, V.B.; Zanini, M.; Zattera, A.J.; Baldasso, C. Sorption capacity of hydrophobic cellulose cryogels silanized by two different methods. *Cellulose* **2017**, *24*, 3421–3431. [CrossRef]

International Journal of *Molecular Sciences*

MDPI

Review

Cyclodextrin-Containing Hydrogels: A Review of Preparation Method, Drug Delivery, and Degradation Behavior

Jiayue Liu [1,†], Bingren Tian [2,†], Yumei Liu [2,*] and Jian-Bo Wan [1,*]

1 State Key Laboratory of Quality Research in Chinese Medicine, Institute of Chinese Medical Sciences, University of Macau, Macao 999078, China; liujiayue0331@163.com
2 School of Chemical Engineering and Technology, Xinjiang University, Urumqi 830046, China; tianbingren1@163.com
* Correspondence: xjdxlym@xju.edu.cn (Y.L.); jbwan@um.edu.mo (J.-B.W.); Tel.: +853-88224680 (J.-B.W)
† These authors contributed equally to this work and should be considered co-first authors.

Abstract: Hydrogels possess porous structures, which are widely applied in the field of materials and biomedicine. As a natural oligosaccharide, cyclodextrin (CD) has shown remarkable application prospects in the synthesis and utilization of hydrogels. CD can be incorporated into hydrogels to form chemically or physically cross-linked networks. Furthermore, the unique cavity structure of CD makes it an ideal vehicle for the delivery of active ingredients into target tissues. This review describes useful methods to prepare CD-containing hydrogels. In addition, the potential biomedical applications of CD-containing hydrogels are reviewed. The release and degradation process of CD-containing hydrogels under different conditions are discussed. Finally, the current challenges and future research directions on CD-containing hydrogels are presented.

Keywords: cyclodextrin; hydrogel; preparation; release; degradation

Citation: Liu, J.; Tian, B.; Liu, Y.; Wan, J.-B. Cyclodextrin-Containing Hydrogels: A Review of Preparation Method, Drug Delivery, and Degradation Behavior. *Int. J. Mol. Sci.* **2021**, *22*, 13516. https://doi.org/10.3390/ijms222413516

Academic Editor: Yury A. Skorik

Received: 6 November 2021
Accepted: 14 December 2021
Published: 16 December 2021

1. Introduction

Many studies in the material science field have explored diverse materials that could improve drug delivery, thus ensuring effective treatment of various diseases [1,2]. In the past few decades, numerous materials have been synthesized and applied in drug delivery systems [3,4]. Particularly, several clinical trials have explored the development of hydrogels and their potential applications [5]. The physical properties of hydrogels are biocompatible to organisms. Therefore, they have become indispensable materials in several biomedical applications [6]. Hydrophilic groups present in the hydrogel network enable the hydrogel to interact with water molecules in the surrounding environment. As a result, the hydrogel can absorb large amounts of water to maintain its structure and viscoelasticity. The volume changes of hydrogels can be controlled by changing parameters such as composite molecules and cross-linking density in different environments. Hydrogels are mainly synthesized in the water phase, therefore the introduction of cross-linking points throughout the hydrogel network is important in preventing the hydrogel from dissolving. According to the different cross-linking properties in the hydrogel network, hydrogels are mainly divided into two categories, including physical and chemical cross-linked hydrogels [7,8]. Physically cross-linked hydrogels are formed through physical interactions between the polymers that form hydrogels. On the other hand, chemically cross-linked hydrogels are formed by covalent bonds.

These properties enable the use of hydrogels as macromolecular platforms in drug delivery, wound healing dressings, and implant coatings [9–11]. Previous studies developed controllable, localized drug release systems for hydrogel systems [12,13]. Commonly used methods for drug loading include forming unstable chemical bonds to covalently bind drug molecules to the hydrogel matrix or employing non-covalent methods to encapsulate drug molecules into the hydrogel. Although the hydrogel can effectively control the release time when the drug is loaded through a covalent method, the synthesis process and the loading amount of the drug are not satisfactory [14,15]. On the contrary, physically trapping drug molecules into the hydrogel is simpler, and results in a hydrogel with a higher drug loading capacity. Drugs are mainly loaded to hydrogels through hydrophobic, electrostatic, or hydrogen bond interactions. However, most hydrogels are hydrophilic in nature, therefore, loading some hydrophobic drugs is not effective. Bringing hydrophobic regions into the hydrogel network can increase drug loading capacity and reduce the burst effect of the drug at the initial stage of entering the body [16–18].

Cyclodextrin (CD) is a cyclic oligosaccharide consisting of glucopyranoside units linked through α-1,4 glycosidic bonds obtained [19]. Structural analysis shows that CD is characterized by external hydrophilicity and internal hydrophobicity. Due to its unique cavity structure, hydrophobic molecules can be loaded to form an inclusion complex in dynamic equilibrium [19]. Because of several hydroxyl groups existing in the external structure, CD could form physically cross-linked hydrogel through intermolecular forces. In addition, CD can be connected to form a chemically cross-linked hydrogel network through covalent bonds. Several studies have explored these two different types of CD-containing hydrogels [2,20].

Although CD-containing hydrogels are still in the basic research stage, they have prominent advantages in different applications [21–26] and remarkable potential in biomedical applications, thus improving human health (Table 1). This review is aimed to discuss and summarize the development of CD-containing hydrogels in drug delivery. The traits of hydrogels are classified and discussed based on the different preparation methods. Furthermore, the potential applications of these hydrogels are summarized. In addition, the release rate of drugs and degradation of hydrogels are explored. Further, we summarize prospects for the future development of CD-containing hydrogels based on previous research findings.

Table 1. Comparison of cyclodextrin-containing hydrogels with other types of hydrogels.

Material for Forming Hydrogel	Cyclodextrin	Chitosan	Cellulose	Alginic Acid	Gum Arabic	Polyacryl Amide	Polyvinyl Alcohol
Source	Starch	Chitin	Plant	Alga	Acacia trees	Acrylonitrile	Vinyl acetate
Connection type	α-1, 4-glycosidic bond	β-1, 4-glycosidic bond	β-1, 4-glycosidic bond	1, 4-glycosidic bond	-	-	-
Techniques	Radical polymerization; Click reaction; Nucleophilic addition/ substitution	Photo- polymerization; Thermal polymerization	Chemical crosslinking; Free-radical polymerization; Grafting; Freeze-thaw	Enzymatically crosslinking; Chemical crosslinking	Photo-induced radical polymerization	Radiation-induced	Freeze-thaw
Kinds of drug delivery	Hydrophobic drug	Small molecules; peptides; proteins	Small molecules; peptides; proteins	Traditional low-molecular-weight drugs and macromolecules	Small molecules; proteins	Small molecules; peptides; proteins	Small molecules; peptides; proteins
Clinic trial	Yes	Yes	Yes	Yes	No	Yes	Yes
Ref.	[27–29]	[30,31]	[32,33]	[34,35]	[36]	[37,38]	[39]

2. Preparation Methods of CD-Containing Hydrogels

Studies on the use of the unique cavity of CD to encapsulate and deliver drugs are ongoing [40–42]. CD hydrogels can be divided into two groups including physically crosslinked hydrogels and chemically cross-linked hydrogels (Figure 1).

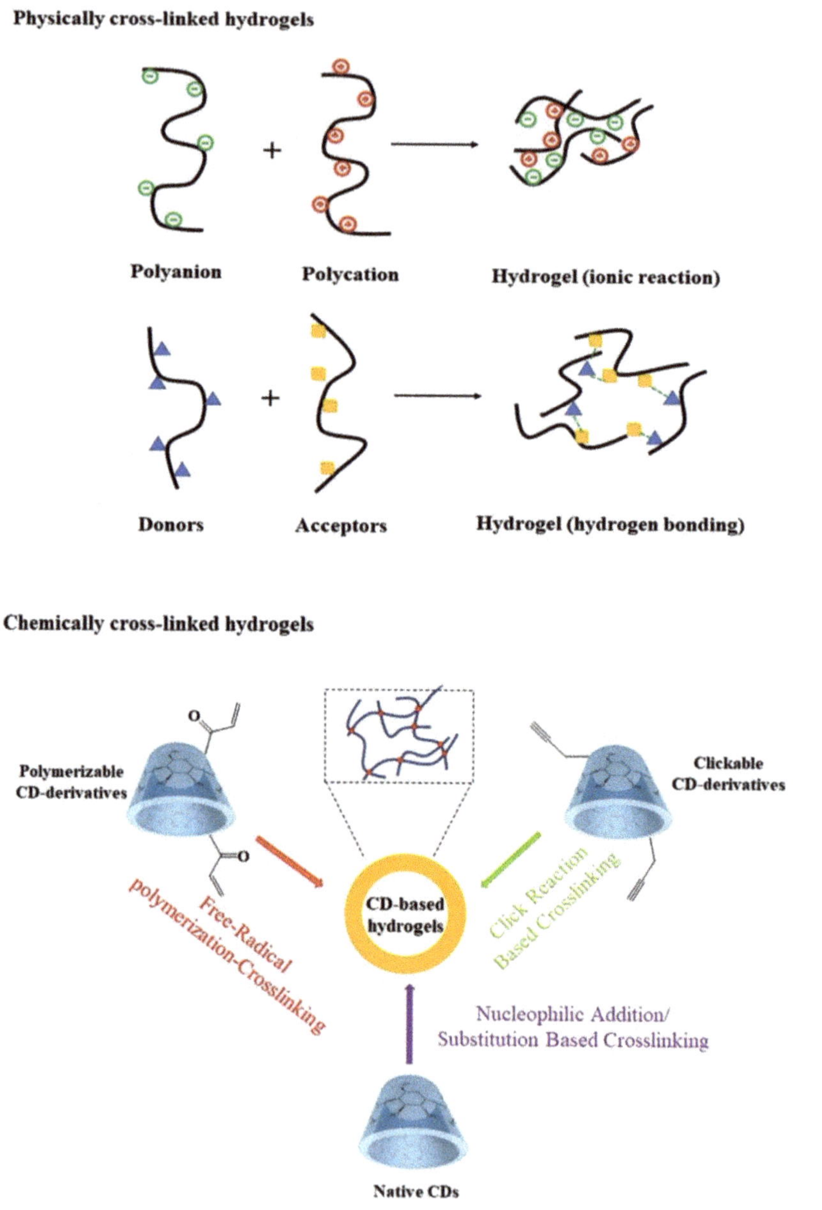

Figure 1. Schematic representation of common methodologies for preparation of CD-containing hydrogels.

Table 2 summarizes some examples of the preparation methods of physically cross-linked or chemically cross-linked cyclodextrin-containing hydrogels. Although physical hydrogels are non-toxic, they still have some shortcomings, such as low mechanical strength, and their pore size cannot be easily adjusted [43]. Physically cross-linked hydrogels are stable enough to prevent them from dissolving in water [44,45]. The methods involved in chemical cross-linking include free radical polymerization cross-linking-based methods; nucleophilic addition/substitution-based methods; cross-linking methods based on 'click' reactions and incorporation of CDs through post-gelation attachment [2]. The chemical activity produces permanent hydrogels through the covalent interaction of polymer and crosslinker functional groups. Polymerization into a hydrogel network produces fine-tuned hydrogels through chain growth, addition, or condensation reactions [46,47].

Table 2. The preparation methods of cyclodextrin-containing hydrogels.

Types	Matrix	Preparation Methods	Characteristic	Ref.
Physically cross-linked cyclodextrin-containing hydrogels	Chitosan	Casting method	Bilayer hydrogels	[48]
	Chitosan	Freeze-thaw cycling method	pH sensitivity	[49]
	Chitosan	Freezing method	Thermosensitive; Shortly gelation time (3 min or less)	[50]
	Chitosan/Poly(Vinyl Alcohol)	Dry at room temperature in vacuo	pH-specific release behavior	[51]
	Gellan gum	Gelation at room temperature	Biocompatible material	[52]
	Hydroxypropyl methylcellulose	Dispersion method	Benefit for skin	[53]
	Nanocellulose	30 min with autoclaving (121 °C, 103 kPa)	Sustained release	[54]
	poly (vinyl alcohol)	Freezing drying method	Long-term release	[55]
	Soy soluble polysaccharide	Reduced pressure and stored in a desiccator	3D-nanocomposites, superabsorbent, malleable, bioadhesive	[56]
	Xanthan	Freezing drying method	Long-term release	[57]
Chemically cross-linked cyclodextrin-containing hydrogels	4-arm-Polyethylene glycol-Succinimidyl Glutarate	Nucleophilic substitution-based method	Improved the therapeutic effect	[58]
	Agarose gel	Nucleophilic substitution-based method	Low gelling temperature for controlled drug delivery	[59]
	Carboxymethyl cellulose	Free radical polymerization crosslinking-based method	pH-responsive behaviour	[60]
	Carboxymethyl cellulose	Nucleophilic substitution-based method	Biocompatible, capable of controlling the release for a long duration	[61]
	Chitosan	Nucleophilic substitution-based method	Local antibiotic release	[62]
	Nanocellulose	Nucleophilic substitution-based method	Cell compatibility, non-cytotoxicity	[63]
	Polyvinyl alcohol	Nucleophilic substitution-based method	Good strength, elasticity, WVP, and swelling ability	[64]
	Poly(*N*-isopropylacrylamide)	Free radical polymerization crosslinking-based method	Thermoresponsive	[65]
	Poly(2-hydroxyethyl methacrylate)	Nucleophilic substitution-based method	Sustained drug delivery	[66]
	Sodium hyaluronan	Nucleophilic substitution-based method	Controlled release	[67]

In addition, hydroxyl groups occur at different positions in the CD molecule, thus they have different effects on the formed derivatives [68]. CD formed from glucopyranose has two types of hydroxyl groups, one is the primary hydroxyl group at the 6- position and the other is the secondary hydroxyl group at the 2- and 3- positions [69]. On the other hand, the primary hydroxyl group outside the cyclodextrin is free to move. In addition, the acidic and basic properties of the three hydroxyl groups are significantly different. The hydroxyl group at the 6- position is basic; the hydroxyl group at the 2- position is acidic (pKa = 12.1), whereas the 3- hydroxyl group is not easily modified. Therefore, the chemical modification process of CD is affected by the nucleophilicity of the hydroxyl group and the modification reagent. Under normal reaction conditions, the 6-position hydroxyl group is the most active and easily participates in the reaction after attack by electrophiles. Electrophiles can also attack other positions, including the less popular hydroxyl groups (2- and 3-). More than 40% of the compounds have low water-solubility properties. In addition, some compounds have poor stability or poor taste [70–72]. CD has become the ideal material to solve solubility, light stability, and poor taste limitations. Different CDs have different sizes of hydrophobic cavities, so it is important to choose specific CD derivatives based on the molecular size of the target molecule [73]. Owing to the low solubility of parent CD, its further application in medicine was limited. The preparation of water-soluble derivatives is important for improving the application of CD in drug delivery systems. Therefore, different CD-based derivatives have been developed to improve water solubility and functionality of natural CD.

3. The Promising Application of CD-Containing Hydrogels for Drug Delivery

In this section, we classify and discuss different types of CD-containing hydrogels based on the preparation method (Table 3). The release behavior of drugs after the introduction of different CD into hydrogels will be explored. Besides, the potential application of the prepared physical/chemical CD-containing hydrogel in the medical field will be discussed.

Table 3. Comparison of the properties of CD-containing hydrogels loaded with drugs [74–79].

Property	Physical Cross-Linked Hydrogel	Chemical Cross-Linked Hydrogel
Size of guest molecules	Small molecules (lipophilic)	Small molecules (lipophilic)
Drug loading strategies	Encapsulation	Encapsulation
Drug release speed	Can be controlled	Can be controlled
Drug release possible mechanisms	External stimulus; competition of external molecules	External stimulus; competition of external molecule
Duration times	Hours to days	Days to months
Drug delivery characteristic	High drug loading effectivity; low chance of drug deactivation	High drug loading effectivity; low chance of drug deactivation
Potential application	Drug delivery systems, injectable, wound dressings	Transdermal drug delivery, injectable, implantable, oral/ophthalmic drug carrier
Advantages	Non-toxic; cross-linking is reversible	Strong mechanical strength; the pore size can be adjusted; the variety of synthesis methods; difficult to degrade
Disadvantages	Low mechanical strength; difficult to adjust the pore size	Potentially toxic; no cross-linking is reversible

3.1. *Physically Cross-Linked CD-Containing Hydrogels*

The physically cross-linked hydrogel formed by intermolecular interaction does not cause damage to the environment during the preparation process [49,80]. Therefore, physically cross-linked hydrogels have wide applications in the medical field. This section mainly summarizes and analyzes potential applications of physically cross-linked CD-containing hydrogels loaded with different drugs in different medical fields (Table 4). Because in the process of preparation, the physically cross-linked CD-containing hydrogels don't involve the chemical cross-linking agents and chemical solvents, so physical gelatinization and drug encapsulation simultaneously have important research and application value, especially in the aspect of oral administration medicine, ocular delivery system, wound dressing materials, local drug delivery system to treat cancer, etc.

Table 4. Potential application of physically cross-linked CD-containing hydrogels.

No.	Drug	Formation Materials	Hydrogel State	Type of Cells	Summary	Potential Application	Ref.
1	Berberine hydrochloride	β-CD; Bacterial cellulose;	Nano-particle	*S. aureus; P. aeruginosa; E. coli*	The ultra-fine network of bacterial cellulose resulted in different release characteristics of berberine hydrochloride. The drug-loaded hydrogel had a good antibacterial effect as revealed by in vitro experiments.	Oral administration medicine	[54]
2	Chlorhexidine	β-CD; NaCl; $NaHCO_3$; $CaCl_2$	Contact lenses	*E. coli*	β-CD in eye drops significantly enhanced the delivery of chlorhexidine into the cornea.	Ocular delivery	[81]
3	Coumestrol	Hydroxypropyl-β-CD; methylcellu-lose	Not mentioned	Animals	Hydrogel has high efficacy in wound healing when compared to Dersani, with 50% wound healing achieved within a shorter period compared to this positive control.	Wound dressing materials	[53]
4	Curcumin	Hydroxypropyl-β-CD; silver nanoparticles; bacterial cellulose	Film	*P. aeruginosa; S. aureus; C. aureus*; Panc 1, U251, MSTO	The nano-silver particles loaded into the bacterial cellulose hydrogel showed high cytocompatibility and therapeutic effects against three common wound infection pathogens.	Wound dressing materials	[82]
5	Curcumin	β-CD; Polyvinyl alcohol	Film	Glioblastoma cell line C6; melanoma cell line B16F10; astrocyte cells	The hydrogel controlled the release of curcumin (48 h, 85% release). The polymer membrane had higher cytotoxicity than curcumin. The drug-loaded hydrogel showed prolonged cytotoxic effects (up to 96 h) at a lower concentration (50 μg/mL).	Local drug delivery system to treat cancer	[83]
6	Curcumin	2-hydroxypropyl-β-CD; sodium alginate; chitosan	Film	*E. coli; S. aureus*; NCTC clone 929 cells; NHDF cells	High concentration of crosslinking agent concentration improved the mechanical properties of the hydrogel and decreased the hygroscopicity, water swelling, and weight loss. In addition, hydrogel showed a slow-release effect ($t > 50$ h). Curcumin-loaded double-layer hydrogel effectively treated *E. coli* and *S. aureus*. The double-layer hydrogel was not toxic to NCTC clone 929 cells and normal human dermal fibroblasts.	Wound dressing materials	[48]
7	Gallic acid	Hydroxypropyl-β-CD; bacterial cellulose; poly (vinyl alcohol)	Not mentioned	Not mentioned	The swelling properties during encapsulation were inferior. The release profile of the complex was slower compared with gallic acid.	Pharmaceutical and cosmetic products	[55]
8	Honey bee propolis extract	β-CD; κ-Carrageenan	Not mentioned	*S. aureus; P. aeruginosa; Aspergillus Flavus; Candida albicans*	Higher active compound concentration ensures sustained in vitro release.	Wound dressing	[80]
9	Levofloxacin; methotrexate	Hydroxypropyl-β-CD; xanthan gum	Film	*E. coli; S. aureus*	The hydrogel loaded with the methotrexate showed a well-controlled release profile ($t > 600$ min). The hydrogel loaded with levofloxacin had a good antibacterial effect.	Drug delivery system	[57]

Table 4. *Cont.*

No.	Drug	Formation Materials	Hydrogel State	Type of Cells	Summary	Potential Application	Ref.
10	Red thyme oil	γ-CD; polyvinyl alcohol; chitosan; clinoptilolite	Film	L929 cells	Hydrogels with clinoptilolite contained characteristics such as compressed structure, improved mechanical properties, decreased swelling values, and reduced release rate of the drug. In addition, prepared hydrogels were low-toxic based on MTT assay.	Drug delivery systems and wound dressings	[84]
11	Thyme oil	Methyl-β-CD; hydroxypropyl-β-CD; γ-CD; chitosan; polyvinyl alcohol	Film	*E. coli*; *S. aureus*	The water vapor transmission rate of the hydrogel was appropriate for application in wound dressing. The swelling degree of hydrogel loaded with thyme oil varied with the pH. The hydrogels containing γ-CD had good antibacterial activity.	Wound dressings	[49, 85]

Sajeesh et al. used chitosan, methacrylic acid, and polyethylene glycol to prepare hydrogels, and then added insulin-loaded methyl-β-CD into the hydrogels [86]. Final hydrogel particles were formed through the interaction between CD and the hydrogel matrix. The encapsulation efficiency of the inclusion compound and insulin encapsulated by hydrogel microparticles were evaluated. No significant differences were observed in the encapsulation efficiency of the two systems. Insulin concentrations of 0.5 and 1 mg/mL showed an encapsulation efficiency of the inclusion compound preparation of 87% and 82%, respectively. The efficiency of the non-encapsulated system was 90% and 85% for insulin concentration at 0.5 and 1 mg/mL, respectively. In addition, in vitro drug release experiment showed that the inclusion compound encapsulated by the hydrogel exhibited pH responsiveness. At pH = 1.2 and 7.4, the amount of insulin released under more acidic conditions was significantly lower compared with that under neutral conditions over the same period. The effect of oral delivery of insulin by CD hydrogel inclusion compound microparticles was studied using streptozotocin-induced diabetic rats. The experimental results exhibited that insulin loading, and release characteristics of the hydrogel matrix were not affected by the complexation of CD. In addition, CD compound insulin coated by hydrogel particles effectively reduced blood sugar levels in diabetic animals.

To deliver oral hypoglycemic drugs into the body, Okubo et al. employed the host-guest interaction between hydrophobically modified hydroxypropyl methylcellulose and CD to develop and prepare heat-responsive injectable drug sustained-release hydrogel. The hydrogel was prepared by mixing the CD inclusion compound containing insulin with cellulose. Due to the interaction between the stearyl group of cellulose and the β-CD cavity, the hydrogel underwent a thermal gelation reaction near the human body temperature. The newly prepared hydrogel was effective for up to 24 h after subcutaneous administration of mice. Notably, pharmacokinetic experiment results showed that the hydrogel released insulin which reduced the blood sugar levels [87].

Drugs can be delivered into the human intestinal tract through oral administration to achieve high efficacy [88]. The solubility of berberine hydrochloride could be enhanced by β-CD (25 °C, 11.41 mM). CD-loaded berberine hydrochloride-containing bacterial cellulose hydrogel was prepared by physical adsorption, and its drug loading (34%) was higher compared with that of hydrogels prepared without CD (17.2%). In vitro, drug release experiments revealed that the hydrogel achieved sustained drug release (t > 70 h) under different pH conditions (pH = 1.2, 6.8, and 12.1) of the gastrointestinal fluid. In vitro antibacterial experiments exhibited that hydrogels had better antibacterial effects (*E. coli, S. aureus*, and *P. aeruginosa*), thus laying the foundation for oral drug delivery [54].

To speed up the healing of wounds on the skin surface and prevent bacterial infections, some wound dressings are prepared through the physical synthesis method [89,90]. Sodium alginate and chitosan are mixed to form a hydrogel matrix, and then β-CD inclusion compound containing curcumin is added as an active ingredient. It is reported that sodium alginate and chitosan adsorb each other through electrostatic interaction. The mechanical properties of hydrogel materials are significantly improved by an increase in calcium chloride content. In the active ingredient release study, the hydrogel displayed a sustained release effect (t > 48 h), which was not affected by the addition of calcium chloride. The curcumin-loaded hydrogel showed a good inhibitory effect on the growth of *E. coli* (73.95%) and *S. aureus* (71.59%). Toxicity studies displayed that the hydrogel was non-toxic to NCTC clone 929 cells and normal human dermal fibroblasts [48]. A similar study used hydroxypropyl-β-CD as a drug carrier for loading curcumin and added silver nanoparticles with antibacterial activity to the hydrogels. These hydrogels showed broad-spectrum antibacterial activity and antioxidant properties and can be used for wound dressing. In addition, the hydrogels showed good cell compatibility with different cell lines, including Panc 1 (human pancreatic ductal adenocarcinoma), U251 (Human brain glioma U251 cell line), and MSTO (human mesothelioma). Its high moisture content and good transparency further promote its application potential for the treatment of chronic wounds [82].

Because most people do not pay much attention to hygiene, the risk of pathogenic microorganisms coming into contact with the body is high. The eye, an organ that is directly in contact with the external environment, often suffers from keratitis due to infection by pathogens [91]. If keratitis is left untreated, it may cause permanent vision damage. Eye drops are the preferred method of drug delivery for the timely treatment of infections [92]. However, when eyes are stimulated by the outside environment, the number of blinks and secretion of tears will increase, thus most of the eye drops do not reach the infected areas. Therefore, the frequency of administration of eye drops is increased to increase efficacy. Hewitt et al. explored the possibility of adding drugs to contact lenses. They selected pig eyes as the research object and prepared a hydrogel with β-CD loaded with chlorhexidine. Antibacterial analysis showed that the drug-containing contact lenses delivered a high amount of chlorhexidine to the cornea within 24 h. Although the contact lens loaded with chlorhexidine β-CD failed to improve the drug delivery effect, it was able to deliver the drug to the cornea. In addition, β-CD hindered drug release in the hydrogel matrix. Continuous irrigation with simulated tear fluid can significantly reduce the amount of drug delivered to the cornea. Chlorhexidine retains antibacterial activity in all methods of administration. The hydrogel contact lens injected with chlorhexidine showed a significantly higher effect on the cornea, whether it was used multiple times or once compared with eye drops. Therefore, this method can be used to reduce the number of administrations thus improving patient tolerance degree [81].

3.2. Chemically Cross-Linked CD-Containing Hydrogels

The chemically cross-linked CD-containing hydrogels are not easily degraded in the external environment, thus reducing the number of hydrogels used during the application, ultimately mitigating side effects by the hydrogel drug delivery system (Table 5). Chemical methods can be used to prepare chemically cross-linked CD-containing hydrogels which possess much higher stability than the physically cross-linked CD-containing hydrogels, comparatively speaking. Therefore, the application scope of chemically cross-linked CD-containing hydrogels has been extended to some extent by changing the properties of hydrogels through different chemical reactions., especially in the aspect of injectable nanocarriers, cancer therapy, transdermal drug delivery, tissue engineering, regenerative medicine, wound healing, oral drug delivery, etc.

Table 5. Potential application of chemically cross-linked CD-containing hydrogels.

No.	Drug	Formation Materials	Hydrogel State	Types of Cell	Summary	Potential Application	Ref.
1	5-Fluorouracil	β-CD; N-vinylcaprolactam; *N*, *N*′-methylene bisacrylamide	Nanogel	Human colon cancer cell lines (HCT 116); MRC-5 normal cells	The hydrogel had the best drug loading (659.7 mg/g) after controlling the feeding ratio. The drug release curve showed that the hydrogel could continue to release drugs for up to 30 h; especially in the intestinal juice with pH = 7.4, the 5-fluorouracil drug molecules contained therein were not completely released; and the maximum release rate was 68%.	Implantable hydrogels	[93]
2	Coumarin	β-CD; alginate; calcium homopoly-L-guluronate	Supramolecular hydrogel	RAW 264.7 cells; T. cruzi cells	The lowest release of substituted amidocoumarins from the hydrogels occurred at pH = 1.2 whereas the maximum release (34%) was observed at pH 8.0.	Biomedicine	[94]
3	Curcumin	β-CD; epiclon	Nanosponge	Non-tumorigenic human breast; invasive mouse breast cell lines (4T1)	The high degree of cross-linking led to the formation of mesoporous having high specific surface area and high loading capacity. Nanosponge showed no toxicity against MCF 10A and 4T1 cells as normal and cancerous cells, respectively.	Cancer therapy	[95]
4	Curcumin	Carboxymethyl-β-CD; gelatin; methacrylic anhydride	Microneedle arrays	B16F10 melanoma cell	The inclusion complex of curcumin maintained 90% of the initial concentration. Besides, the hydrogel could enhance the drug loading and adjust release. In vivo study showed that hydrogel had good biocompatibility and degradability.	Transdermal drug delivery	[96]
5	Dexamethasone	β-CD; low-acyl gellan gum; EDC	Injectable hydrogel	NIH/3T3 mouse embryo fibroblast	After drug loading, the gel-forming temperature of the modified hydrogel was reduced and the mechanical properties are improved. Hydrogel had a high affinity and release rate for drugs. In vivo studies had shown that the drug-loaded hydrogel improved the anti-inflammatory response.	Tissue engineering and regenerative medicine	[52]
6	Dexamethasone	β-CD; sodium hyaluronate	Delivery hydrogel	3T3 cells	The novel hydrogels significantly improved the therapeutic effect of dexamethasone in burn wound healing.	Wound healing	[58]
7	Dexibuprofen	β-CD; acrylic acid; methylene bisacrylamide	Nanosponges	Not mentioned	The solubility of ibuprofen in the hydrogel was increased 6.3 times. In vitro release experiments demonstrated that the drug release rate of β-CD nanosponges reached 89% within 30 min under the condition of pH = 6.8.	Oral administration of lipophilic drugs	[97]
8	Diclofenac sodium	β-CD; sodium hyaluronan; EDC;	Contact lens materials	*S. aureus*; 3T3 fibroblasts	The hydrogel not only reduced the adsorption of tearing proteins due to electrostatic mutual repulsion but also improved encapsulation capacity and sustainable release of diclofenac (t > 72 h). In vitro cell viability analysis displayed that all hydrogels were non-toxic to 3T3 mouse fibroblasts.	Ophthalmic diseases	[66]

Table 5. *Cont.*

No.	Drug	Formation Materials	Hydrogel State	Types of Cell	Summary	Potential Application	Ref.
9	Doxorubicin	β-CD; 2-ethyl-2-oxazoline; aminopropyl-triethoxy silane; $FeCl_2.4H_2O$; $FeCl_3.6H_2O$	Magnetic nanohydrogel	MCF7 cells	The magnetic nanohydrogel had a good drug loading rate (74%) and encapsulation rate (81%). Under acidic conditions (pH = 5.3), adding a small amount of GSH (10 mM) increased the release value (89.21%). The magnetic nanohydrogel had good cell compatibility even at high concentrations (10 mg/mL).	Implantable hydrogels	[98]
10	Doxorubicin	β-CD; agarose	Injectable hydrogel	Human embryonic kidney 239 cells; HeLa cells	The hydrogel was able to easily and uniformly load a drug at 30 °C. The drug-loaded hydrogel maintained the drug's anti-cancer activity. In addition, the hydrogels did not exhibit toxicity toward the HEK-293 and HeLa cells.	Injectable hydrogel	[59]
11	Doxorubicin	β-CD; hyaluronic acid; bis(4-nitrophenyl) carbonate	Injectable hydrogel	Human colorectal cancer cells HCT-116	Rheological tests showed that this hydrogel could be easily prepared and used on a schedule compatible with normal operating room procedures. In vitro experiments showed that the unique physical and chemical properties of the hydrogel ensured the sustained release of anticancer drugs (t > 32 d) and prevented the growth of colorectal cancer micelles under 3D culture conditions.	Device for localized chemotherapy of solid tumors	[99]
12	Doxorubicin; curcumin	β-CD; multiwalled carbon nanotubes; maleic anhydride; folic acid; hexamethylene diisocyanate	Nanocarrier	Not mentioned	This injectable hydrogel exhibited pH/thermo response and exerted a deleterious effect on the tumor. A sustained release of the two drugs was observed over a period of 30 h. The release rate of doxorubicin reached 90% under tumor microenvironmental conditions, and the release rate of curcumin reached 85% under high temperature and physiological pH conditions.	Injectable nanocarriers	[100]
13	Doxorubicin	β-CD; tetronic; adamantane	Injectable shear-thinning hydrogels	HeLa cell	The hydrogels showed shear-thinning behaviors, rapid recovery properties, pH-responsive properties, and long-term release of the hydrophobic drug.	Embolic material	[101]
14	Insulin	Carboxymethyl β-CD; carboxymethyl chitosan	Microparticles	Caco-2 cells	The insulin was loaded into the hydrogel, and the results of the drug release experiment found that the insulin was successfully retained in the stomach environment and slowly released after passing through the intestine. In vitro studies had shown that the hydrogel particles exhibited non-cytotoxicity and were mainly transported in the Caco-2 cell monolayer through paracellular pathways.	Oral drug delivery	[102]

Table 5. *Cont.*

No.	Drug	Formation Materials	Hydrogel State	Types of Cell	Summary	Potential Application	Ref.
15	Vitamin E	β-CD; soy soluble polysaccharides; galacturonic acid	Core-shell bio-nanomaterials hydrogel	Not mentioned	The hydrogel exhibited significant swelling adsorption and sustained release (t > 230 h) for the release of vitamin E in vitro. The encapsulation efficiency and drug loadings were 79.10% and 16.04%, respectively. In addition, after oral administration of the vitamin E-loaded hydrogel in rats, the vitamin E level in the plasma continued to increase within 12 h.	Oral drug carrier	[56]

Several studies are currently exploring controlled drug delivery systems based on CD-containing hydrogels [103–106]. Xia et al. used new hesperidin-copper (II) (NH-Cu (II)) as the model drug, then added it into a hydrogel composed of carboxymethyl cellulose (CMC), cellulose nanocrystals (CNC), and hydroxypropyl-β-CD. Citric acid was used as a cross-linking agent to prepare a natural hydrogel film which exhibited controllable swelling behavior. The different dynamic behaviors of NH-Cu (II) in the hydrogel film were then explored. Drug release studies showed that the hydrogel had a sustained release effect at different temperatures. Furthermore, it had different swelling behavior under different pH and different ion concentrations. The swelling kinetics followed the Fick diffusion and Schott second-order kinetic model. In addition, the addition of CNC into hydrogel film changed the mechanical properties, thermal stability at high temperatures, swelling rate, salt sensitivity, and pH sensitivity of the hydrogel film in different solutions. Moreover, CNC greatly improved the loading and encapsulation efficiency of the hydrogel film. The addition of 4% CNC showed an optimal loading efficiency of 753.75 mg/g and a cumulative release rate of 85.08%. The hydrogel membrane showed good cell compatibility and was non-cytotoxic, thus it can be used as a potential drug delivery and controlled release system for wound dressing [63].

Targeted delivery of anti-cancer drugs is one of the most effective treatment methods for tumors [107]. Therefore, direct injection of the drug-containing hydrogel at the tumor area to maximize drug concentration improves the efficacy of the drug. Hyaluronic acid derivative and functionalized CD can be linked by a covalent bond to prepare an injectable hydrogel, and doxorubicin is successfully loaded in the CD cavity. In vitro release experiments of the drug-loaded hydrogels displayed a sustained-release effect. In this case, the hydrophobic interaction between doxorubicin and CD cavity resulted in drug retention in the hydrogel, thereby slowing its diffusion through the three-dimensional network. Notably, after 32 days of incubation, only 50% of the drug load was released in the medium without a significant burst effect. In addition, the hydrogel reduced the size of solid tumors in mice. Moreover, histological analysis of the heart of treated mice displayed showed no cardiotoxicity after treatment of the tumor with the drug-loaded hydrogel. These results implied that this drug-loaded hydrogel is an effective biomedical device to locally treat unresectable solid tumors or for preventing regeneration of residual tumors and inhibiting disease recurrence [99].

Subcutaneous administration of drugs has some side effects. For example, when treating diabetes, people prefer injecting insulin under the skin. However, daily insulin injection is associated with adverse effects, such as hypoglycemia, allergies, and peripheral hyperinsulinemia [102]. Therefore, an oral insulin delivery system can be used to avoid these side effects. A previous study reports oral hydrogen comprising CD and chitosan as raw materials and a water-soluble carbodiimide as cross-linking agent. SEM (scanning electron microscope), FTIR (Fourier transform infrared spectroscopy), XRD (X-Ray diffraction), and swelling experiments indicated the hydrogel had a porous structure. Insulin release behavior was shown to be triggered by in vitro pH. Notably, insulin was successfully retained in the stomach environment and slowly released after passing through the intestine. The stability of insulin secondary structure was studied by circular dichroism and fluorescence spectrophotometry. Analysis experiment revealed no significant difference in secondary structure between native insulin and released insulin. Furthermore, the hydrogel particles exhibit non-cytotoxicity and were mainly transported in the Caco-2 cell monolayer through paracellular pathways. Different insulin-loaded hydrogel microparticles were applied to diabetic mice to evaluate the effectiveness of hydrogel sustained-release microparticles in delivering insulin in the body. Insulin-loaded hydrogel particles significantly and continuously (6–12 h) reduced blood sugar levels in diabetic mice compared with subcutaneous injection. In summary, these findings indicate that hydrogel is a promising oral drug carrier to achieve sustained release [102].

Previous studies explored the use of chemically cross-linked hydrogels and compared them with physically cross-linked hydrogels for the treatment of eye diseases [108]. Cyclosporine is a commonly used drug for the treatment of various immune-mediated ocular surface diseases [109]. However, the poor water solubility of cyclosporine limits its application. Only two topical formulations of cyclosporine have been approved for the treatment of dry eye syndrome, Resis (Allergan, USA). An anionic emulsion (0.5 mg/mL cyclosporine) was approved by the FDA in 2002, and Ikervis was recently approved in Europe (Santen, Tampere, Finland), a cationic nanoemulsion containing 1 mg/mL cyclosporine [110–112]. Sodium hyaluronate and hydroxypropyl-β-CD are used as raw materials to prepare a chemically cross-linked hydrogel. Cyclosporin is loaded into the hydrogel by dipping to avoid degradation of cyclosporin during the cross-linking reaction. Interestingly, changing the weight ratio of the raw materials in the hydrogel adjusts the swelling and the rate of drug release. The swelling plays a key role in the feasibility of the drug penetrating through the sclera and accumulating into the eye tissue. In vitro drug release, experiments show that the hydrogel has a release time of up to 8 h. Therefore, chemically cross-linked hydrogels are effective systems for drug release to the ocular surface, however, the effectiveness should be confirmed in preclinical studies [67].

The polymerization system of the hydrogel obtained by CD polymerization and swelling has very few pores. Khalid et al. used β-CD, acrylic acid, and acrylamide to prepare β-CD nanosponges for loading dexibuprofen. Water solubility experiments showed that the water solubility of dexibuprofen was increased by 6.3 times after loading it to the hydrogel. The particle size of this material was 275.1 ± 28.5 nm. The swelling index of the nanosponge under weakly acidic conditions (pH = 6.8) was 3, and up to 89% dexibuprofen was successfully released within 30 min under this condition. An acute oral toxicity study using rats showed no toxicity-related conditions, death, adverse clinical symptoms and had no toxicological changes in hematology, clinical biochemistry, and histology. Therefore, the β-CD nanosponges prepared through the optimized condensation method may be superior compared with other β-CD nanoformulations, thus they improve oral administration of lipophilic drugs such as dexibuprofen [97].

4. Simulated Degradation Behavior for CD-Containing Hydrogels

The degradation process of hydrogel is one of the key indicators of its safety, and it is also the basis for studying in vivo decomposition of the hydrogel [113–115]. In vitro degradation behavior is studied through changes of the hydrogel under conditions such as simulated body fluids, different types of biological enzymes, and different acids and bases (Figure 2).

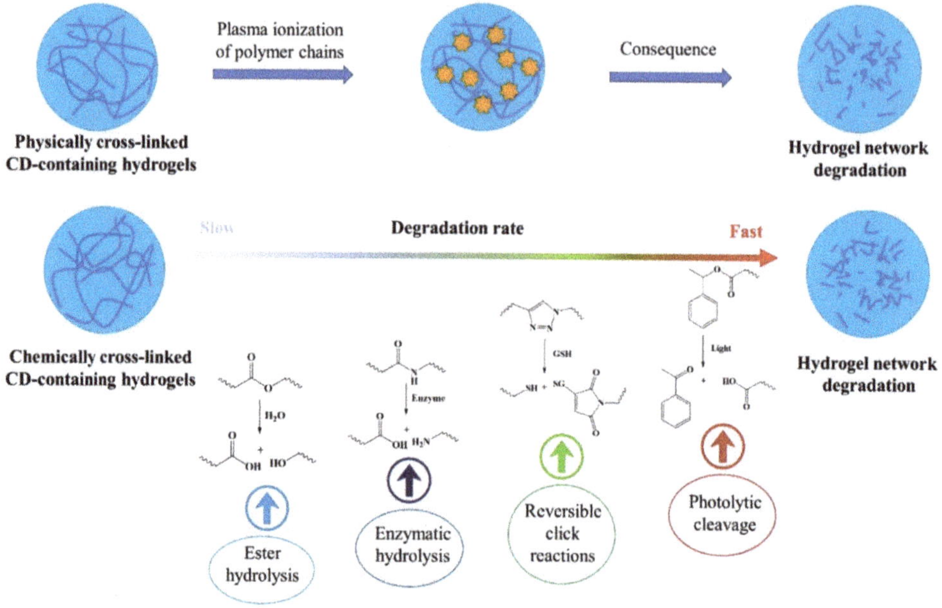

Figure 2. Simulated degradation behavior for CD-containing hydrogels.

A previous study placed CD-based hydrogel loaded with gellan gum in a phosphate buffer solution (pH = 7.4), and the weight loss of the hydrogel was measured at 1, 7, 14, 21 days. Analysis showed that the weight loss rate of the hydrogel was low with the content of CD and that the weight loss rate of the modified hydrogel was higher compared with hydrogels without CD. This phenomenon can be attributed to the reduction of cross-linking sites in the modified hydrogel that lead to more unassociated chains in the gel network. In addition, hydrogels without CD had a slower degradation rate compared with those with CD, implying that the presence of CD in the matrix may increase the decomposition rate of the hydrogel [52]. Exposure of different hydrogels to the external environment, such as a phosphate buffer (pH = 7.4), showed that hydrogels with large cross-linking concentrations required longer hydrolysis time. Unfortunately, the higher cross-linking degree of the hydrogels caused more resistance to degradation in collagenase solution in PBS buffer. The cross-linking could limit the accessibility of enzymes to the cleavage sites of the hydrogels, and prevent the enzymes from penetrating the bulk of the material [116]. The degradation mechanism of these hydrogels mainly resulted from the breaking of cross-linking sites, ester bonds, hydrazone bonds, or steric hindrance through chemical hydrolysis. The polymer backbone was then hydrolyzed and broken into lower molecular weight and soluble fragments [117–123].

Degradation of hydrogel under pure water conditions was simulated, and the degradation rate of gallic acid-loaded hydrogel was explored to determine the environmental safety of bio-based hydrogels. Analysis showed that degradation of the hydrogel was relatively fast in the initial stages, however, it remained stable within 8–360 h. This degradation behavior can be attributed to the effect of water penetration in the bacterial cellulose hydrogel network which resulted in the good degradation behavior of hydrogel. Moreover, the aromatic ring (the presence of gallic acid and CD), slightly slows the degradation ability of the hydrogel. The aromatic ring may delay the degradation of the hydrogel [55].

The ability of hydrogels to degrade in the presence of organisms is important for drug release. A previous study evaluated the degradation of CD-based hydrogel using *Penicillium*, *Aspergillus niger*, and mushroom. Analysis showed that the gel changed to a liquid after six days of degradation. The degradation rate of the hydrogel increased gradually in

Penicillium. In *Penicillium*, the degradation rate was 6.11% after three days, whereas after 21 days the degradation rate was 21.49%. In *Aspergillus niger*, the degradation efficiency of the hydrogel increases rapidly with time. The degradation rate was 15.9% at three days and 68.7% at 21 days. In addition, the degradation effect of mushrooms was higher compared with that of other strains. The experimentally prepared hydrogel was biodegradable and had a good degradation rate in mushrooms. Therefore, even if the hydrogel is disposed to the natural environment, the threat to the environment is minimal [124]. A different study used lysozyme to degrade the cyclic oligosaccharide backbone of a hydrogel by enzymatically hydrolyzing the glycosidic bonds on the hexasaccharide ring. The weight of the hydrogel decreased, indicating that the hydrogel was biodegradable. In addition, analysis of the infrared spectra of the degradation products (obtained on day 14 and day 28) showed that the C-O-C tensile peak intensity of the degradation products decreased after 14 days, whereas the C-O-C tensile peak disappeared after 28 days of degradation. At the same time, analysis of the day 14 and day 28 SEM images of the degraded hydrogel showed that the porous network morphology of the hydrogel was deformed and degraded, showing the degradability property of the hydrogel [125].

When preparing CD-loaded hydrogels, materials that are easily degraded in vivo should be selected as preparation materials. A poly[(R)-3-hydroxybutyrate] fragment was introduced into a hydrogel containing CD and analysis showed that the copolymer produced was biodegradable under physiological conditions. In addition, the hydrogel formulation was bioabsorbable after administration and was able to dissociate into its components [126]. However, materials used in some heat-sensitive hydrogels are not biodegradable [65]. A previous study explored a hydrogel that could be hydrolyzed by α-amylase after the introduction of CD. Moreover, the degradation rate increases when part of the poly-*N*-propylacrylamide molecular chain is replaced by maleic anhydride-β-CD (MAH-β-CD), mainly due to an increase in the hydrolysis of β-CD by α-amylase. Analysis of ^{1}H NMR spectrum shows β-CD reacts with at least 2 maleic anhydride (MAH) molecules on average. Therefore, MAH-β-CD is a better cross-linking agent in the hydrogel network. At 37 °C, all hydrogels are in a semi-swelled state, and the speed at which enzymes enter the internal network of the hydrogel slows down [127].

When the degradation behavior of hydrogels is studied in an organism, the hydrogel is placed on the organism and then changes of the hydrogel at different times are observed [128]. Using mice as the model to verify the degradation of hydrogel, the degradation rate of the hydrogel was found to be faster under irradiation than without irradiation, indicating that the injectable hydrogel could be rapidly decomposed by photothermal treatment. Moreover, the fluorescence decay rate of the hydrogel was faster after irradiation than without irradiation. On the 4th day, the relative fluorescence signal intensity of the irradiated hydrogel group was about 60%, and the relative fluorescence signal intensity of the non-irradiated hydrogel group was close to 80%. On the 8th day, the relative fluorescence signal intensity of the non-irradiated hydrogel group was only about 40%, whereas the relative fluorescence signal intensity of the irradiated hydrogel group was about 20%. Therefore, laser irradiation can accelerate the decomposition of the hydrogel nanocomposite system. Light response assisted fluorescence imaging allows visualization of the disassembly of medical biomaterials. The near-infrared light-responsive supramolecular nanocomposite system is widely used to explore the fluorescent imaging tracking process of biomedical materials. The visualization of degradable supramolecular hydrogels on demand helps to eliminate carriers. The controllable non-invasive supramolecular nanocomposite is a promising biomedical material that allows instant stimulation and sustainable monitoring [129]. In another experiment on the degradation of hydrogels in mice during transdermal administration, Zhou et al. discovered that long-term biodegradation of hydrogels administered transdermally in mice may be caused by cross-linked hydrogel side chains. The existence of the cyclodextrin cavity made it difficult for the biological enzymes in the body to contact the hydrogel [96].

5. Conclusions and Perspectives

As a host molecule that can form host-guest inclusion complexes, CD has great application potential in the fields of medicine and materials science. Promisingly, the administration of CD-containing hydrogels can avoid the annoying side effects of traditional drug administration. In recent years, several studies have shown that hydrogel systems composed of CDs can effectively deliver drugs into target tissues. Especially, many physical and chemical methods have been successfully developed for preparing CD hydrogels.

Presently, the research on CD-containing hydrogel mainly focused on the field of biomedicine. In this review, different methods to design and prepare CD hydrogels are discussed. In addition, the potential applications of drug-loaded CD hydrogels are summarized. The data reviewed here show that introduction of CD improves the properties of the hydrogel and the drug release time of the hydrogels. Interestingly, some CD-containing hydrogels based on light and heat response have been successfully synthesized. This review broadens our understanding of the development trends in the application of CD-containing hydrogels, which lays the foundation for future clinical research. At the same time, looking forward to the future medical frontiers, we believe that an increasing number of emerging technologies will be introduced to guide the preparation of CD-containing hydrogels.

Future studies should design drug-loaded CD-containing hydrogels which can withstand factors in the external environment and internal environment in the body to guarantee excellent drug-release dynamics and drug-loading capacity. Moreover, it is of great importance to establish novel methods and tools for monitoring the metabolism of drugs in the body when drug-loaded CD-containing hydrogels enter the body. Meanwhile, attention should be paid to the safety issues associated with hydrogel materials inside the body. How the CD-containing hydrogels are degraded in the body, how they change, and whether they will interact with the body should be closely analyzed. Gratifyingly, more smart stimuli-responsive hydrogels based on CD for drug delivery will be designed and prepared, which can respond to different external and interior environmental stimuli to release drugs. In the future, we can actively take a page from cutting-edge technologies to prepare the CD-containing hydrogels for precision medicine that can accurately load appropriate doses of drugs. In addition, we should have a deep insight into the basic principles of drug loading and release of CD-containing hydrogels in the aspect of site-targeting, release conditions, dosing intervals for optimal treatment to different diseases. If the above-mentioned problems are fully solved, CD-containing hydrogels are expected to be more widely applied in clinical practice.

Author Contributions: All authors drafted the article and critically modified the important knowledge content, endorsing the final version. Study conception and design: Y.L. and J.-B.W. Literature and data collection: J.L. and B.T. Drafting the article or revising it critically for important intellectual content: Y.L. and J.-B.W. Final approval of the version of the article to be published: J.L., B.T., Y.L. and J.-B.W. All authors have read and agreed to the published version of the manuscript.

Funding: This study was supported by the National Natural Science Foundation of China (31660490), and the Enterprise Cooperation Project (CP/003/2018).

Institutional Review Board Statement: Not applicable.

Informed Consent Statement: Not applicable.

Data Availability Statement: Not applicable.

Conflicts of Interest: The authors declare no conflict of interest.

Abbreviation

FTIR	Fourier transform infrared spectroscopy
NMR	nuclear magnetic resonance
SEM	scanning electron microscope
XRD	X-Ray diffraction
S. aureus	*Staphylococcus aureus*
P. aeruginosa	*Pseudomonas aeruginosa*
E. coli	*Escherichia coli*
C. aureus	*Staphylococcus aureus enterotoxin C*
Panc 1	Human pancreatic cancer cell
U251	Human brain glioma U251 cell line
MSTO	human mesothelioma
NHD	Fskin fibroblast cells
L929 cells	L929 mouse fibroblast cells

References

1. Champeau, M.; Heinze, D.A.; Viana, T.N.; de Souza, E.R.; Chinellato, A.C.; Titotto, S. 4D printing of hydrogels: A review. *Adv. Funct. Mater.* **2020**, *30*, 1910606. [CrossRef]
2. Arslan, M.; Sanyal, R.; Sanyal, A. Cyclodextrin embedded covalently crosslinked networks: Synthesis and applications of hydrogels with nano-containers. *Polym. Chem.* **2020**, *11*, 615–629. [CrossRef]
3. Tang, W.; Zou, C.; Da, C.; Cao, Y.; Peng, H. A review on the recent development of cyclodextrin-based materials used in oilfield applications. *Carbohydr. Polym.* **2020**, *240*, 116321. [CrossRef]
4. Wang, W.; Yang, Z.; Zhang, A.; Yang, S. Water retention and fertilizer slow release integrated superabsorbent synthesized from millet straw and applied in agriculture. *Ind. Crop. Prod.* **2020**, *160*, 113126. [CrossRef]
5. Mandal, A.; Clegg, J.R.; Anselmo, A.C.; Mitragotri, S. Hydrogels in the clinic. *Bioeng. Transl. Med.* **2020**, *5*, e10158. [CrossRef] [PubMed]
6. Hosseini, M.S.; Nabid, M.R. Synthesis of chemically cross-linked hydrogel films based on basil seed (*Ocimum basilicum* L.) mucilage for wound dressing drug delivery applications. *Int. J. Biol. Macromol.* **2020**, *163*, 336–347. [CrossRef]
7. Cai, T.; Huo, S.; Wang, T.; Sun, W.; Tong, Z. Self-healable tough supramolecular hydrogels crosslinked by poly-cyclodextrin through host-guest interaction. *Carbohydr. Polym.* **2018**, *193*, 54–61. [CrossRef] [PubMed]
8. Tan, S.; Ladewig, K.; Fu, Q.; Blencowe, A.; Qiao, G.G. Cyclodextrin-based supramolecular assemblies and hydrogels: Recent advances and future perspectives. *Macromol. Rapid Commun.* **2014**, *35*, 1166–1184. [CrossRef] [PubMed]
9. Cheng, W.; Wang, M.; Chen, M.; Niu, W.; Li, Y.; Wang, Y.; Luo, M.; Xie, C.; Leng, T.; Lei, B. Injectable Antibacterial Antiinflammatory Molecular Hybrid Hydrogel Dressing for Rapid MDRB-Infected Wound Repair and Therapy. *Chem. Eng. J.* **2021**, *409*, 128140. [CrossRef]
10. Peng, L.; Chang, L.; Si, M.; Lin, J.; Wei, Y.; Wang, S.; Liu, H.; Han, B.; Jiang, L. Hydrogel-coated dental device with adhesion-inhibiting and colony-suppressing properties. *ACS Appl. Mater. Interfaces* **2020**, *12*, 9718–9725. [CrossRef]
11. Singh, B.; Kumar, A. Synthesis and characterization of alginate and sterculia gum based hydrogel for brain drug delivery applications. *Int. J. Biol. Macromol.* **2020**, *148*, 248–257. [CrossRef]
12. Liu, Y.; Du, J.; Peng, P.; Cheng, R.; Lin, J.; Xu, C.; Yang, H.; Cui, W.; Mao, H.; Li, Y.; et al. Regulation of the inflammatory cycle by a controllable release hydrogel for eliminating postoperative inflammation after discectomy. *Bioact. Mater.* **2020**, *6*, 146–157. [CrossRef] [PubMed]
13. Qu, J.; Liang, Y.; Shi, M.; Guo, B.; Gao, Y.; Yin, Z. Biocompatible conductive hydrogels based on dextran and aniline trimer as electro-responsive drug delivery system for localized drug release. *Int. J. Biol. Macromol.* **2019**, *140*, 255–264. [CrossRef]
14. Dreiss, C.A. Hydrogel design strategies for drug delivery. *Curr. Opin. Colloid Interface Sci.* **2020**, *48*, 1–17. [CrossRef]
15. Peers, S.; Montembault, A.; Ladavière, C. Chitosan hydrogels for sustained drug delivery. *J. Control. Release* **2020**, *326*, 150–163. [CrossRef] [PubMed]
16. Krukiewicz, K.; Zak, J.K. Biomaterial-based regional chemotherapy: Local anticancer drug delivery to enhance chemotherapy and minimize its side-effects. *Mater. Sci. Eng. C* **2016**, *62*, 927–942. [CrossRef]
17. Li, J.; Mooney, D.J. Designing hydrogels for controlled drug delivery. *Nat. Rev. Mater.* **2016**, *1*, 1–17. [CrossRef]
18. Torres-Luna, C.; Fan, X.; Domszy, R.; Hu, N.; Wang, N.S.; Yang, A. Hydrogel-based ocular drug delivery systems for hydrophobic drugs. *Eur. J. Pharm. Sci.* **2020**, *154*, 105503. [CrossRef]
19. Szejtli, J. Introduction and general overview of cyclodextrin chemistry. *Chem. Rev.* **1998**, *98*, 1743–1754. [CrossRef] [PubMed]
20. Concheiro, A.; Alvarez-Lorenzo, C. Chemically cross-linked and grafted cyclodextrin hydrogels: From nanostructures to drug-eluting medical devices. *Adv. Drug Deliv. Rev.* **2013**, *65*, 1188–1203. [CrossRef] [PubMed]
21. Morin-Crini, N.; Crini, G. Environmental applications of water-insoluble β-cyclodextrin–epichlorohydrin polymers. *Prog. Polym. Sci.* **2013**, *38*, 344–368. [CrossRef]

22. Crini, G.; Fourmentin, S.; Fenyvesi, É.; Torri, G.; Fourmentin, M.; Morin-Crini, N. Fundamentals and applications of cyclodextrins. In *Cyclodextrin Fundamentals, Reactivity and Analysis*; Springer: Cham, Switzerland, 2018; pp. 1–55.
23. Crini, G. Cyclodextrin–epichlorohydrin polymers synthesis, characterization and applications to wastewater treatment: A review. *Enviro. Chem. Lett.* **2021**, *19*, 2383–2403. [CrossRef]
24. Cova, T.F.; Murtinho, D.; Pais, A.A.; Valente, A.J. Combining cellulose and cyclodextrins: Fascinating designs for materials and pharmaceutics. *Front. Chem.* **2018**, *6*, 271. [CrossRef]
25. Cova, T.F.; Murtinho, D.; Aguado, R.; Pais, A.A.; Valente, A.J. Cyclodextrin polymers and cyclodextrin-containing polysaccharides for water remediation. *Polysaccharides* **2021**, *2*, 16–38. [CrossRef]
26. Pinho, E. Cyclodextrins-based hydrogel. In *Plant and Algal Hydrogels for Drug Delivery and Regenerative Medicine*; Woodhead Publishing: Sawston, UK, 2021; pp. 113–141.
27. Machín, R.; Isasi, J.R.; Vélaz, I. β-Cyclodextrin hydrogels as potential drug delivery systems. *Carbohydr. Polym.* **2012**, *87*, 2024–2030. [CrossRef]
28. Liu, G.; Yuan, Q.; Hollett, G.; Zhao, W.; Kang, Y.; Wu, J. Cyclodextrin-based host–guest supramolecular hydrogel and its application in biomedical fields. *Polym. Chem.* **2018**, *9*, 3436–3449. [CrossRef]
29. Tian, B.; Liu, Y.; Liu, J. Smart stimuli-responsive drug delivery systems based on cyclodextrin: A review. *Carbohydr. Polym.* **2021**, *251*, 116871. [CrossRef] [PubMed]
30. Hamedi, H.; Moradi, S.; Hudson, S.M.; Tonelli, A.E. Chitosan based hydrogels and their applications for drug delivery in wound dressings: A review. *Carbohydr. Polym.* **2018**, *199*, 445–460. [CrossRef]
31. Tian, B.; Hua, S.; Tian, Y.; Liu, J. Chemical and physical chitosan hydrogels as prospective carriers for drug delivery: A review. *J. Mater. Chem. B* **2020**, *8*, 10050–10064. [CrossRef] [PubMed]
32. Zainal, S.H.; Mohd, N.H.; Suhaili, N.; Anuar, F.H.; Lazim, A.M.; Othaman, R. Preparation of cellulose-based hydrogel: A review. *J. Mater. Res. Technol.* **2021**, *10*, 935–952. [CrossRef]
33. Chang, C.; Zhang, L. Cellulose-based hydrogels: Present status and application prospects. *Carbohydr. Polym.* **2011**, *84*, 40–53. [CrossRef]
34. Guo, X.; Wang, Y.; Qin, Y.; Shen, P.; Peng, Q. Structures, properties and application of alginic acid: A review. *Int. J. Biol. Macromol.* **2020**, *162*, 618–628. [CrossRef] [PubMed]
35. Hoffman, A.S. Hydrogels for biomedical applications. *Adv. Drug Deliver. Rev.* **2012**, *64*, 18–23. [CrossRef]
36. Patel, S.; Goyal, A. Applications of natural polymer gum arabic: A review. *Int. J. Food Prop.* **2015**, *18*, 986–998. [CrossRef]
37. Sennakesavan, G.; Mostakhdemin, M.; Dkhar, L.K.; Seyfoddin, A.; Fatihhi, S.J. Acrylic acid/acrylamide based hydrogels and its properties-A review. *Polym. Degrad. Stabil.* **2020**, *180*, 109308. [CrossRef]
38. Rodríguez-Rodríguez, R.; Espinosa-Andrews, H.; Velasquillo-Martínez, C.; García-Carvajal, Z.Y. Composite hydrogels based on gelatin, chitosan and polyvinyl alcohol to biomedical applications: A review. *Int. J. Polym. Mater. Polym. Biomater.* **2020**, *69*, 1–20. [CrossRef]
39. Timofejeva, A.; D'Este, M.; Loca, D. Calcium phosphate/polyvinyl alcohol composite hydrogels: A review on the freeze-thawing synthesis approach and applications in regenerative medicine. *Eur. Polym. J.* **2017**, *95*, 547–565. [CrossRef]
40. Liu, Q.; Yang, D.; Shang, T.; Guo, L.; Yang, B.; Xu, X. Chain conformation transition induced host-guest assembly between triple helical curdlan and β-CD for drug delivery. *Biomater. Sci.* **2020**, *8*, 1638–1648. [CrossRef]
41. Tian, B.; Hua, S.; Liu, J. Cyclodextrin-based delivery systems for chemotherapeutic anticancer drugs: A review. *Carbohydr. Polym.* **2020**, *232*, 115805. [CrossRef]
42. Tian, B.; Liu, Y.; Liu, J. Cyclodextrin as a magic switch in covalent and non-covalent anticancer drug release systems. *Carbohydr. Polym.* **2020**, *242*, 116401. [CrossRef] [PubMed]
43. Cooper, R.C.; Yang, H. Hydrogel-based ocular drug delivery systems: Emerging fabrication strategies, applications, and bench-to-bedside manufacturing considerations. *J. Control. Release* **2019**, *306*, 29–39. [CrossRef]
44. Gunathilake, T.M.S.U.; Ching, Y.C.; Chuah, C.H.; Abd Rahman, N.; Nai-Shang, L. Recent advances in celluloses and their hybrids for stimuli-responsive drug delivery. *Int. J. Biol. Macromol.* **2020**, *158*, 670–688. [CrossRef]
45. Soppimath, K.S.; Aminabhavi, T.M.; Dave, A.M.; Kumbar, S.G.; Rudzinski, W.E. Stimulus-responsive "smart" hydrogels as novel drug delivery systems. *Drug Dev. Ind. Pharm.* **2002**, *28*, 957–974. [CrossRef] [PubMed]
46. Mignon, A.; De Belie, N.; Dubruel, P.; Van Vlierberghe, S. Superabsorbent polymers: A review on the characteristics and applications of synthetic, polysaccharide-based, semi-synthetic and 'smart'derivatives. *Eur. Polym. J.* **2019**, *117*, 165–178. [CrossRef]
47. Zhang, K.; Feng, Q.; Fang, Z.; Gu, L.; Bian, L. Structurally Dynamic Hydrogels for Biomedical Applications: Pursuing a Fine Balance between Macroscopic Stability and Microscopic Dynamics. *Chem. Rev.* **2021**, *121*, 11149–11193. [CrossRef] [PubMed]
48. Kiti, K.; Suwantong, O. Bilayer wound dressing based on sodium alginate incorporated with curcumin-β-cyclodextrin inclusion complex/chitosan hydrogel. *Int. J. Biol. Macromol.* **2020**, *164*, 4113–4124. [CrossRef] [PubMed]
49. Moradi, S.; Barati, A.; Tonelli, A.E.; Hamedi, H. Chitosan-based hydrogels loading with thyme oil cyclodextrin inclusion compounds: From preparation to characterization. *Eur. Polym. J.* **2020**, *122*, 109303. [CrossRef]
50. Barragán, C.A.R.; Balleza, E.R.M.; García-Uriostegui, L.; Ortega, J.A.A.; Toríz, G.; Delgado, E. Rheological characterization of new thermosensitive hydrogels formed by chitosan, glycerophosphate, and phosphorylated β-cyclodextrin. *Carbohydr. Polym.* **2018**, *201*, 471–481. [CrossRef]

51. Das, S.; Subuddhi, U. Cyclodextrin mediated controlled release of naproxen from pH-sensitive chitosan/poly (vinyl alcohol) hydrogels for colon targeted delivery. *Ind. Eng. Chem. Res.* **2013**, *52*, 14192–14200. [CrossRef]
52. Choi, J.H.; Park, A.; Lee, W.; Youn, J.; Rim, M.A.; Kim, W.; Kim, N.; Song, J.E.; Khang, G. Preparation and characterization of an injectable dexamethasone-cyclodextrin complexes-loaded gellan gum hydrogel for cartilage tissue engineering. *J. Control. Release* **2020**, *327*, 747–765. [CrossRef]
53. Bianchi, S.E.; Machado, B.E.; da Silva, M.G.; da Silva, M.M.; Dal Bosco, L.; Marques, M.S.; Horn, A.P.; Persich, L.; Geller, F.C.; Argenta, D.; et al. Coumestrol/hydroxypropyl-β-cyclodextrin association incorporated in hydroxypropyl methylcellulose hydrogel exhibits wound healing effect: In vitro and in vivo study. *Eur. J. Pharm. Sci.* **2018**, *119*, 179–188. [CrossRef] [PubMed]
54. Xiao, L.; Poudel, A.J.; Huang, L.; Wang, Y.; Abdalla, A.M.; Yang, G. Nanocellulose hyperfine network achieves sustained release of berberine hydrochloride solubilized with β-cyclodextrin for potential anti-infection oral administration. *Int. J. Biol. Macromol.* **2020**, *153*, 633–640. [CrossRef] [PubMed]
55. Chunshom, N.; Chuysinuan, P.; Thanyacharoen, T.; Techasakul, S.; Ummartyotin, S. Development of gallic acid/cyclodextrin inclusion complex in freeze-dried bacterial cellulose and poly (vinyl alcohol) hydrogel: Controlled-release characteristic and antioxidant properties. *Mater. Chem. Phys.* **2019**, *232*, 294–300. [CrossRef]
56. Eid, M.; Sobhy, R.; Zhou, P.; Wei, X.; Wu, D.; Li, B. β-cyclodextrin-soy soluble polysaccharide based core-shell bionanocomposites hydrogel for vitamin E swelling controlled delivery. *Food Hydrocoll.* **2020**, *104*, 105751. [CrossRef]
57. Zhang, B.; Wang, J.; Li, Z.; Ma, M.; Jia, S.; Li, X. Use of hydroxypropyl β-cyclodextrin as a dual functional component in xanthan hydrogel for sustained drug release and antibacterial activity. *Colloid Surf. A* **2020**, *587*, 124368. [CrossRef]
58. Yu, B.; Zhan, A.; Liu, Q.; Ye, H.; Huang, X.; Shu, Y.; Yang, Y.; Liu, H. A designed supramolecular cross-linking hydrogel for the direct, convenient, and efficient administration of hydrophobic drugs. *Int. J. Pharm.* **2020**, *578*, 119075. [CrossRef]
59. Kim, C.; Jeong, D.; Kim, S.; Kim, Y.; Jung, S. Cyclodextrin functionalized agarose gel with low gelling temperature for controlled drug delivery systems. *Carbohydr. Polym.* **2019**, *222*, 115011. [CrossRef]
60. Malik, N.S.; Ahmad, M.; Minhas, M.U. Cross-linked β-cyclodextrin and carboxymethyl cellulose hydrogels for controlled drug delivery of acyclovir. *PLoS ONE* **2017**, *12*, e0172727. [CrossRef] [PubMed]
61. Ghorpade, V.S.; Yadav, A.V.; Dias, R.J. Citric acid crosslinked β-cyclodextrin/carboxymethylcellulose hydrogel films for controlled delivery of poorly soluble drugs. *Carbohydr. Polym.* **2017**, *164*, 339–348. [CrossRef]
62. Amiel, A.G.; Palomino-Durand, C.; Maton, M.; Lopez, M.; Cazaux, F.; Chai, F.; Neut, C.; Foligné, B.; Martel, B.; Blanchemain, N. Designed sponges based on chitosan and cyclodextrin polymer for a local release of ciprofloxacin in diabetic foot infections. *Int. J. Pharm.* **2020**, *587*, 119677. [CrossRef] [PubMed]
63. Xia, N.; Wan, W.; Zhu, S.; Liu, Q. Preparation of crystalline nanocellulose/hydroxypropyl β-cyclodextrin/carboxymethyl cellulose polyelectrolyte complexes and their controlled release of neohesperidin-copper (II) in vitro. *Int. J. Biol. Macromol.* **2020**, *163*, 1518–1528. [CrossRef]
64. Pooresmaeil, M.; Namazi, H. Preparation and characterization of polyvinyl alcohol/β-cyclodextrin/GO-Ag nanocomposite with improved antibacterial and strength properties. *Polym. Adv. Technol.* **2019**, *30*, 447–456. [CrossRef]
65. Song, X.; Zhang, Z.; Zhu, J.; Wen, Y.; Zhao, F.; Lei, L.; Phan-Thien, N.; Khoo, B.C.; Li, J. Thermoresponsive hydrogel induced by dual supramolecular assemblies and its controlled release property for enhanced anticancer drug delivery. *Biomacromolecules* **2020**, *21*, 1516–1527. [CrossRef]
66. Li, R.; Guan, X.; Lin, X.; Guan, P.; Zhang, X.; Rao, Z.; Du, L.; Zhao, J.; Rong, J.; Zhao, J. Poly (2-hydroxyethyl methacrylate)/β-cyclodextrin-hyaluronan contact lens with tear protein adsorption resistance and sustained drug delivery for ophthalmic diseases. *Acta Biomater.* **2020**, *110*, 105–118. [CrossRef] [PubMed]
67. Grimaudo, M.A.; Nicoli, S.; Santi, P.; Concheiro, A.; Alvarez-Lorenzo, C. Cyclosporine-loaded cross-linked inserts of sodium hyaluronan and hydroxypropyl-β-cyclodextrin for ocular administration. *Carbohydr. Polym.* **2018**, *201*, 308–316. [CrossRef] [PubMed]
68. Davis, M.E.; Brewster, M.E. Cyclodextrin-based pharmaceutics: Past, present and future. *Nat. Rev. Drug Discov.* **2004**, *3*, 1023–1035. [CrossRef]
69. Loftsson, T.; Duchêne, D. Cyclodextrins and their pharmaceutical applications. *Int. J. Pharm.* **2007**, *329*, 1–11. [CrossRef]
70. Öztürk-Atar, K.; Kaplan, M.; Çalış, S. Development and evaluation of polymeric micelle containing tablet formulation for poorly water-soluble drug: Tamoxifen citrate. *Drug Dev. Ind. Pharm.* **2020**, *46*, 1695–1704. [CrossRef]
71. Tian, B.; Xiao, D.; Hei, T.; Ping, R.; Hua, S.; Liu, J. The application and prospects of cyclodextrin inclusion complexes and polymers in the food industry: A review. *Polym. Int.* **2020**, *69*, 597–603. [CrossRef]
72. Wankar, J.; Kotla, N.G.; Gera, S.; Rasala, S.; Pandit, A.; Rochev, Y.A. Recent advances in host-guest self-assembled cyclodextrin carriers: Implications for responsive drug delivery and biomedical engineering. *Adv. Funct. Mater.* **2020**, *30*, 1909049. [CrossRef]
73. Tejashri, G.; Amrita, B.; Darshana, J. Cyclodextrin based nanosponges for pharmaceutical use: A review. *Acta Pharm.* **2013**, *63*, 335–358. [CrossRef]
74. Liu, Z.; Ye, L.; Xi, J.; Wang, J.; Feng, Z.G. Cyclodextrin Polymers: Structure, Synthesis, and Use as Drug Carriers. *Prog. Polym. Sci.* **2021**, *118*, 101408. [CrossRef]
75. Zhang, M.; Wang, J.; Jin, Z. Supramolecular hydrogel formation between chitosan and hydroxypropyl β-cyclodextrin via Diels-Alder reaction and its drug delivery. *Int. J. Biol. Macromol.* **2018**, *114*, 381–391. [CrossRef]

76. Gami, P.; Kundu, D.; Seera, S.D.K.; Banerjee, T. Chemically crosslinked xylan–β-Cyclodextrin hydrogel for the in vitro delivery of curcumin and 5-Fluorouracil. *Int. J. Biol. Macromol.* **2020**, *158*, 18–31. [CrossRef] [PubMed]
77. Blanco-Fernandez, B.; Lopez-Viota, M.; Concheiro, A.; Alvarez-Lorenzo, C. Synergistic performance of cyclodextrin–agar hydrogels for ciprofloxacin delivery and antimicrobial effect. *Carbohydr. Polym.* **2011**, *85*, 765–774. [CrossRef]
78. Otero-Espinar, F.J.; Torres-Labandeira, J.J.; Alvarez-Lorenzo, C.; Blanco-Méndez, J. Cyclodextrins in drug delivery systems. *J. Drug Deliv. Sci. Tec.* **2010**, *20*, 289–301. [CrossRef]
79. Li, J.; Loh, X.J. Cyclodextrin-based supramolecular architectures: Syntheses, structures, and applications for drug and gene delivery. *Adv. Drug Deliver. Rev.* **2008**, *60*, 1000–1017. [CrossRef]
80. Sharaf, S.; El-Naggar, M.E. Wound dressing properties of cationized cotton fabric treated with carrageenan/cyclodextrin hydrogel loaded with honey bee propolis extract. *Int. J. Biol. Macromol.* **2019**, *133*, 583–591. [CrossRef] [PubMed]
81. Hewitt, M.G.; Morrison, P.W.; Boostrom, H.M.; Morgan, S.R.; Fallon, M.; Lewis, P.N. In vitro topical delivery of chlorhexidine to the cornea: Enhancement using drug-loaded contact lenses and β-cyclodextrin complexation, and the importance of simulating tear irrigation. *Mol. Pharm.* **2020**, *17*, 1428–1441. [CrossRef] [PubMed]
82. Gupta, A.; Briffa, S.M.; Swingler, S.; Gibson, H.; Kannappan, V.; Adamus, G.; Kowalczuk, M.; Martin, C.; Radecka, I. Synthesis of silver nanoparticles using curcumin-cyclodextrins loaded into bacterial cellulose-based hydrogels for wound dressing applications. *Biomacromolecules* **2020**, *21*, 1802–1811. [CrossRef] [PubMed]
83. Gularte, M.S.; Quadrado, R.F.; Pedra, N.S.; Soares, M.S.; Bona, N.P.; Spanevello, R.M.; Fajardo, A.R. Preparation, characterization and antitumor activity of a cationic starch-derivative membrane embedded with a β-cyclodextrin/curcumin inclusion complex. *Int. J. Biol. Macromol.* **2020**, *148*, 140–152. [CrossRef] [PubMed]
84. Moradi, S.; Barati, A.; Tonelli, A.E.; Hamedi, H. Effect of clinoptilolite on structure and drug release behavior of chitosan/thyme oil γ-Cyclodextrin inclusion compound hydrogels. *J. Appl. Polym. Sci.* **2021**, *138*, 49822. [CrossRef]
85. Moradi, S.; Barati, A.; Salehi, E.; Tonelli, A.E.; Hamedi, H. Preparation and characterization of chitosan based hydrogels containing cyclodextrin inclusion compounds or nanoemulsions of thyme oil. *Polym. Int.* **2019**, *68*, 1891–1902. [CrossRef]
86. Sajeesh, S.; Bouchemal, K.; Marsaud, V.; Vauthier, C.; Sharma, C.P. Cyclodextrin complexed insulin encapsulated hydrogel microparticles: An oral delivery system for insulin. *J. Control. Release* **2010**, *147*, 377–384. [CrossRef]
87. Okubo, M.; Iohara, D.; Anraku, M.; Higashi, T.; Uekama, K.; Hirayama, F. A thermoresponsive hydrophobically modified hydroxypropylmethylcellulose/cyclodextrin injectable hydrogel for the sustained release of drugs. *Int. J. Pharm.* **2020**, *575*, 118845. [CrossRef]
88. Sarkar, A.; Mackie, A.R. Engineering oral delivery of hydrophobic bioactives in real-world scenarios. *Curr. Opin. Colloid Interface Sci.* **2020**, *48*, 40–52. [CrossRef]
89. Xiang, J.; Shen, L.; Hong, Y. Status and future scope of hydrogels in wound healing: Synthesis, materials and evaluation. *Eur. Polym. J.* **2020**, *130*, 109609. [CrossRef]
90. Zhang, L.; Liu, M.; Zhang, Y.; Pei, R. Recent progress of highly adhesive hydrogels as wound dressings. *Biomacromolecules* **2020**, *21*, 3966–3983. [CrossRef] [PubMed]
91. Klotz, S.A.; Penn, C.C.; Negvesky, G.J.; Butrus, S.I. Fungal and parasitic infections of the eye. *Clin. Microbiol. Rev.* **2000**, *13*, 662–685. [CrossRef]
92. Jumelle, C.; Gholizadeh, S.; Annabi, N.; Dana, R. Advances and limitations of drug delivery systems formulated as eye drops. *J. Control. Release* **2020**, *321*, 1–22. [CrossRef]
93. El-Zeiny, H.M.; Abukhadra, M.R.; Sayed, O.M.; Osman, A.H. Ahmed, S.A. Insight into novel β-cyclodextrin-grafted-poly (N-vinylcaprolactam) nanogel structures as advanced carriers for 5-fluorouracil: Equilibrium behavior and pharmacokinetic modeling. *Colloid Surf. A* **2020**, *586*, 124197. [CrossRef]
94. Moncada-Basualto, M.; Matsuhiro, B.; Mansilla, A.; Lapier, M.; Maya, J.D.; Olea-Azar, C. Supramolecular hydrogels of β-cyclodextrin linked to calcium homopoly-l-guluronate for release of coumarins with trypanocidal activity. *Carbohydr. Polym.* **2019**, *204*, 170–181. [CrossRef]
95. Gholibegloo, E.; Mortezazadeh, T.; Salehian, F.; Ramazani, A.; Amanlou, M.; Khoobi, M. Improved curcumin loading, release, solubility and toxicity by tuning the molar ratio of cross-linker to β-cyclodextrin. *Carbohydr. Polym.* **2019**, *213*, 70–78. [CrossRef]
96. Zhou, X.; Luo, Z.; Baidya, A.; Kim, H.J.; Wang, C.; Jiang, X.; Qu, M.; Zhu, J.; Ren, L.; Vajhadin, F.; et al. Biodegradable β-Cyclodextrin Conjugated Gelatin Methacryloyl Microneedle for Delivery of Water-Insoluble Drug. *Adv. Healthc. Mater.* **2020**, *9*, 2000527. [CrossRef]
97. Khalid, Q.; Ahmad, M.; Minhas, M.U.; Batool, F.; Malik, N.S.; Rehman, M. Novel β-cyclodextrin nanosponges by chain growth condensation for solubility enhancement of dexibuprofen: Characterization and acute oral toxicity studies. *J. Drug Deliv. Sci. Technol.* **2021**, *61*, 102089. [CrossRef]
98. Soleimani, K.; Arkan, E.; Derakhshankhah, H.; Haghshenas, B.; Jahanban-Esfahlan, R.; Jaymand, M. A novel bioreducible and pH-responsive magnetic nanohydrogel based on β-cyclodextrin for chemo/hyperthermia therapy of cancer. *Carbohydr. Polym.* **2020**, *252*, 117229. [CrossRef]
99. Fiorica, C.; Palumbo, F.S.; Pitarresi, G.; Puleio, R.; Condorelli, L.; Collura, G.; Giammona, G. A hyaluronic acid/cyclodextrin based injectable hydrogel for local doxorubicin delivery to solid tumors. *Int. J. Pharm.* **2020**, *589*, 119879. [CrossRef] [PubMed]

100. Das, M.; Nariya, P.; Joshi, A.; Vohra, A.; Devkar, R.; Seshadri, S.; Thakore, S. Carbon nanotube embedded cyclodextrin polymer derived injectable nanocarrier: A multiple faceted platform for stimulation of multi-drug resistance reversal. *Carbohydr. Polym.* **2020**, *247*, 116751. [CrossRef] [PubMed]
101. Lee, H.J.; Le, P.T.; Kwon, H.J.; Park, K.D. Supramolecular assembly of tetronic–adamantane and poly (β-cyclodextrin) as injectable shear-thinning hydrogels. *J. Mat. Chem. B* **2019**, *7*, 3374–3382. [CrossRef]
102. Yang, Y.; Liu, Y.; Chen, S.; Cheong, K.L.; Teng, B. Carboxymethyl β-cyclodextrin grafted carboxymethyl chitosan hydrogel-based microparticles for oral insulin delivery. *Carbohydr. Polym.* **2020**, *246*, 116617. [CrossRef] [PubMed]
103. Hossen, S.; Hossain, M.K.; Basher, M.K.; Mia, M.N.H.; Rahman, M.T.; Uddin, M.J. Smart nanocarrier-based drug delivery systems for cancer therapy and toxicity studies: A review. *J. Adv. Res.* **2019**, *15*, 1–18. [CrossRef]
104. Luo, J.; Shi, X.; Li, L.; Tan, Z.; Feng, F.; Li, J.; Pang, M.; Wang, X.; He, L. An injectable and self-healing hydrogel with controlled release of curcumin to repair spinal cord injury. *Bioact. Mater.* **2021**, *6*, 4816–4829. [CrossRef] [PubMed]
105. Gull, N.; Khan, S.M.; Butt, O.M.; Islam, A.; Shah, A.; Jabeen, S.; Khan, S.U.; Khan, A.; Khan, R.U.; Butt, M.T.Z. Inflammation targeted chitosan-based hydrogel for controlled release of diclofenac sodium. *Int. J. Biol. Macromol.* **2020**, *162*, 175–187. [CrossRef]
106. Ata, S.; Rasool, A.; Islam, A.; Bibi, I.; Rizwan, M.; Azeem, M.K.; Iqbal, M. Loading of Cefixime to pH sensitive chitosan based hydrogel and investigation of controlled release kinetics. *Int. J. Biol. Macromol.* **2020**, *155*, 1236–1244. [CrossRef]
107. Jeong, K.; Yu, Y.J.; You, J.Y.; Rhee, W.J.; Kim, J.A. Exosome-mediated microRNA-497 delivery for anti-cancer therapy in a microfluidic 3D lung cancer model. *Lab. Chip* **2020**, *20*, 548–557. [CrossRef]
108. Yu, Y.; Xu, S.; Yu, S.; Li, J.; Tan, G.; Li, S.; Pan, W. A Hybrid Genipin-Cross-Linked Hydrogel/Nanostructured Lipid Carrier for Ocular Drug Delivery: Cellular, ex Vivo, and in Vivo Evaluation. *ACS Biomater. Sci. Eng.* **2020**, *6*, 1543–1552. [CrossRef]
109. Utine, C.A.; Stern, M.; Akpek, E.K. Clinical review: Topical ophthalmic use of cyclosporin A. *Ocul. Immunol. Inflamm.* **2010**, *18*, 352–361. [CrossRef] [PubMed]
110. Periman, L.M.; Mah, F.S.; Karpecki, P.M. A review of the mechanism of action of cyclosporine A: The role of cyclosporine a in dry eye disease and recent formulation developments. *Clin. Ophthalmol.* **2020**, *14*, 4187. [CrossRef] [PubMed]
111. Singh, M.; Bharadwaj, S.; Lee, K.E.; Kang, S.G. Therapeutic nanoemulsions in ophthalmic drug administration: Concept in formulations and characterization techniques for ocular drug delivery. *J. Control. Release* **2020**, *328*, 895–916. [CrossRef]
112. Campos, P.M.; Petrilli, R.; Lopez, R.F. The prominence of the dosage form design to treat ocular diseases. *Int. J. Pharm.* **2020**, *586*, 119577. [CrossRef]
113. Pradhan, S.; Keller, K.A.; Sperduto, J.L.; Slater, J.H. Fundamentals of laser-based hydrogel degradation and applications in cell and tissue engineering. Adv. *Healthc. Mater.* **2017**, *6*, 1700681. [CrossRef]
114. Tondera, C.; Hauser, S.; Krüger-Genge, A.; Jung, F.; Neffe, A.T.; Lendlein, A.; Klopfleisch, R.; Steinbach, J.; Neuber, C.; Pietzsch, J. Gelatin-based hydrogel degradation and tissue interaction in vivo: Insights from multimodal preclinical imaging in immunocompetent nude mice. *Theranostics* **2016**, *6*, 2114. [CrossRef] [PubMed]
115. Yang, R.; Liu, X.; Ren, Y.; Xue, W.; Liu, S.; Wang, P.; Zhao, M.; Xu, H.; Chi, B. Injectable adaptive self-healing hyaluronic acid/poly (γ-glutamic acid) hydrogel for cutaneous wound healing. *Acta Biomater.* **2021**, *127*, 102–115. [CrossRef]
116. Liu, C.; Zhang, Z.; Liu, X.; Ni, X.; Li, J. Gelatin-based hydrogels with β-cyclodextrin as a dual functional component for enhanced drug loading and controlled release. *RSC Adv.* **2013**, *3*, 25041–25049. [CrossRef]
117. Li, J. Self-assembled supramolecular hydrogels based on polymer–cyclodextrin inclusion complexes for drug delivery. *NPG Asia Mater.* **2010**, *2*, 112–118. [CrossRef]
118. Jalalvandi, E.; Cabral, J.; Hanton, L.R.; Moratti, S.C. Cyclodextrin-polyhydrazine degradable gels for hydrophobic drug delivery. *Mater. Sci. Eng. C* **2016**, *69*, 144–153. [CrossRef] [PubMed]
119. Kersani, D.; Mougin, J.; Lopez, M.; Degoutin, S.; Tabary, N.; Cazaux, F.; Janus, L.; Maton, M.; Chai, F.; Sobocinski, J.; et al. Stent coating by electrospinning with chitosan/poly-cyclodextrin based nanofibers loaded with simvastatin for restenosis prevention. *Eur. J. Pharm. Biopharm.* **2020**, *150*, 156–167. [CrossRef]
120. Yang, N.; Wang, Y.; Zhang, Q.; Chen, L.; Zhao, Y. In situ formation of poly (thiolated chitosan-co-alkylated β-cyclodextrin) hydrogels using click cross-linking for sustained drug release. *J. Mater. Sci.* **2019**, *54*, 1677–1691. [CrossRef]
121. Sheng, J.; Wang, Y.; Xiong, L.; Luo, Q.; Li, X.; Shen, Z.; Zhu, W. Injectable doxorubicin-loaded hydrogels based on dendron-like β-cyclodextrin-poly (ethylene glycol) conjugates. *Polym. Chem.* **2017**, *8*, 1680–1688. [CrossRef]
122. Van de Manakker, F.; Braeckmans, K.; Morabit, N.E.; De Smedt, S.C.; van Nostrum, C.F.; Hennink, W.E. Protein-Release Behavior of Self-Assembled PEG-β-Cyclodextrin/PEG-Cholesterol Hydrogels. *Adv. Funct. Mater.* **2009**, *19*, 2992–3001. [CrossRef]
123. Yu, J.; Fan, H.; Huang, J.; Chen, J. Fabrication and evaluation of reduction-sensitive supramolecular hydrogel based on cyclodextrin/polymer inclusion for injectable drug-carrier application. *Soft Matter* **2011**, *7*, 7386–7394. [CrossRef]
124. Huang, Z.; Liu, S.; Zhang, B.; Wu, Q. Preparation and swelling behavior of a novel self-assembled β-cyclodextrin/acrylic acid/sodium alginate hydrogel. *Carbohydr. Polym.* **2014**, *113*, 430–437. [CrossRef] [PubMed]
125. Roy, A.; Maity, P.P.; Dhara, S.; Pal, S. Biocompatible, stimuli-responsive hydrogel of chemically crosslinked β-cyclodextrin as amoxicillin carrier. *J. Appl. Polym. Sci.* **2018**, *135*, 45939. [CrossRef]
126. Li, J.; Li, X.; Ni, X.; Wang, X.; Li, H.; Leong, K.W. Self-assembled supramolecular hydrogels formed by biodegradable PEO-PHB-PEO triblock copolymers and α-cyclodextrin for controlled drug delivery. *Biomaterials* **2006**, *27*, 4132–4140. [CrossRef] [PubMed]

127. Wang, Y.; Yang, N.; Wang, D.; He, Y.; Chen, L.; Zhao, Y. Poly (MAH-β-cyclodextrin-co-NIPAAm) hydrogels with drug hosting and thermo/pH-sensitive for controlled drug release. *Polym. Degrad. Stabil.* **2018**, *147*, 123–131. [CrossRef]
128. Yoon, S.J.; Hyun, H.; Lee, D.W.; Yang, D.H. Visible light-cured glycol chitosan hydrogel containing a beta-cyclodextrin-curcumin inclusion complex improves wound healing in vivo. *Molecules* **2017**, *22*, 1513. [CrossRef]
129. Liang, J.; Dong, X.; Wei, C.; Ma, G.; Liu, T.; Kong, D.; Lv, F. A visible and controllable porphyrin-poly (ethylene glycol)/α-cyclodextrin hydrogel nanocomposites system for photo response. *Carbohydr. Polym.* **2017**, *175*, 440–449. [CrossRef]

International Journal of
Molecular Sciences

MDPI

Article

Role of Polymer Concentration and Crosslinking Density on Release Rates of Small Molecule Drugs

Francesca Briggs †, Daryn Browne † and Prashanth Asuri *

Department of Bioengineering, Santa Clara University, Santa Clara, CA 95053, USA; fbriggs@scu.edu (F.B.); dbrowne@scu.edu (D.B.)
* Correspondence: asurip@scu.edu; Tel.: +1-408-551-3005
† These authors contributed equally to this work.

Abstract: Over the past few years, researchers have demonstrated the use of hydrogels to design drug delivery platforms that offer a variety of benefits, including but not limited to longer circulation times, reduced drug degradation, and improved targeting. Furthermore, a variety of strategies have been explored to develop stimulus-responsive hydrogels to design smart drug delivery platforms that can release drugs to specific target areas and at predetermined rates. However, only a few studies have focused on exploring how innate hydrogel properties can be optimized and modulated to tailor drug dosage and release rates. Here, we investigated the individual and combined roles of polymer concentration and crosslinking density (controlled using both chemical and nanoparticle-mediated physical crosslinking) on drug delivery rates. These experiments indicated a strong correlation between the aforementioned hydrogel properties and drug release rates. Importantly, they also revealed the existence of a saturation point in the ability to control drug release rates through a combination of chemical and physical crosslinkers. Collectively, our analyses describe how different hydrogel properties affect drug release rates and lay the foundation to develop drug delivery platforms that can be programmed to release a variety of bioactive payloads at defined rates.

Keywords: controlled drug delivery; hydrogel properties; polymer concentration; crosslinking density

Citation: Briggs, F.; Browne, D.; Asuri, P. Role of Polymer Concentration and Crosslinking Density on Release Rates of Small Molecule Drugs. *Int. J. Mol. Sci.* **2022**, *23*, 4118. https://doi.org/10.3390/ijms23084118

Academic Editor: Yury A. Skorik

Received: 23 March 2022
Accepted: 6 April 2022
Published: 8 April 2022

1. Introduction

Conventional drug delivery systems are often limited by poor targeting and circulation times [1–4]. Recent research has demonstrated the use of controlled drug delivery systems based on nanoparticles, liposomes, hydrogels, and membranes, amongst others, to address these issues [1,4,5]. In principle, controlled drug delivery can control when and how bioactive molecules are made available to cells and tissues, thereby enhancing their efficacy and reducing their required dosage and toxicity [2,4]. Hydrogels, in particular, have received much attention given their low toxicity and excellent biocompatibility, tunable physical properties (e.g., porosity, crosslinking density) that can be leveraged to modulate drug loads and release rates, and their ability to stabilize and protect labile biomolecules from degradation [6–9]. Furthermore, hydrogels can be prepared with a wide variety of additional physical properties, including bioinertness, biodegradability, and bioresorbability to meet the needs of specific applications [8,10,11]. They can also be formed into any desired shape or size, providing a flexible platform to meet the requirements of various drug delivery routes and systems [6,12,13]. Finally, hydrogels can be designed to respond to various biological stimuli (both internal and external), and such stimulus-responsive hydrogels have demonstrated wide applicability for the delivery of both small molecule and macromolecule drugs [8,14,15].

Currently, the majority of the studies focus on the stimulus responsiveness of hydrogels to develop a wide variety of smart systems that exhibit tailorable drug delivery properties in response to one or more (dual-responsive) stimuli, such as pH, temperature, light, enzymes, etc. [16–19]. Fewer studies focus on manipulating and studying the role of inherent hydrogel properties (i.e., monomer concentration, crosslinking density) to tailor drug delivery amounts and rates [20–26]. Developing a strong understanding of how innate hydrogel properties influence their ability to deliver drugs will serve to provide an additional design feature in the development of more efficacious and responsive drug delivery platforms. In this paper, we studied the role of polymer and crosslinker concentrations on the release profiles of the cancer drug, 5-fluorouracil (5-FU), using polyacrylamide (pAAm) hydrogels as the model system. pAAm hydrogel was chosen as the model as it is well characterized, commercially available, and routinely used in a wide variety of scientific applications, including for drug delivery [27–30]. These studies and results from additional supplementary experiments (using nanoparticle-based crosslinkers and double-network hydrogels) demonstrated a strong role for both polymer concentration and crosslinking density on drug delivery rates. The differences in the drug delivery rates translated into statistically significant differences in cell viability when U-87 glioblastoma cells were exposed to the different hydrogel-5-FU formulations.

2. Results

2.1. Influence of Polymer Concentration on Drug Release Rates

First, we explored the role of polymer concentration on the release rates for different concentrations of 5-FU in pAAm hydrogels (Figure 1). These experiments clearly revealed reduced drug release rates with increasing pAAm concentration. It is important to note that the differences in drug release rates are quite significant; after 24 h only, <40% of the drug had been released from 10% of the pAAm hydrogels, whereas ca. 70% of the drug had been released from 2.5% pAAm hydrogels. Furthermore, the differences in slopes for the initial and later stages of drug release for the different hydrogel samples—9, 5.6, and 1.5 for 2.5, 5, and 10% pAAm, respectively—indicated increased burst releases for lower concentrations of the hydrogel. Previous studies have indicated that higher drug loadings often result in increased burst release [31]; we were curious, therefore, to study if the polymer concentration had an effect on the burst release at higher drug loadings. Consistent with the drug release profiles for lower loadings, we observed reduced drug release rates and burst releases at higher pAAm concentrations (Figure 1b). The differences in slopes for the initial and later stages of drug release for the higher drug loadings were 9.5, 9.0, and 2.5 for 2.5%, 5%, and 10% pAAm, respectively.

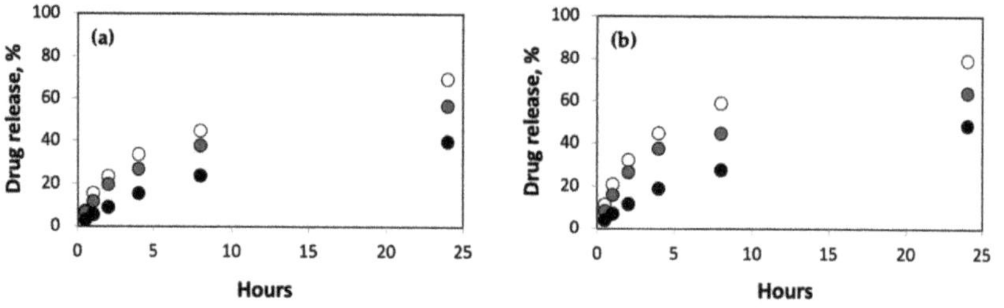

Figure 1. Drug release (%) from hydrogels prepared using 2.5% (white circles), 5% (grey circles), and 10% (black circles) AAm for (**a**) normal drug loading (1:8 *w*/*w* 5-FU:pAAm) and (**b**) high drug loading (1:2 *w*/*w* 5-FU:pAAm). Data shown are the mean of triplicate measurements with a standard error of <15%.

2.2. Influence of Crosslinking Density on Drug Release Rates

Next, we studied the effect of crosslinking density on the rate of drug release from pAAm hydrogels of varying polymer concentrations. The crosslinking density was modified by changing the concentration of the chemical crosslinker, N,N′-methylenebis-acrylamide (Bis). Irrespective of the polymer concentration, we observed a decrease in drug release rates for pAAM hydrogels crosslinked using higher concentrations of Bis (Figure 2). Previous studies have indicated that the swellability of hydrogels is an important parameter in controlling drug release rates, with lower swelling ratios leading to slower release profiles [32–34]. To confirm this correlation, we compared the swellability of the different hydrogel formulations (Figure 3). These results clearly confirmed a significant negative correlation between crosslinking density and swellability. Taken together, results from Figures 2 and 3 support previously established hypotheses that lower drug release rates for hydrogels prepared using a higher crosslinking ratio may be due to decreased swellability.

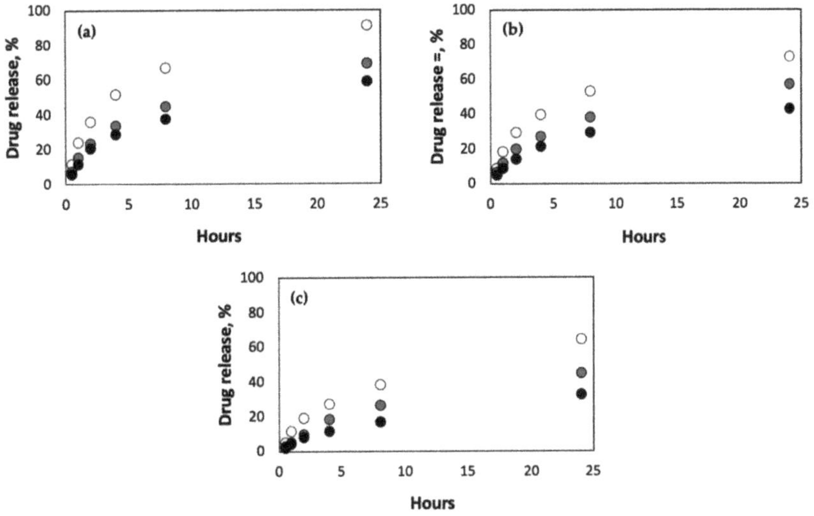

Figure 2. Drug release (%) from hydrogels prepared using (**a**) 2.5% AAm, (**b**) 5% AAm, and (**c**) 10% AAm, and different concentrations of Bis: 0.0625% (white circles), 0.25% (grey circles) and 1% (black circles). Data shown are the mean of triplicate measurements with standard error of <15%.

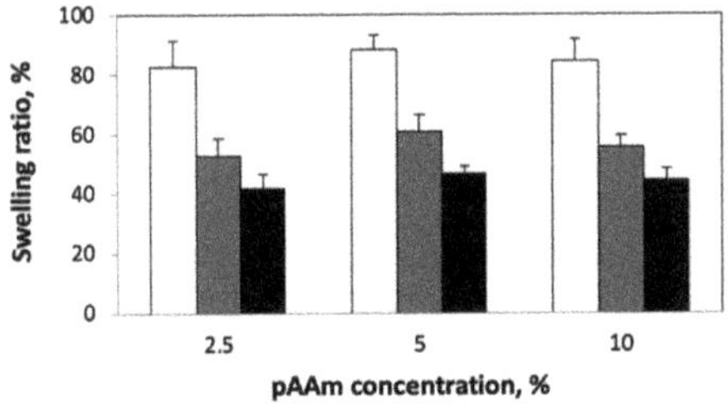

Figure 3. Swelling ratios (%) of hydrogels after 24 h incubation in pH 7.2 buffer and at 37 °C, prepared using different concentrations of AAm Bis: 0.0625% (white), 0.25% (grey) and 1% (black). Data shown are the mean and standard error of triplicate measurements.

2.3. *Nanoparticle-Mediated Increases in Crosslinking Density Affects Drug Release Rates*

To confirm the role of crosslinking density on swellability and drug release rates, we performed additional experiments using pAAm hydrogels incorporating silica nanoparticles (SiNPs). Previous investigations, including our own research, have demonstrated that nanoparticles may facilitate the formation of non-covalent or pseudo crosslinks within a hydrogel network and thereby contribute to the overall crosslinking density [35–39]. Figure 4 compares the drug release rates and swellability for 5% pAAm prepared using different concentrations of SiNPs. The data indicated that both drug release rates and swellability decreased with increasing nanoparticle concentration; furthermore, they exhibited similar saturation behaviors (between 2 and 3% SiNPs). To offer additional support to the data that suggested that nanoparticle-mediated increases in crosslinking density can affect drug release rates, we repeated the drug release studies over a range of Bis and SiNP concentrations (Table 1). Unsurprisingly, both Bis and SiNP concentrations played a significant role on the drug release rates. Interestingly, we observed a decreased impact of SiNPs on the drug release rates at higher Bis concentrations. Collectively, these results (i.e., data presented in Figure 4 and Table 1) suggest the existence of a saturation point in the ability to control drug release rates by modulating the crosslinking density of the hydrogel network.

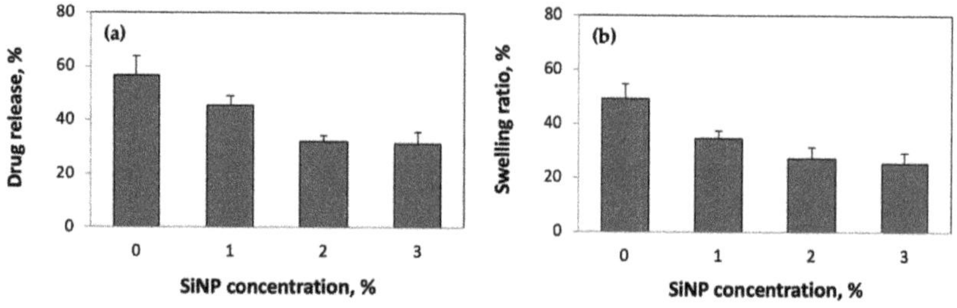

Figure 4. (**a**) Drug release (%) and (**b**) swelling ratios (%) for 5% pAAm hydrogels prepared using different concentrations of SiNPs after 24 h incubation in pH 7.2 buffer and at 37 °C. Data shown are the mean and standard error of triplicate measurements.

Table 1. Drug release (%) from 5% pAAm hydrogels prepared using different concentrations of Bis and SiNPs after 24 h incubation in pH 7.2 buffer and at 37 °C.

Bis, %	0% SiNPs	2% SiNPs
0.0625	76.1 ± 9	35.4 ± 3
0.25	56.6 ± 5	31.2 ± 3
1.0	42.3 ± 5	30.3 ± 2

2.4. Drug Release Rates in Double-Network Hydrogels

To further validate the aforementioned observations that suggest the ability to control drug release rates by modulating the hydrogel crosslinking density, we repeated the nanoparticle studies using double-network (DN) hydrogels. In a previous study, we demonstrated that nanoparticle-mediated enhancements in the mechanical properties of DN hydrogels were strongly dependent on the extent to which SiNPs interact with the individual networks and contribute to the overall crosslinking density of the hydrogel network [40]. We were intrigued to test if these observations may be extended to drug release, or in other words, do differences in the extent of nanoparticle-mediated increases in crosslinking within DN hydrogels also impact drug release rates. Table 2 compares the drug release rates for DN hydrogels prepared using either alginate or agarose as the second network and incorporating or not incorporating SiNPs. The data reveals two important takeaways: (i) the addition of a second polymer network led to reduced drug release rates, and (ii) reduced drug release rates for DN hydrogels prepared using SiNPs relative to neat hydrogels incorporating SiNPs. Furthermore, the observed effects on SiNPs on drug release rates were consistent with the observations made for the mechanical properties of DN hydrogels due to the addition of nanoparticles [40]. We observed reduced decreases in release rates for pAAM-agarose DN hydrogels compared to pAAM-alginate DN hydrogels. We attribute these differences in SiNP mediated effects on drug release rates (and mechanical properties) to differences in the extent to which the nanoparticles may interact with the polymer chains. While it has been reported that alginate may non-covalently associate with silica nanoparticle surface [41], there are no published works (to the best of our knowledge) demonstrating positive interactions between agarose and silica nanoparticles.

Table 2. Drug release (%) from double-network hydrogels prepared using different concentrations of SiNPs after 24 h incubation in pH 7.2 buffer and at 37 °C.

Sample	0% SiNPs	2% SiNPs
5% pAAm	56.6 ± 5	31.2 ± 3
5% pAAm–1% alginate	43.2 ± 3	9.5 ± 1
5% pAAm–1% agarose	40.4 ± 5	30.3 ± 4

2.5. Differences in Drug Release Rates Translates to Differences in Cytotoxicity

Finally, we performed experiments to confirm that the differences in drug release rates were biologically relevant, i.e., differences in drug release rates from the different hydrogel formulations translate to differences in compound-mediated cell viability. For this, we exposed human U-87 glioblastoma cells cultured on tissue-culture polystyrene surfaces to pAAm hydrogels, prepared using different polymer and crosslinker (Bis) concentrations, containing 5-FU. Figure 5 compares the percentage viability for U-87 cells exposed to 5-FU released from different formulations of pAAm hydrogels. These results clearly demonstrate significant differences in cell viability and a strong correlation between differences in drug release rates (mediated by differences in either polymer or crosslinker concentrations) and cell viability. We observe lower cell death for U-87 cells exposed to hydrogel formulations prepared using higher concentrations of pAAm, i.e., conditions that led to reduced drug release rates (Figure 5a). Similarly, increased crosslinking densities (achieved by increasing the concentration of Bis) led to reduced 5-FU release rates and decreased cell death (Figure 5b).

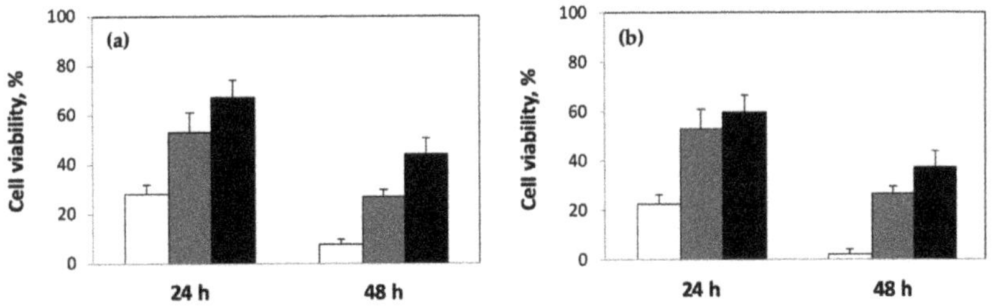

Figure 5. Viability of U-87 cells (%) exposed to 5-FU released from pAAm hydrogels prepared using (**a**) different concentrations of AAm: 2.5% (white), 5% (grey), and 10% (black) and 0.25% Bis, or (**b**) 5% AAm and different concentrations of Bis: 0.0625% (white), 0.25% (grey) and 1% (black). Data shown are the mean and standard error of triplicate measurements.

3. Discussion

Research over the past few years has clearly demonstrated the advantages of controlled drug delivery over conventional delivery platforms, including increased stability and bioavailability, improved and more reliable therapeutic effects, and reduced occurrence and intensity of adverse effects [1,4,42]. Hydrogels hold enormous potential for controlled drug delivery owing to their tunable physical properties, including porosity and degradability, and ability to respond to a variety of chemical and biological stimuli [6,8,43–46]. Furthermore, they are nontoxic and biocompatible, can be engineered to deliver drugs with a range of chemical properties, and possess the ability to protect labile drugs from degradation [44,46–49]. Combined, these properties enable hydrogel-based platforms to serve as excellent candidates for the controlled delivery of various therapeutic agents, including small molecule and macromolecular drugs. Hydrogel properties may be manipulated using a range of physical and chemical strategies to tailor these properties for controlled drug delivery [21,25,50]. In this study, our primary objective was to investigate if we could leverage our current understanding of manipulating the properties of hydrogels to develop drug delivery platforms with tunable release properties. Specifically, we wished to study the role of polymer and crosslinker concentrations on drug release profiles and develop an understanding of how these variables influence the ability of hydrogels to deliver drugs in a more responsive manner.

Our results provided strong evidence in support of the influence of polymer concentration and crosslinking density on drug release rates. Although others have reported similar results (i.e., negative correlations between polymer and/or crosslinker concentrations and drug release rates), we believe a more systematic approach, as performed in this work, is warranted before these properties may be used in combination with the stimulus-responsive properties as design features to develop more efficacious and flexible drug delivery platforms. To this effect, we were also intrigued to explore the correlation between nanoparticle-mediated enhancements in mechanical properties of hydrogels and the ability of nanoparticles to influence drug release properties of hydrogels. Previous investigations have clearly indicated that the ability of nanoparticles to improve hydrogel elastic modulus stems from their ability to increase the crosslinking density of polymer networks—an attribute that has also been shown to influence drug release rates from polymers [35–39]. Therefore, perhaps, it is not surprising that we observe a strong correlation between nanoparticle-mediated effects on hydrogel modulus and drug release rates. Our results indicated that at higher concentrations of nanoparticles, the hydrogel nanocomposite formulations exhibited a slower rate of drug release, possibly due to the increase in crosslinking density. We believe our work will provide a foundation for future studies that aim to combine one or more properties of hydrogels to develop drug delivery platforms with orthogonal design features for manipulating and controlling the release of small molecule drugs. These future studies and models may also enable the development and use of tunable hydrogel platforms, including polymeric nanocarriers and nanogels [51,52], for the delivery of a wide range of therapeutics, including peptide, protein, and nucleic acid payloads.

4. Materials and Methods

4.1. Materials

Materials for the preparation of the hydrogels, acrylamide (AAm, 40% *w/v*), N,N′-methylenebisacrylamide (Bis, 2% *w/v*), alginic acid (sodium salt, low viscosity), ammonium persulfate (APS), N,N,N′,N′-tetramethylethylenediamine (TEMED), and calcium chloride ($CaCl_2$) were purchased from Sigma Aldrich (Saint Louis, MO, USA), and agarose (low melting) was purchased from Thermo Fisher Scientific (Waltham, MA, USA), and used as received. Tris-HCl buffer (pH 7.2) was purchased from Life Technologies (Carlsbad, CA, USA) and binzil silica nanoparticle colloid solution with a mean particle size of 4 nm was a gift from AkzoNobel Pulp and Performance Chemicals Inc. (Marietta, GA, USA). 5-FU was purchased from Sigma (Saint Louis, MO, USA). Human U-87 glioblastoma cells were obtained from ATCC (Manassas, VA, USA), Dulbecco's modified Eagle medium (DMEM) from Mediatech (Manassas, VA, USA), fetal bovine serum and penicillin-streptomycin from Invitrogen (Carlsbad, CA, USA), and sodium pyruvate, MEM non-essential amino acids and GlutaMax from Life Technologies (Carlsbad, CA, USA). WST cell proliferation assay kit was purchased from Dojindo Molecular Technologies (Rockville, MD, USA).

4.2. Preparation of Hydrogel Samples

All hydrogel samples for drug release studies were prepared using an acrylic mold (1.6 mm thick and 6.5 mm in radius) at room temperature as previously described [37]. The polymerization reactions were performed between parallel plates of the mold to minimize exposure to air as oxygen inhibits the free radical polymerization reaction for pAAm. For pAAm samples, the monomer (AAm), crosslinker (Bis), and 5-FU stocks were diluted to their desired concentrations in pH 7.2, 250 mM Tris–HCl buffer, followed by the addition of TEMED (0.1% of the final reaction volume) and 10% *w*/*v* APS solution (1% of the final reaction volume). Final concentration of 5-FU was 1:8 *w*/*w* drug:polymer and 1:2 *w*/*w* drug:polymer for normal and high drug loadings, respectively. DN hydrogel samples composed of pAAm and alginate were prepared by first dissolving alginate in Tris–HCl buffer at room temperature. The alginate stock solutions were then diluted to the desired concentrations and added to the pAAm/5-FU reaction mixture prior to the addition of APS and TEMED, followed by the addition of 100 mM $CaCl_2$ in Tris–HCl buffer to crosslink the alginate. For DN hydrogels composed of pAAm and agarose, agarose stocks were prepared by first dissolving agarose in Tris–HCl buffer at 70 °C. The pAAm reaction mixture was then warmed to 37 °C, before the addition of agarose stock solutions at 37 °C and subsequent addition of APS and TEMED. For nanocomposite hydrogels, various amounts of silica nanoparticles were added to the reaction mixture prior to the addition of APS and TEMED (and $CaCl_2$ for hydrogels made using alginate).

4.3. Measurement of Drug Release Rates

Drug release rates for 5-FU were determined by incubating the hydrogel-5-FU samples in pH 7.2, 100 mM Tris–HCl buffer, at 37 °C. At specific time intervals, the absorbance of the releasate solution was measured at 266 nm (Tecan Infinite 200 PRO, Tecan, Switzerland). The experiments were performed in triplicate, and the average drug release rates were calculated as:

$$\text{Drug release, \%} = \frac{\text{Drug released from the hydrogel}}{\text{Total drug in the hydrogel}} \times 100 \quad (1)$$

4.4. Measurement of Swelling Rates

To measure the swelling properties of the hydrogel samples, the hydrogel disks containing no 5-FU were prepared as described above, wiped with tissue paper to remove any excess water, before weighing to determine the initial weight (W_{0h}). The samples were then immersed in pH 7.2, 100 mM Tris–HCl buffer for 24 h at 37 °C. Their final weights were recorded (W_{24h}) after first blotting excess buffer with tissue paper. The experiments were performed in triplicate, and the average swelling ratios were calculated as:

$$\text{Swelling ratio, \%} = \frac{W_{24h} - W_{0h}}{W_{0h}} \times 100 \quad (2)$$

4.5. Cell Maintenance and Measurement of Cytotoxic Responses

U-87 cells were maintained in DMEM supplemented with 15% fetal bovine serum, sodium pyruvate, MEM non-essential amino acids, GlutaMax, and 1% penicillin–streptomycin at 37 °C in a 5% CO_2 humidified environment. Cell culture media was changed every other day and cells were passaged every 4–5 days using 0.25% trypsin/EDTA. For cell toxicity studies, U-87 cells were seeded into 96-well flat-bottom plates at a cell density of 8000–10,000 cells/well and allowed to proliferate for 24 h before exposure to the hydrogel disks containing or not containing (control) 5-FU. Cell toxic responses to the drug released from the hydrogel disks were measured by exposing the cells to the hydrogel-5-FU samples for predetermined periods of time and quantified using the commercially available, formazan-based WST assay, as per manufacturer's instructions. Briefly, 20 mL of the WST solution for every 100 mL of culture media was added directly to the wells (after removing the hydrogel-5-FU samples) and the cells were incubated for up to 4 h in a humidified incubator at 37 °C and 5% CO_2. 80 mL of the WST-media solution was transferred from each well into a well of a new 96-well plate in order to avoid any background absorbance, and the absorbance was measured at 570 nm (Tecan Infinite 200 PRO, Tecan, Switzerland).

4.6. Statistical Analysis

Average and standard error was calculated using Microsoft Excel (v. 16.54) and the standard error was presented in the form of error bars in the graphs.

Author Contributions: P.A. conceived and designed the experiments; P.A. collected the data; F.B., D.B. and P.A. analyzed the data and wrote the paper. All authors have read and agreed to the published version of the manuscript.

Funding: This research received no external funding.

Institutional Review Board Statement: Not applicable.

Informed Consent Statement: Not applicable.

Data Availability Statement: Not applicable.

Acknowledgments: This work was supported by the School of Engineering at Santa Clara University.

Conflicts of Interest: The authors declare no conflict of interest.

References

1. Park, K. The Controlled Drug Delivery Systems: Past Forward and Future Back. *J. Control. Release* **2014**, *190*, 3–8. [CrossRef]
2. Senapati, S.; Mahanta, A.K.; Kumar, S.; Maiti, P. Controlled Drug Delivery Vehicles for Cancer Treatment and Their Performance. *Signal Transduct. Target. Ther.* **2018**, *3*, 7. [CrossRef]
3. Lorscheider, M.; Gaudin, A.; Nakhlé, J.; Veiman, K.-L.; Richard, J.; Chassaing, C. Challenges and Opportunities in the Delivery of Cancer Therapeutics: Update on Recent Progress. *Ther. Deliv.* **2021**, *12*, 55–76. [CrossRef]
4. Adepu, S.; Ramakrishna, S. Controlled Drug Delivery Systems: Current Status and Future Directions. *Molecules* **2021**, *26*, 5905. [CrossRef] [PubMed]
5. Yun, Y.; Lee, B.K.; Park, K. Controlled Drug Delivery Systems: The next 30 Years. *Front. Chem. Sci. Eng.* **2014**, *8*, 276–279. [CrossRef]
6. Li, J.; Mooney, D.J. Designing Hydrogels for Controlled Drug Delivery. *Nat. Rev. Mater.* **2016**, *1*, 16071. [CrossRef] [PubMed]
7. Thambi, T.; Li, Y.; Lee, D.S. Injectable Hydrogels for Sustained Release of Therapeutic Agents. *J. Control. Release* **2017**, *267*, 57–66. [CrossRef]
8. Sun, Z.; Song, C.; Wang, C.; Hu, Y.; Wu, J. Hydrogel-Based Controlled Drug Delivery for Cancer Treatment: A Review. *Mol. Pharm.* **2020**, *17*, 373–391. [CrossRef]
9. Jacob, S.; Nair, A.B.; Shah, J.; Sreeharsha, N.; Gupta, S.; Shinu, P. Emerging Role of Hydrogels in Drug Delivery Systems, Tissue Engineering and Wound Management. *Pharmaceutics* **2021**, *13*, 357. [CrossRef]
10. Zhang, Y.; Huang, Y. Rational Design of Smart Hydrogels for Biomedical Applications. *Front. Chem.* **2021**, *8*, 615665. [CrossRef]
11. Cao, H.; Duan, L.; Zhang, Y.; Cao, J.; Zhang, K. Current Hydrogel Advances in Physicochemical and Biological Response-Driven Biomedical Application Diversity. *Signal Transduct. Target. Ther.* **2021**, *6*, 426. [CrossRef]
12. Gonçalves, C.; Pereira, P.; Gama, M. Self-Assembled Hydrogel Nanoparticles for Drug Delivery Applications. *Materials* **2010**, *3*, 1420–1460. [CrossRef]
13. Neto, A.I.; Demir, K.; Popova, A.A.; Oliveira, M.B.; Mano, J.F.; Levkin, P.A. Fabrication of Hydrogel Particles of Defined Shapes Using Superhydrophobic-Hydrophilic Micropatterns. *Adv. Mater.* **2016**, *28*, 7613–7619. [CrossRef]
14. Sood, N.; Bhardwaj, A.; Mehta, S.; Mehta, A. Stimuli-Responsive Hydrogels in Drug Delivery and Tissue Engineering. *Drug Deliv.* **2016**, *23*, 758–780. [CrossRef]
15. Kasiński, A.; Zielińska-Pisklak, M.; Oledzka, E.; Sobczak, M. Smart Hydrogels—Synthetic Stimuli-Responsive Antitumor Drug Release Systems. *Int. J. Nanomed.* **2020**, *15*, 4541–4572. [CrossRef]
16. Lin, X.; Ma, Q.; Su, J.; Wang, C.; Kankala, R.K.; Zeng, M.; Lin, H.; Zhou, S.-F. Dual-Responsive Alginate Hydrogels for Controlled Release of Therapeutics. *Molecules* **2019**, *24*, 2089. [CrossRef]
17. Chatterjee, S.; Hui, P.C.; Kan, C.; Wang, W. Dual-Responsive (PH/Temperature) Pluronic F-127 Hydrogel Drug Delivery System for Textile-Based Transdermal Therapy. *Sci. Rep.* **2019**, *9*, 11658. [CrossRef]
18. Laurano, R.; Boffito, M.; Abrami, M.; Grassi, M.; Zoso, A.; Chiono, V.; Ciardelli, G. Dual Stimuli-Responsive Polyurethane-Based Hydrogels as Smart Drug Delivery Carriers for the Advanced Treatment of Chronic Skin Wounds. *Bioact. Mater.* **2021**, *6*, 3013–3024. [CrossRef]
19. Obireddy, S.R.; Lai, W.-F. Multi-Component Hydrogel Beads Incorporated with Reduced Graphene Oxide for PH-Responsive and Controlled Co-Delivery of Multiple Agents. *Pharmaceutics* **2021**, *13*, 313. [CrossRef]
20. Cury, B.S.F.; de Castro, A.D.; Klein, S.I.; Evangelista, R.C. Influence of Phosphated Cross-Linked High Amylose on in Vitro Release of Different Drugs. *Carbohydr. Polym.* **2009**, *78*, 789–793. [CrossRef]
21. Roig-Roig, F.; Solans, C.; Esquena, J.; García-Celma, M.J. Preparation, Characterization, and Release Properties of Hydrogels Based on Hyaluronan for Pharmaceutical and Biomedical Use. *J. Appl. Polym. Sci.* **2013**, *130*, 1377–1382. [CrossRef]
22. Bertz, A.; Wöhl-Bruhn, S.; Miethe, S.; Tiersch, B.; Koetz, J.; Hust, M.; Bunjes, H.; Menzel, H. Encapsulation of Proteins in Hydrogel Carrier Systems for Controlled Drug Delivery: Influence of Network Structure and Drug Size on Release Rate. *J. Biotechnol.* **2013**, *163*, 243–249. [CrossRef]

23. Akhtar, M.F.; Ranjha, N.M.; Hanif, M. Effect of Ethylene Glycol Dimethacrylate on Swelling and on Metformin Hydrochloride Release Behavior of Chemically Crosslinked PH–Sensitive Acrylic Acid–Polyvinyl Alcohol Hydrogel. *DARU J. Pharm. Sci.* **2015**, *23*, 41. [CrossRef]
24. Agarwal, S.; Murthy, R.S.R. Effect of Different Polymer Concentration on Drug Release Rate and Physicochemical Properties of Mucoadhesive Gastroretentive Tablets. *Indian J. Pharm. Sci.* **2015**, *77*, 705–714. [CrossRef]
25. Kopač, T.; Ručigaj, A.; Krajnc, M. The Mutual Effect of the Crosslinker and Biopolymer Concentration on the Desired Hydrogel Properties. *Int. J. Biol. Macromol.* **2020**, *159*, 557–569. [CrossRef]
26. Lai, W.-F.; Huang, E.; Lui, K.-H. Alginate-based Complex Fibers with the Janus Morphology for Controlled Release of Co-delivered Drugs. *Asian J. Pharm. Sci.* **2021**, *16*, 77–85. [CrossRef]
27. Telegeev, G.; Kutsevol, N.; Chumachenko, V.; Naumenko, A.; Telegeeva, P.; Filipchenko, S.; Harahuts, Y. Dextran-Polyacrylamide as Matrices for Creation of Anticancer Nanocomposite. *Int. J. Polym. Sci.* **2017**, *2017*, e4929857. [CrossRef]
28. Bashir, S.; Teo, Y.Y.; Naeem, S.; Ramesh, S.; Ramesh, K. PH Responsive N-Succinyl Chitosan/Poly (Acrylamide-Co-Acrylic Acid) Hydrogels and in Vitro Release of 5-Fluorouracil. *PLoS ONE* **2017**, *12*, e0179250. [CrossRef]
29. Qiao, Z.; Tran, L.; Parks, J.; Zhao, Y.; Hai, N.; Zhong, Y.; Ji, H. Highly Stretchable Gelatin-polyacrylamide Hydrogel for Potential Transdermal Drug Release. *Nano Sel.* **2021**, *2*, 107–115. [CrossRef]
30. Torres-Figueroa, A.V.; Pérez-Martínez, C.J.; del Castillo-Castro, T.; Bolado-Martínez, E.; Corella-Madueño, M.A.G.; García-Alegría, A.M.; Lara-Ceniceros, T.E.; Armenta-Villegas, L. Composite Hydrogel of Poly(Acrylamide) and Starch as Potential System for Controlled Release of Amoxicillin and Inhibition of Bacterial Growth. *J. Chem.* **2020**, *2020*, e5860487. [CrossRef]
31. Chou, S.-F.; Carson, D.; Woodrow, K.A. Current Strategies for Sustaining Drug Release from Electrospun Nanofibers. *J. Control. Release* **2015**, *220*, 584–591. [CrossRef]
32. Kou, J.H.; Fleisher, D.; Amidon, G.L. Modeling Drug Release from Dynamically Swelling Poly(Hydroxyethyl Methacrylate-Co-Methacrylic Acid) Hydrogels. *J. Control. Release* **1990**, *12*, 241–250. [CrossRef]
33. Kim, S.W.; Bae, Y.H.; Okano, T. Hydrogels: Swelling, Drug Loading, and Release. *Pharm. Res.* **1992**, *9*, 283–290. [CrossRef]
34. Hegab, R.A.; Pardue, S.; Shen, X.; Kevil, C.; Peppas, N.A.; Caldorera-Moore, M.E. Effect of Network Mesh Size and Swelling to the Drug Delivery from PH Responsive Hydrogels. *J. Appl. Polym. Sci.* **2020**, *137*, 48767. [CrossRef]
35. Xia, L.-W.; Xie, R.; Ju, X.-J.; Wang, W.; Chen, Q.; Chu, L.-Y. Nano-Structured Smart Hydrogels with Rapid Response and High Elasticity. *Nat. Commun.* **2013**, *4*, 2226. [CrossRef]
36. Rose, S.; Prevoteau, A.; Elzière, P.; Hourdet, D.; Marcellan, A.; Leibler, L. Nanoparticle Solutions as Adhesives for Gels and Biological Tissues. *Nature* **2014**, *505*, 382–385. [CrossRef]
37. Zaragoza, J.; Babhadiashar, N.; O'Brien, V.; Chang, A.; Blanco, M.; Zabalegui, A.; Lee, H.; Asuri, P. Experimental Investigation of Mechanical and Thermal Properties of Silica Nanoparticle-Reinforced Poly(Acrylamide) Nanocomposite Hydrogels. *PLoS ONE* **2015**, *10*, e0136293. [CrossRef]
38. Adibnia, V.; Taghavi, S.M.; Hill, R.J. Roles of Chemical and Physical Crosslinking on the Rheological Properties of Silica-Doped Polyacrylamide Hydrogels. *Rheol. Acta* **2017**, *56*, 123–134. [CrossRef]
39. Zaragoza, J.; Fukuoka, S.; Kraus, M.; Thomin, J.; Asuri, P. Exploring the Role of Nanoparticles in Enhancing Mechanical Properties of Hydrogel Nanocomposites. *Nanomaterials* **2018**, *8*, 882. [CrossRef]
40. Chang, A.; Babhadiashar, N.; Barrett-Catton, E.; Asuri, P. Role of Nanoparticle–Polymer Interactions on the Development of Double-Network Hydrogel Nanocomposites with High Mechanical Strength. *Polymers* **2020**, *12*, 470. [CrossRef]
41. Theodorakis, N.; Saravanou, S.-F.; Kouli, N.-P.; Iatridi, Z.; Tsitsilianis, C. PH/Thermo-Responsive Grafted Alginate-Based SiO2 Hybrid Nanocarrier/Hydrogel Drug Delivery Systems. *Polymers* **2021**, *13*, 1228. [CrossRef]
42. Langer, R. Drug Delivery and Targeting. *Nature* **1998**, *392*, 5–10.
43. Pillai, O.; Panchagnula, R. Polymers in Drug Delivery. *Curr. Opin. Chem. Biol.* **2001**, *5*, 447–451. [CrossRef]
44. McKenzie, M.; Betts, D.; Suh, A.; Bui, K.; Kim, L.D.; Cho, H. Hydrogel-Based Drug Delivery Systems for Poorly Water-Soluble Drugs. *Molecules* **2015**, *20*, 20397–20408. [CrossRef]
45. Narayanaswamy, R.; Torchilin, V.P. Hydrogels and Their Applications in Targeted Drug Delivery. *Molecules* **2019**, *24*, 603. [CrossRef]
46. Rizzo, F.; Kehr, N.S. Recent Advances in Injectable Hydrogels for Controlled and Local Drug Delivery. *Adv. Healthc. Mater.* **2021**, *10*, e2001341. [CrossRef]
47. Roorda, W.E.; Boddé, H.E.; De Boer, A.G.; Bouwstra, J.A.; Junginer, H.E. Synthetic Hydrogels as Drug Delivery Systems. *Pharm. Weekbl. Sci.* **1986**, *8*, 165–189. [CrossRef]
48. Hoare, T.R.; Kohane, D.S. Hydrogels in Drug Delivery: Progress and Challenges. *Polymer* **2008**, *49*, 1993–2007. [CrossRef]
49. Trombino, S.; Cassano, R. Special Issue on Designing Hydrogels for Controlled Drug Delivery: Guest Editors' Introduction. *Pharmaceutics* **2020**, *12*, 57. [CrossRef] [PubMed]
50. Das, R.; Bhattacharjee, C. 23—Hydrogel Nanocomposite for Controlled Drug Release. In *Applications of Nanocomposite Materials in Drug Delivery*; Asiri, A.M., Mohammad, A., Eds.; Woodhead Publishing Series in Biomaterials; Woodhead Publishing: Sawston, UK, 2018; pp. 575–588, ISBN 978-0-12-813741-3.
51. Sharma, A.; Garg, T.; Aman, A.; Panchal, K.; Sharma, R.; Kumar, S.; Markandeywar, T. Nanogel—an Advanced Drug Delivery Tool: Current and Future. *Artif. Cells Nanomed. Biotechnol.* **2016**, *44*, 165–177. [CrossRef] [PubMed]
52. Das, S.S.; Bharadwaj, P.; Bilal, M.; Barani, M.; Rahdar, A.; Taboada, P.; Bungau, S.; Kyzas, G.Z. Stimuli-Responsive Polymeric Nanocarriers for Drug Delivery, Imaging, and Theragnosis. *Polymers* **2020**, *12*, 1397. [CrossRef]

International Journal of
Molecular Sciences

MDPI

Article

Hyaluronan/Diethylaminoethyl Chitosan Polyelectrolyte Complexes as Carriers for Improved Colistin Delivery

Natallia V. Dubashynskaya [1], Sergei V. Raik [1], Yaroslav A. Dubrovskii [2,3,4], Elena V. Demyanova [5], Elena S. Shcherbakova [5], Daria N. Poshina [1], Anna Y. Shasherina [3], Yuri A. Anufrikov [3] and Yury A. Skorik [1,*]

1 Institute of Macromolecular Compounds, Russian Academy of Sciences, Bolshoi VO 31, 199004 St. Petersburg, Russia; dubashinskaya@gmail.com (N.V.D.); raiksv@gmail.com (S.V.R.); poschin@yandex.ru (D.N.P.)
2 Almazov National Medical Research Centre, Akkuratova 2, 197341 St. Petersburg, Russia; dubrovskiy.ya@gmail.com
3 Institute of Chemistry, St. Petersburg State University, Universitetskii 26, Peterhof, 198504 St. Petersburg, Russia; annakirillova980@mail.ru (A.Y.S.); anufrikov_yuri@mail.ru (Y.A.A.)
4 Research and Training Center of Molecular and Cellular Technologies, St. Petersburg State Chemical Pharmaceutical University, Prof. Popova 14, 197376 St. Petersburg, Russia
5 State Research Institute of Highly Pure Biopreparations, Pudozhsakya 7, 197110 St. Petersburg, Russia; lenna_22@mail.ru (E.V.D.); elenka.shcherbakova@ya.ru (E.S.S.)
* Correspondence: yury_skorik@mail.ru

Citation: Dubashynskaya, N.V.; Raik, S.V.; Dubrovskii, Y.A.; Demyanova, E.V.; Shcherbakova, E.S.; Poshina, D.N.; Shasherina, A.Y.; Anufrikov, Y.A.; Skorik, Y.A. Hyaluronan/Diethylaminoethyl Chitosan Polyelectrolyte Complexes as Carriers for Improved Colistin Delivery. *Int. J. Mol. Sci.* **2021**, *22*, 8381. https://doi.org/10.3390/ijms22168381

Academic Editor: Inmaculada Aranaz Corral

Received: 11 July 2021
Accepted: 30 July 2021
Published: 4 August 2021

Publisher's Note: MDPI stays neutral with regard to jurisdictional claims in published maps and institutional affiliations.

Abstract: Improving the therapeutic characteristics of antibiotics is an effective strategy for controlling the growth of multidrug-resistant Gram-negative microorganisms. The purpose of this study was to develop a colistin (CT) delivery system based on hyaluronic acid (HA) and the water-soluble cationic chitosan derivative, diethylaminoethyl chitosan (DEAECS). The CT delivery system was a polyelectrolyte complex (PEC) obtained by interpolymeric interactions between the HA polyanion and the DEAECS polycation, with simultaneous inclusion of positively charged CT molecules into the resulting complex. The developed PEC had a hydrodynamic diameter of 210–250 nm and a negative surface charge (ζ-potential = −19 mV); the encapsulation and loading efficiencies were 100 and 16.7%, respectively. The developed CT delivery systems were characterized by modified release (30–40% and 85–90% of CT released in 15 and 60 min, respectively) compared to pure CT (100% CT released in 15 min). In vitro experiments showed that the encapsulation of CT in polysaccharide carriers did not reduce its antimicrobial activity, as the minimum inhibitory concentrations against *Pseudomonas aeruginosa* of both encapsulated CT and pure CT were 1 μg/mL.

Keywords: colistin; polymyxin; hyaluronic acid; diethylaminoethyl chitosan; polyelectrolyte complexes; drug delivery system; ESKAPE pathogens; antimicrobial activity

1. Introduction

Nanoparticles (NPs) based on natural and modified polysaccharides are well established as potential platforms for the treatment of severe infectious diseases. These drug delivery systems protect active pharmaceutical substances from untimely destruction in the body, while also providing controlled release, overcoming biological barriers, and targeting delivery of active agents to infection sites, thereby reducing drug doses and side effects [1–5]. In addition, natural polysaccharides (hyaluronic acid (HA), alginic acid, starch, and dextrin) and semi-synthetic chitin derivatives, including chitosan and its derivatives (e.g., diethylaminoethyl chitosan (DEAECS), succinyl chitosan, and glutaryl chitosan), are biocompatible, biodegradable, and non-toxic, while simultaneously providing suitable drug loading efficiency [6–9].

One potential drug that could benefit from complexation with NPs is the antimicrobial drug colistin (CT). CT is a mixture of the cyclic polypeptides colistin A (polymyxin E1) and colistin B (polymyxin E2), which differ in their fatty acids [10]. CT is currently used as a last-reserve drug against multidrug-resistant infections, including nosocomial pneumonia, caused by Gram-negative bacteria (e.g., *Pseudomonas aeruginosa, Klebsiella pneumoniae, Escherichia coli*). Severe systemic infections require administration of CT by injection; however, because the small CT molecules are quickly eliminated from systemic circulation, high drug doses are required. Further disadvantages of CT therapy are its side effects of neurotoxicity and nephrotoxicity [11–13]. These disadvantages can be reduced by using nanotechnology-based drugs [3,4,14].

Polyelectrolytes are widely used in biomedicine due to the variety of their conformations and nanostructures, as well as their molecular interactions with biomedical materials. For example, by varying the molar ratio between two oppositely charged polyelectrolytes, it is possible to obtain different types of supramolecular systems (such as host–guest complexes and interpolyelectrolyte complexes) through cooperative electrostatic interactions. Such systems are capable of specifically interacting with a third component (the drug being administered) [15,16].

As a cationic peptide antibiotic, CT has the ability to interact with anionic molecules, including anionic polymers, and to form negatively charged NPs, and the resulting complexes can be used as carriers for the targeted delivery of CT [6,17–20]. For example, our research group [21] studied the formation of CT delivery systems formed by complexation of the drug with HA via electrostatic interaction. The resulting particles had a negative surface charge (−19 mV) and ranged in size from 200 nm to 1 μm, depending on the ratio of HA/CT and the molecular weight (MW) of HA. The encapsulation efficiency (EE) was 30–100%, and the loading efficiency (LE) was 20–67%. The antibiotic release was 45% and 85% in 15 and 60 min, respectively (pH 7.4). The minimum inhibitory concentrations (MICs) of both encapsulated CT and pure CT were 1 μg/mL at pH 7.4 (against *Pseudomonas aeruginosa*).

The positively charged CT can also be loaded into a system consisting of a polycation and a polyanion. For example, Yasar et al. [6] developed polyelectrolyte complexes (PECs) based on anionic starch (MW > 100,000) and oligochitosan (MW 5000) for the delivery of CT and tobramycin. These PECs had a size of 170–380 nm, and a surface charge from −17 mV to −30 mV; the EE was 97–99% and the LE was 17% (for CT) and 3% (for tobramycin). The drug release into phosphate-buffered saline (PBS) at pH 7.4 reached 20–40% in 16 h. The antimicrobial activity of the loaded antibiotics against *E. coli* and *P. aeruginosa* was similar to that of the corresponding free drugs. Deacon et al. [19] fabricated NPs of tobramycin with alginate and alginate/chitosan and found that the presence of low-molecular-weight chitosan increased the colloidal stability of the NPs. These NPs had the following parameters: a size of 500–540 nm, ζ-potential of −25 to −28 mV, EE of 20–45%, and a tobramycin release (PBS, pH 7.4) of 45% in the first 90 min, which increased to 80% after 48 h. The antimicrobial activity of the NPs against *P. aeruginosa* was equivalent to that of unencapsulated tobramycin (MIC 0.625 mg/L). Balmayor et al. [20] produced an injectable biodegradable starch/chitosan delivery depot system for the sustained release of gentamicin for the treatment of bone infections. The obtained microparticles, prepared by a reductive alkylation crosslinking method, had a spherical shape, a size range of 80–150 μm, EE of 50–60%, and LE of 30%. The in vitro release profile (PBS, pH 7.4) was 50–70% in 24 h, followed by a sustained release for 30 days. In addition, a bacterial inhibition test on *Staphylococcus aureus* showed 70–100% of the antimicrobial activity of free gentamicin.

In summary, the delivery of vital antibiotics (polypeptides, macrolides, aminoglycosides, and fluoroquinolones) can be improved by encapsulating these drugs in carriers based on polysaccharide PECs. In this case, the antibiotic release is regulated by modification of the polymer MW, the component ratio, the particle size, and/or the preparation method.

We previously studied polyelectrolyte interactions between HA and DEAECS [7] and showed that mixing HA (MW of 950,000) with DEAEC (degree of substitution (DS) of 26, 55, 85, and 113%) resulted in the formation of spherical NPs whose size and charge depended both on the ratio of the polymers and on the mixing order. The most stable PEC was obtained by mixing HA and DEAEC at ratios of 1:5 and 1.7:5. The hydrodynamic diameter of these particles was 120–300 nm, and their surface charge was −10 mV to −23 mV. The structure–sensitivity ratio (Rg/Rh), determined by comparing the gyration radius (Rg) obtained by static light scattering and the hydrodynamic radius (Rh) obtained by dynamic light scattering (DLS), depended on the DS of the DEAECS. The use of chitosan with a DS of 25% and 55% gave an Rg/Rh of 1.1, whereas chitosan with high DS (85% and 113%) gave an Rg/Rh of 0.7. Spherically symmetric objects have an Rg/Rh of about 1, while monodisperse spheres have values of 0.7–0.8.

DEAECS [22–25] and HA [26–29] have attractive biomedical properties, as both are biodegradable, biocompatible, and non-toxic, water-soluble polymers. Chitosan and its water-soluble derivatives are enzymatically biodegraded in body fluids and tissues by lysozyme and are mainly distributed in the kidneys and urine, where they are further destructed by enzymes and excreted; this prevents their accumulation in the body [30–32]. HA has a high affinity for stabilin-2 and CD44 receptors (expressed at inflammation sites and on the surface of immunocompetent cells, such as T- and B-lymphocytes and macrophages) [33–38] and for toll-like receptor 4 (TLR4), which is associated with protection against Gram-negative bacteria [39,40]. The aim of the present work was to develop a suitable drug delivery system for improved CT delivery based on HA-DEAECS PECs. The CT delivery systems were expected to show an initial burst release of CT to provide a therapeutic dose in the systemic circulation at 0.5–1 h, followed by a sustained release to maintain an essential dose and antimicrobial activity for at least 10–12 h.

We determined the optimal conditions for the formation of suitable complexes by studying parameters that affect the particle size and surface charge, as well as the EE and the LE. The parameters included: (i) the MW of HA, (ii) the mass ratio of HA and CT, and (iii) the addition of strong polycation DEAECS. We hypothesized that the inclusion of CT in the HA-based polymer system would result in prolonged and targeted action due to the high affinity of HA for CD44 receptors, while DEAECS would increase the stability of the complex and allow controlled CT release. The advantages of these polymer carriers are expected to improve CT delivery and antimicrobial activity, while reducing CT side effects.

2. Results and Discussion

2.1. Preparation and Characterization of HA-DEAECS PECs

The first step in our study was the preparation and characterization of drug-free HA-DEAECS PECs (Scheme 1A). This step had two purposes: one was to study the effects of MW and DS on the characteristics of the resulting particles to allow later comparisons with CT-loaded NPs, and the other was to obtain drug-free particles for use in biological experiments to exclude the influence of NPs on antimicrobial activity and toxicity. We used polyanion HA with different MWs to have better control of the NP size, and we used polycation DEAECS with different DS to have better control of the strength of the formed PEC and the surface charge. Both HA and DEAECS are water-soluble, biocompatible, biodegradable, and non-toxic, and their potential to serve as drug delivery vehicles is enhanced by their simple preparation procedure. HA is an alternating copolymer of D-glucuronic acid and N-acetyl-D-glucosamine, linked via alternating β-(1→4) and β-(1→3) glycosidic bonds. DEAECS is an alkylated derivative of chitosan. As shown previously, both amine and hydroxyl groups are substituted, and 0–15% of the diethylaminoethyl groups are alkylated, forming quaternary ammonium groups [7].

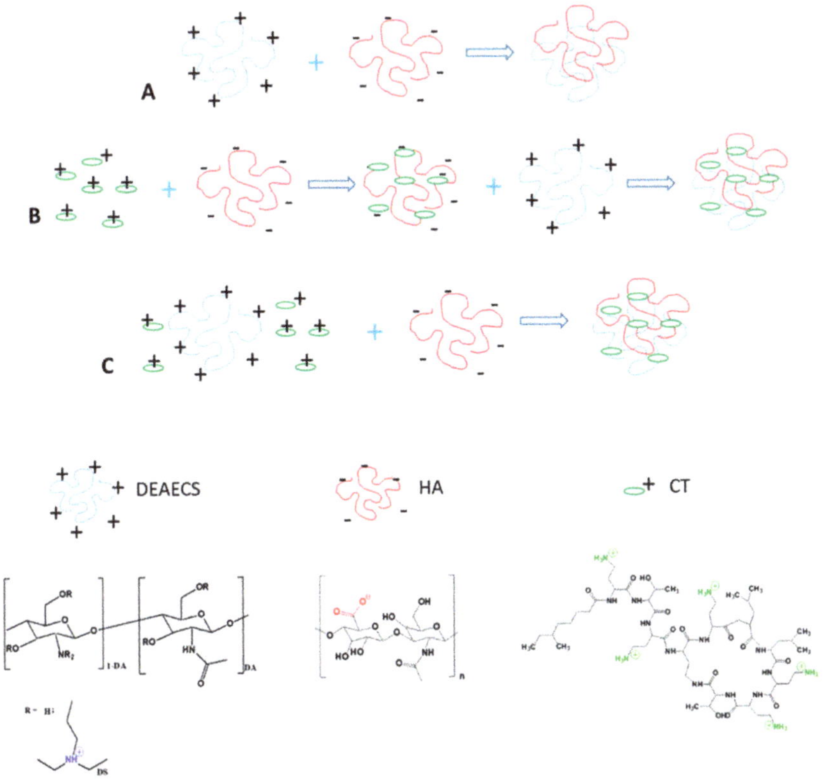

Scheme 1. Preparation of HA-DEAECS PEC (**A**) and CT-HA-DEAECS PEC (**B**,**C**).

The hydrodynamic diameter and ζ-potential of the resulting PEC are presented in Table 1. The HA-DEAECS PECs were obtained with a negative surface charge, as evidenced by a ζ-potential of around −32 to −35 mV. The hydrodynamic diameter of the PECs varied from 400 to 2000 nm, and this depended heavily on the MW of the HA. The obtained data led to the choice of HA_{54} for further experiments.

Table 1. Formation parameters and properties of HA-DEAECS and CT-HA-DEAECS PECs (mean ± standard deviation, n = 3).

No.	Formulation (Weight Ratio)	Procedure	2Rh (nm)	ζ-Potential (mV)	PDI	EE (%)	LE (%)
Formulation 1	HA_{54}:$DEAECS_{84}$ (5:1)	Scheme 1A	446 ± 136	−32.8 ± 0.2	0.11	-	-
Formulation 2	HA_{750}:$DEAECS_{84}$ (5:1)		1862 ± 442	−35.0 ± 0.8	0.09	-	-
Formulation 3	HA_{54}:$DEAECS_{45}$ (5:1)		604 ± 158	−31.7 ± 0.6	0.10	-	-
Formulation 4	HA_{54}:$DEAECS_{64}$ (5:1)		520 ± 76	−32.0 ± 0.6	0.06	-	-
Formulation 5	CT:HA_{54}:$DEAECS_{84}$ (1:5:1)	Scheme 1C	244 ± 66	−18.8 ± 0.2	0.10	100	16.7
Formulation 6	CT:HA_{54}:$DEAECS_{84}$ (2:5:1)		284 ± 134	−12.2 ± 0.6	0.24	98.5	32.8
Formulation 7	CT:HA_{54}:$DEAECS_{84}$ (4:5:1)		950 ± 248	−2.6 ± 0.2	0.10	47.8	31.8
Formulation 8	CT:HA_{54}:$DEAECS_{84}$ (6:5:1)		1104 ± 300	−1.9 ± 0.9	0.10	36.9	36.9
Formulation 9	CT:HA_{750}:$DEAECS_{84}$ (1:5:1)		612 ± 104	−24.4 ± 0.4	0.07	100	16.7
Formulation 10	CT:HA_{54}:$DEAECS_{45}$ (1:5:1)		210 ± 72	−19.0 ± 0.2	0.14	100	16.7
Formulation 11	CT:HA_{54}:$DEAECS_{64}$ (1:5:1)		210 ± 96	−18.9 ± 0.9	0.23	100	16.7
Formulation 12	CT:HA_{54}:$DEAECS_{84}$ (1:5:2)		210 ± 38	−14.9 ± 0.6	0.07	100	14.3
Formulation 13	CT:HA_{54}:$DEAECS_{84}$ (1:5:2)	Scheme 1B	210 ± 68	−13.5 ± 0.4	0.12	100	14.3

The thermodynamic stability of the HA-DEAECS PECs was studied using isothermal titration calorimetry (ITC). The ITC data (Figure 1) are represented by two fragments: the first is the PEC formation with an endothermal effect. Points for molar ratios from 0 to 1 were fitted by a one-site independent binding model, and the corresponding thermodynamic parameters were Kd = 3.49×10^{-5}; n = 0.521; $\Delta H = -0.207$ kJ/mol; $\Delta G = -25.44$ kJ/mol; and $\Delta S = 84.64$ J/mol$\times$K. Therefore, not surprisingly, the HA-DEAECS PEC formation is an entropy-driven process.

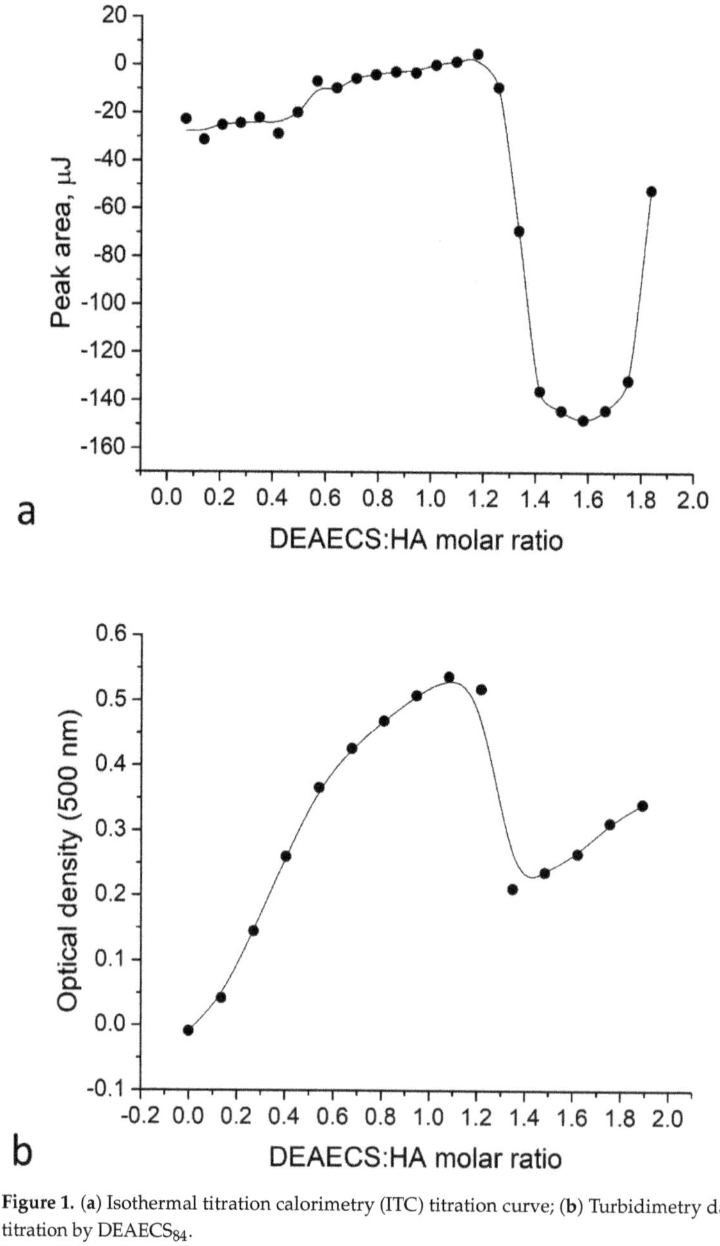

Figure 1. (**a**) Isothermal titration calorimetry (ITC) titration curve; (**b**) Turbidimetry data for HA_{54} titration by $DEAECS_{84}$.

The heat effect at a molar ratio of 1.2–2 could not be explained by the ionic interactions of HA and DEAECS. Turbidimetry data (Figure 1b) showed that a molar ratio of 1.2–1.4 led to a rapid decrease in turbidity. This reflects an endothermic effect on the ITC curve and suggests that this effect represents a disaggregation at an excessive amount of titrant.

Consequently, the HA-DEAECS PEC preparation procedure represents a simple method for the formulation of stable anionic particles of different sizes.

2.2. Preparation, Optimization, and Characterization of CT-HA-DEAECS PECs

CT is a cyclic heptapeptide with a tripeptide side chain that bears the covalently attached fatty acid. It is positively charged at a neutral pH due to the presence of amine functional groups in the structure. The positively charged CT molecules are able to interact with polyanionic HA molecules and form PECs. In addition, the CT molecule contains a lipophilic fatty acyl tail and a hydrophilic head group; therefore, CT can form micellar structures with anionic polymers that are further stabilized by the addition of chitosan and chitosan derivatives [6,17,21]. The use of this type of polyelectrolyte particle as a systemic circulation drug carrier requires that the particles have a size below 300 nm and a negative charge below −15 mV to ensure stability. We investigated the possibility of using HA and DEAECS to load CT by studying the influence of various factors on the size, ζ-potential, EE, and LE. The obtained results are presented in Table 1.

One point to note is that the introduction of CT into the HA-DEAECS (5:1) PECs reduced their size from 400–600 nm to 210–244 nm (HA_{54}) and from 1800 to 600 nm (HA_{750}), regardless of the DS of DEAECS. These results likely reflect the high efficiency of CT binding to HA. The ζ-potential of the CT-loaded PEC decreased from −(31–32) mV to −(18–19) mV (HA_{54}) and from −35 mV to −24 mV (HA_{750}) compared to CT-free particles.

Further experiments were aimed at examining the effect of DEAECS addition on the properties of the resulted PEC. As presented in Table 1, the addition of DEAECS did not significantly change the particle size. The surface charge was significantly reduced only for the CT:HA:DEAECS ratios of 2:5:1 and 4:5:1. We used the CT:HA:DEAECS ratio of 1:5:1 to study the effect of DS, the amount of $DEAECS_{84}$, and the order of addition of the components to the HA. The obtained data show that these factors did not affect the particle size, whereas a 2-fold increase in the $DEAECS_{84}$ amount in the CT-HA-DEAECS PEC changed the ζ-potential from −19 to −15 mV. The addition of $DEAECS_{84}$ to the CT-HA nanosuspension changed the ζ-potential from −19 to −13 mV. The EE and LE for CT-HA-$DEAECS_{84}$ depended primarily on the amount of the loaded CT; the best values were 100% (EE) and 16.7% (LE) for the CT:HA:DEAECS ratio of 1:5:1 (Table 1). The PDI values for most formulations were in the range of 0.1–0.2; PDI values of 0.2 and below are generally considered acceptable for nanocarrier drug delivery systems [41].

Based on the optimization results, formulations 5, 10, 11, 12, and 13 were selected for further study to test the stabilizing effect of DEAECS on CT release and on CT antimicrobial activity.

The morphology of the PECs was visualized using AFM (Figure 2). The resulting particles had a spherical shape with a size ranging from 10 to 500 nm, with most at 50–100 nm. The Formulation 1 particles were surrounded by non-complexed macromolecules. The particle size corresponds to the size estimated using DLS, taking into consideration that AFM measurements were carried out on dry samples that reduced the particle size compared to the hydrated complexes in suspension.

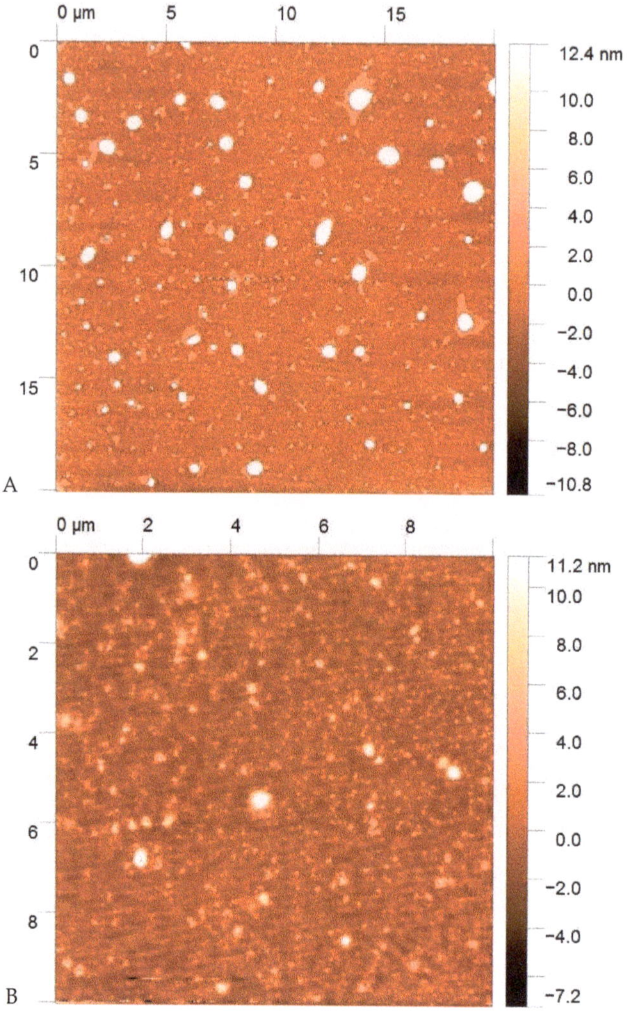

Figure 2. Atomic force microscopy (AFM) images of Formulation 1 (**A**) and Formulation 5 (**B**).

2.3. *In Vitro Release of CT from CT-HA-DEAECS PECs*

The study of release kinetics into PBS (pH 7.4) simulated the pattern of CT behavior in the systemic bloodstream. As shown in Figure 3, the CT release was 30–40% in 0.25 h, with 85–90% of the CT being released in 1–2 h for all three tested formulations (formulations 5, 12 and 13). Thus, when loaded into the HA-DEAECS PEC via polyelectrolyte interactions, CT was rapidly released due to the high ionic strength of the release medium. A 2-fold increase in DEAECS amount (formulations 12 and 13) stabilized the PEC and slowed this CT release, but for only 1 h. After the first hour, the release from all test samples was approximately the same [6,19].

The proposed modification of CT release from polymeric PEC compared to the release profile of free CT will allow targeted drug delivery to the sites of bacterial inflammation due to the high affinity of HA for the relevant receptors (CD44, stabilin-2, and TLR4 [33–40]. However, the initial rapid CT release is also necessary to ensure high antibiotic doses in the first hours after administration, since the bactericidal effect of CT is mainly realized via the carpet model of insertion, and it is dose-dependent [42–45].

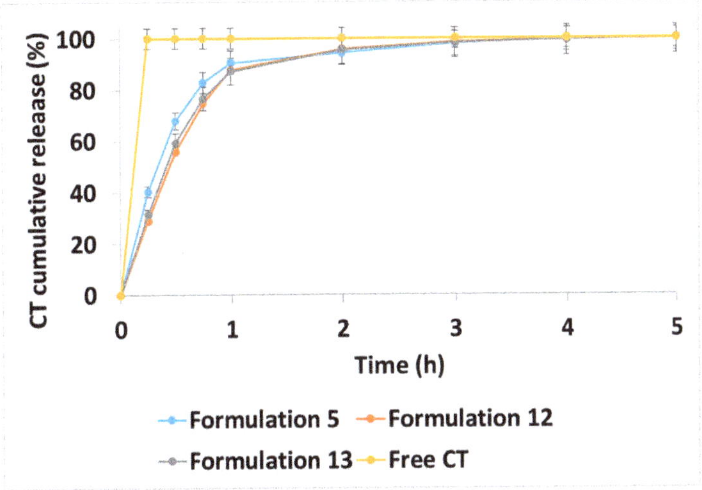

Figure 3. Colistin (CT) release kinetics from different CT-HA-DEAECS PECs formulations at 37 °C into phosphate-buffered saline (PBS). Free CT corresponds to an equivalent concentration of the CT in the PEC. Error bars indicate standard deviations (n = 3).

2.4. Antimicrobial Activity of CT-HA-DEAECS PECs

The MIC for *Pseudomonas aeruginosa* was determined at a concentration of 1 × 10^7 CFU/mL (Figure 4). The CT-loaded NPs showed comparable antimicrobial effects to those of unencapsulated CT. All MICs were 1 μg/mL, which may indicate the preservation of the antibiotic activity in spite of encapsulation in different types of NPs. The blank NPs had no discernible effects on the visible growth of the bacteria.

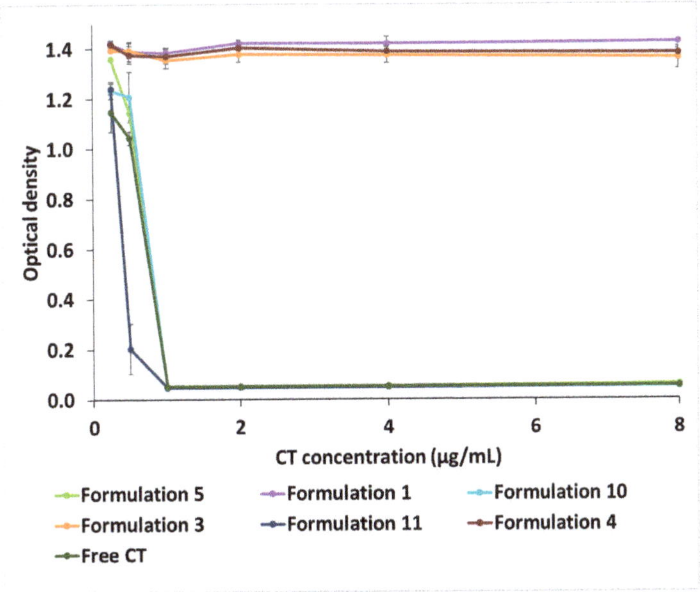

Figure 4. The minimum inhibitory concentrations (MIC) of CT-HA-DEAECS and HA-DEAECS PECs against *Pseudomonas aeruginosa*. Error bars indicate standard deviations (n = 3).

The in vivo test showed that the encapsulated CT was released after 1–2 h in PBS, but it could present different kinetics in the microbial culture medium. Therefore, the study of the antimicrobial activity of developed drug delivery systems confirms that CT remains active and that encapsulated CT is released at a sufficiently rapid rate to maintain the 24 h result.

3. Materials and Methods

3.1. Materials

We used HA with MWs of 54,000 (HA_{54}) and 750,000 (HA_{750}). The MWs were determined by viscometry. The intrinsic viscosity of HA was determined using an Ubbelohde viscometer (Design Bureau Pushchino, Pushchino, Russia) at 30 °C in 0.2 M NaCl. The MW of HA was calculated from the Mark–Houwink equation: $[\eta] = 3.9 \times 10^{-2} \times MW^{0.77}$; $[\eta]$ HA_{54} = 1.7 dL/g and $[\eta]$ HA_{750} = 13.1 dL/g [46,47].

DEAECS with different DS values of 45% ($DEAECS_{45}$), 64% ($DEAECS_{64}$), and 84% ($DEAECS_{84}$), determined by ^{1}H NMR spectroscopy [7], were synthesized using a previously described method [7]. The starting material was crab shell chitosan with an average MW of 37,000 (determined by viscometry) and a degree of acetylation (DA) of 26% (determined by elemental analysis and ^{1}H NMR spectroscopy) [48].

CT sulfate (MW of 1390) was procured from BetaPharm (Wujiang, Shanghai, China). The contents of CT A (polymyxin E1) and CT B (polymyxin E2) were 31.1 ± 0.4% and 68.9 ± 0.4%, respectively [49].

2-Chloro-*N*,*N*-diethylamine hydrochloride (99%), deuterium oxide (99.9 atom % D), trifluoroacetic acid, and PBS were obtained from Sigma-Aldrich Co. (St. Louis, MO, USA). Other reagents and solvents were obtained from commercial sources and were used without further purification.

3.2. General Methods

The ^{1}H NMR spectra were recorded on an Avance 400 spectrometer (Bruker BioSpin GmbH, Rheinstetten, Germany) at 400 MHz and 70 °C using a zgpr pulse sequence with suppression of residual H_2O. Samples for NMR were dissolved in D_2O and X µL of CF_3COOH (where X is the mass of the polymer) was added to protonate all the amino groups.

Elemental analysis was performed on a Vario EL CHN analyzer (Elementar, Langenselbold, Germany).

The hydrodynamic diameter (2Rh) and the ζ-potential were determined by dynamic and electrophoretic light scattering, respectively, using a Compact-Z instrument (Photocor, Moscow, Russia) with a 659.7 nm He–Ne laser at 25 mV power and a detection angle of 90°. The polydispersity index (PDI) was determined by cumulants' analysis of autocorrelation function using DynaLS software.

The particle morphology was studied using atomic force microscopy with a Smena instrument (NT-MDT, Zelenograd, Russia).

3.3. Liquid Chromatography-Mass Spectrometry (LC-MS) Measurements

LC-MS was performed as previously described [21]. Chromatographic separation was performed using an Elute UHPLC (Bruker Daltonics GmbH, Bremen, Germany) equipped with a Millipore Chromolith Performance/PR-18e, C18 analytical column (100 mm × 2 mm, Merck, Darmstadt, Germany) with Chromolith® RP-18 endcapped 5-3 guard cartridges (Merck, Darmstadt, Germany). Mass spectra were obtained on a Maxis Impact Q-TOF mass spectrometer (Bruker Daltonics GmbH, Bremen, Germany) equipped with an electrospray ionization source (Bruker Daltonics GmbH, Bremen, Germany) and operated in positive ionization mode. Mass spectra were analyzed using DataAnalysis® and TASQ® software (Bruker Daltonics GmbH, Bremen, Germany).

3.4. Isothermal Titration Calorimetry and Turbidimetry

HA_{54} (2.12 mM; moles of monomeric units) and $DEAECS_{84}$ (14.3 mM; moles of N atoms) solutions in 0.15 M NaCl were prepared and degassed under vacuum before the titration. HA_{54} (1 mL) was loaded in a Nano ITC microcalorimeter (TA Instruments, New Castle, DE, USA) and titrated by 24 injections (9.98 μL) of $DEAECS_{84}$ solution. The time interval between injections was 2400 s, and the temperature was 25.0000 ± 0.0001 °C. ITC measurements were performed at the Center for Thermogravimetric and Calorimetric Research of Science Park of St. Petersburg State University.

In parallel, 2 mL of 2.12 mM HA_{54} solution were loaded into a 10 mm quartz cuvette and titrated with 14.3 mM $DEAECS_{84}$ solution. After each 20 μL addition, the optical density (OD) at 500 nm was measured on a UV-1700 spectrometer (Shimadzu, Kyoto, Japan).

3.5. Preparation of HA-DEAECS PECs

Solutions of HA (0.5 mg/mL, pH 6.5) and DEAECS (0.5 mg/mL, pH 5.6–6.2) with a DS of 45%, 64%, and 84% were prepared in ultrapure water.

Method A: HA-DEAECS PECs were obtained by the following procedure [7]: DEAECS solution was added dropwise via a 23G needle into the HA solution (Scheme 1A) at a ratio of 1:5, and the mixture was vortexed for 2 min and then allowed to stand for 2 h for PEC formation. Various PEC formation parameters are presented in Table 1. The resulting NPs were freeze-dried.

3.6. Preparation of CT-HA-DEAECS PECs

Solutions of CT (1 mg/mL, pH 6.9), HA (0.5 mg/mL, pH 6.5), and DEAECS (0.5 mg/mL, pH 5.6–6.2) with DS of 45%, 64%, and 84% were prepared in ultrapure water. CT-HA-DEAECS PECs were obtained by the following procedures.

Method B: CT solution was added to the HA solution, and the DEAECS solution was added to this mixture (Scheme 1B).

Method C: CT was mixed with the DEAECS solution, and this mixture was added to the HA solution (Scheme 1C). All solutions were added dropwise via a 23G needle, and the mixtures were vortexed for 2 min and then allowed to stand for 2 h for forming PECs. Various PEC formation parameters are presented in Table 1. The resulting NPs were freeze-dried.

3.7. Encapsulation and Loading Efficiencies

The EE and LE were determined by measuring the concentration of the non-loaded CT (by an indirect method). The CT-HA-DEAECS PEC suspension was concentrated by ultrafiltration at 4500 rpm through a 10,000 MWCO Vivaspin® Turbo 4 centrifugal concentrator (Sartorius AG, Göttingen, Germany) for 15 min at 20 °C. The amount of encapsulated CT in the PEC was calculated by the difference between the total CT amount used to prepare the PEC and the CT amount in the supernatant. The CT concentration in the supernatant was analyzed by LC-MS. The results were calculated using the following equations:

$$\text{EE (\%)} = \frac{\text{CT mass total} - \text{CT mass in the supernatant}}{\text{CT mass total}} \times 100\%$$

$$\text{LE (\%)} = \frac{\text{CT mass total} - \text{CT mass in the supernatant}}{\text{Polymer mass}} \times 100\%$$

3.8. In Vitro Release of CT from CT-HA-DEAECS PEC

The release medium conditions took into account the FDA recommendations for the dissolution methods of injectable dosage forms. A 1 mg sample of PEC was dissolved in PBS (4 mL, pH 7.4) and incubated at 37 °C. At regular intervals, 4 mL of medium was ultracentrifuged at 4500 rpm using a 10,000 MWCO Vivaspin® Turbo 4 centrifugal concentrator. The amount of CT released into the supernatant was determined by LC-MS.

3.9. Antimicrobial Activity of CT-HA-DEAECS PEC

This assay was realized by the microtiter broth dilution method described by the Clinical and Laboratory Standards Institute [50] using *P. aeruginosa* ATCC 27,853 (Museum of Microbiological Cultures, State Research Institute of Highly Pure Biopreparations, St. Petersburg, Russia).

In a 96-well plate, 125 μL of Mueller-Hinton broth (HiMedia, Mumbai, India) were added. Stock solutions of NPs with antibiotics were prepared by diluting the sample in Mueller-Hinton broth to a maximum 2-fold required CT concentration. Serial dilutions of CT at concentrations from 64 to 0.25 μg/mL were then obtained on the plate. The OD of an overnight *P. aeruginosa* suspension in Mueller-Hinton broth was measured on a UVmini-1240 spectrophotometer (Shimadzu, Kyoto, Japan) at a wavelength of 540 nm, and the suspension was serially diluted 1:100 in Mueller-Hinton broth to give approximately 1×10^7 CFU/mL.

The inoculum was prepared by cultivation in a liquid medium (Mueller-Hinton broth) for 18 h by adding 125 μL of inoculum to the wells of the plate. The plate also contained controls: 100% growth (bacteria only), sterility control (Mueller-Hinton broth only), and blank NPs at equivalent polymer concentrations. The plate was incubated for 24 h at 37 °C, and then the OD was measured on an ELx808 ™ Absorbance Microplate Reader (BioTek Instruments, Winooski, VT, USA) at 630 nm.

4. Conclusions

This study has several main conclusions: (i) Polyanionic HA and polycationic DEAECS in aqueous solutions form stable PECs that are suitable for the inclusion of positively charged CT molecules. The reaction entropy component ($-T\Delta S$) and free energy (ΔG) were negative, indicating a favorable reaction. A negative ΔG corresponds to an entropy-driven process, which indicates strong bonding between the molecules; the Kd in the physiological solution was about 3.49×10^{-5}. (ii) Varying the MW of HA, component ratios, and mixing procedure optimized the PEC preparation method and yielded formulations suitable for intravascular injection (mean size 210–250 nm and surface charge −19 mV). (iii) In vitro release experiments showed that the developed drug delivery systems provided sustained CT release for approximately 1–2 h compared to pure CT. (iv) The drug delivery systems exhibited comparable antimicrobial activity against *P. aeruginosa* to that of pure CT (both had a MIC of 1 μg/mL). This study can be extended to include in vivo biological tests to develop safe and effective antimicrobials for the treatment of infections caused by Gram-negative multidrug-resistant microorganisms.

Author Contributions: Conceptualization, N.V.D. and Y.A.S.; methodology, N.V.D., S.V.R. and Y.A.S.; investigation, N.V.D., S.V.R.,Y.A.D., E.V.D., E.S.S., D.N.P., A.Y.S. and Y.A.A.; writing—original draft preparation, N.V.D. and S.V.R.; writing—review and editing, Y.A.S.; supervision, Y.A.S.; project administration, Y.A.S.; funding acquisition, Y.A.S. All authors have read and agreed to the published version of the manuscript.

Funding: This work was financially supported by the Russian Science Foundation (project 19-73-20157).

Institutional Review Board Statement: Not applicable.

Informed Consent Statement: Not applicable.

Data Availability Statement: The data are contained within the article.

Conflicts of Interest: The authors declare no conflict of interest. The funders had no role in the design of the study; in the collection, analyses, or interpretation of data; in the writing of the manuscript, or in the decision to publish the results.

References

1. Ghosh, B.; Giri, T.K. Recent Advances of Chitosan Nanoparticles as a Carrier for Delivery of Antimicrobial Drugs. In *Polysaccharide Based Nano-Biocarrier in Drug Delivery*; CRC Press: Boca Raton, FL, USA, 2018; pp. 63–79.
2. Parisi, O.I.; Scrivano, L.; Sinicropi, M.S.; Puoci, F. Polymeric nanoparticle constructs as devices for antibacterial therapy. *Curr. Opin. Pharmacol.* **2017**, *36*, 72–77. [CrossRef]
3. Abed, N.; Couvreur, P. Nanocarriers for antibiotics: A promising solution to treat intracellular bacterial infections. *Int. J. Antimicrob. Agents* **2014**, *43*, 485–496. [CrossRef]
4. Kang, B.; Opatz, T.; Landfester, K.; Wurm, F.R. Carbohydrate nanocarriers in biomedical applications: Functionalization and construction. *Chem. Soc. Rev.* **2015**, *44*, 8301–8325. [CrossRef]
5. D'Angelo, I.; Conte, C.; Miro, A.; Quaglia, F.; Ungaro, F. Pulmonary drug delivery: A role for polymeric nanoparticles? *Curr. Top. Med. Chem.* **2015**, *15*, 386–400. [CrossRef]
6. Yasar, H.; Ho, D.-K.; De Rossi, C.; Herrmann, J.; Gordon, S.; Loretz, B.; Lehr, C.-M. Starch-chitosan polyplexes: A versatile carrier system for anti-infectives and gene delivery. *Polymers* **2018**, *10*, 252. [CrossRef]
7. Raik, S.V.; Gasilova, E.R.; Dubashynskaya, N.V.; Dobrodumov, A.V.; Skorik, Y.A. Diethylaminoethyl chitosan–hyaluronic acid polyelectrolyte complexes. *Int. J. Biol. Macromol.* **2020**, *146*, 1161–1168. [CrossRef] [PubMed]
8. Skorik, Y.A.; Kritchenkov, A.S.; Moskalenko, Y.E.; Golyshev, A.A.; Raik, S.V.; Whaley, A.K.; Vasina, L.V.; Sonin, D.L. Synthesis of N-succinyl-and N-glutaryl-chitosan derivatives and their antioxidant, antiplatelet, and anticoagulant activity. *Carbohydr. Polym.* **2017**, *166*, 166–172. [CrossRef]
9. Kashapov, R.; Gaynanova, G.; Gabdrakhmanov, D.; Kuznetsov, D.; Pavlov, R.; Petrov, K.; Zakharova, L.; Sinyashin, O. Self-Assembly of Amphiphilic Compounds as a Versatile Tool for Construction of Nanoscale Drug Carriers. *Int. J. Mol. Sci.* **2020**, *21*, 6961. [CrossRef]
10. Orwa, J.A.; Van Gerven, A.; Roets, E.; Hoogmartens, J. Development and validation of a liquid chromatography method for analysis of colistin sulphate. *Chromatographia* **2000**, *51*, 433–436. [CrossRef]
11. Bialvaei, A.Z.; Samadi Kafil, H. Colistin, mechanisms and prevalence of resistance. *Curr. Med. Res. Opin.* **2015**, *31*, 707–721. [CrossRef] [PubMed]
12. Linden, P.K.; Kusne, S.; Coley, K.; Fontes, P.; Kramer, D.J.; Paterson, D. Use of parenteral colistin for the treatment of serious infection due to antimicrobial-resistant Pseudomonas aeruginosa. *Clin. Infect. Dis.* **2003**, *37*, e154–e160. [CrossRef]
13. Nation, R.L.; Li, J.; Cars, O.; Couet, W.; Dudley, M.N.; Kaye, K.S.; Mouton, J.W.; Paterson, D.L.; Tam, V.H.; Theuretzbacher, U. Consistent global approach on reporting of colistin doses to promote safe and effective use. *Clin. Infect. Dis.* **2014**, *58*, 139–141. [CrossRef]
14. Dubashynskaya, N.V.; Skorik, Y.A. Polymyxin Delivery Systems: Recent Advances and Challenges. *Pharmaceuticals* **2020**, *13*, 83. [CrossRef]
15. Nazarova, A.; Khannanov, A.; Boldyrev, A.; Yakimova, L.; Stoikov, I. Self-Assembling Systems Based on Pillar[5]arenes and Surfactants for Encapsulation of Diagnostic Dye DAPI. *Int. J. Mol. Sci.* **2021**, *22*, 6038. [CrossRef]
16. Papagiannopoulos, A. Current Research on Polyelectrolyte Nanostructures: From Molecular Interactions to Biomedical Applications. *Macromol* **2021**, *1*, 155–172. [CrossRef]
17. Wallace, S.J.; Li, J.; Nation, R.L.; Prankerd, R.J.; Velkov, T.; Boyd, B.J. Self-assembly behavior of colistin and its prodrug colistin methanesulfonate: Implications for solution stability and solubilization. *J. Phys. Chem. B* **2010**, *114*, 4836–4840. [CrossRef]
18. Abouelmagd, S.A.; Ellah, N.H.A.; Amen, O.; Abdelmoez, A.; Mohamed, N.G. Self-assembled tannic acid complexes for pH-responsive delivery of antibiotics: Role of drug-carrier interactions. *Int. J. Pharm.* **2019**, *562*, 76–85. [CrossRef]
19. Deacon, J.; Abdelghany, S.M.; Quinn, D.J.; Schmid, D.; Megaw, J.; Donnelly, R.F.; Jones, D.S.; Kissenpfennig, A.; Elborn, J.S.; Gilmore, B.F. Antimicrobial efficacy of tobramycin polymeric nanoparticles for Pseudomonas aeruginosa infections in cystic fibrosis: Formulation, characterisation and functionalisation with dornase alfa (DNase). *J. Control. Release* **2015**, *198*, 55–61. [CrossRef] [PubMed]
20. Balmayor, E.R.; Baran, E.; Azevedo, H.S.; Reis, R. Injectable biodegradable starch/chitosan delivery system for the sustained release of gentamicin to treat bone infections. *Carbohydr. Polym.* **2012**, *87*, 32–39. [CrossRef]
21. Dubashynskaya, N.V.; Raik, S.V.; Dubrovskii, Y.A.; Shcherbakova, E.S.; Demyanova, E.V.; Shasherina, A.Y.; Anufrikov, Y.A.; Poshina, D.N.; Dobrodumov, A.V.; Skorik, Y.A. Hyaluronan/Colistin Polyelectrolyte Complexes: Promising Antiinfective Drug Delivery Systems. *Int. J. Biol. Macromol.* **2021**, *187*, 157–165. [CrossRef]
22. Raik, S.V.; Andranovits, S.; Petrova, V.A.; Xu, Y.; Lam, J.K.; Morris, G.A.; Brodskaia, A.V.; Casettari, L.; Kritchenkov, A.S.; Skorik, Y.A. Comparative Study of Diethylaminoethyl-Chitosan and Methylglycol-Chitosan as Potential Non-Viral Vectors for Gene Therapy. *Polymers* **2018**, *10*, 442. [CrossRef]
23. da Mata Cunha, O.; Lima, A.M.F.; Assis, O.B.G.; Tiera, M.J.; de Oliveira Tiera, V.A. Amphiphilic diethylaminoethyl chitosan of high molecular weight as an edible film. *Int. J. Biol. Macromol.* **2020**, *164*, 3411–3420. [CrossRef]
24. de Souza, R.; Picola, I.P.D.; Shi, Q.; Petronio, M.S.; Benderdour, M.; Fernandes, J.C.; Lima, A.M.F.; Martins, G.O.; Martinez Junior, A.M.; de Oliveira Tiera, V.A.; et al. Diethylaminoethyl-chitosan as an efficient carrier for siRNA delivery: Improving the condensation process and the nanoparticles properties. *Int. J. Biol. Macromol.* **2018**, *119*, 186–197. [CrossRef]

25. Dias, A.M.; Dos Santos Cabrera, M.P.; Lima, A.M.F.; Taboga, S.R.; Vilamaior, P.S.L.; Tiera, M.J.; de Oliveira Tiera, V.A. Insights on the antifungal activity of amphiphilic derivatives of diethylaminoethyl chitosan against Aspergillus flavus. *Carbohydr. Polym.* **2018**, *196*, 433–444. [CrossRef]
26. Almalik, A.; Karimi, S.; Ouasti, S.; Donno, R.; Wandrey, C.; Day, P.J.; Tirelli, N. Hyaluronic acid (HA) presentation as a tool to modulate and control the receptor-mediated uptake of HA-coated nanoparticles. *Biomaterials* **2013**, *34*, 5369–5380. [CrossRef]
27. Chen, L.; Zheng, Y.; Feng, L.; Liu, Z.; Guo, R.; Zhang, Y. Novel hyaluronic acid coated hydrophobically modified chitosan polyelectrolyte complex for the delivery of doxorubicin. *Int. J. Biol. Macromol.* **2019**, *126*, 254–261. [CrossRef]
28. Tripodo, G.; Trapani, A.; Torre, M.L.; Giammona, G.; Trapani, G.; Mandracchia, D. Hyaluronic acid and its derivatives in drug delivery and imaging: Recent advances and challenges. *Eur. J. Pharm. Biopharm.* **2015**, *97*, 400–416. [CrossRef]
29. Larsen, N.E.; Balazs, E.A. Drug delivery systems using hyaluronan and its derivatives. *Adv. Drug Deliv. Rev.* **1991**, *7*, 279–293. [CrossRef]
30. Onishi, H.; Machida, Y. Biodegradation and distribution of water-soluble chitosan in mice. *Biomaterials* **1999**, *20*, 175–182. [CrossRef]
31. Dong, W.; Han, B.; Feng, Y.; Song, F.; Chang, J.; Jiang, H.; Tang, Y.; Liu, W. Pharmacokinetics and biodegradation mechanisms of a versatile carboxymethyl derivative of chitosan in rats: In vivo and in vitro evaluation. *Biomacromolecules* **2010**, *11*, 1527–1533. [CrossRef]
32. Sonin, D.; Pochkaeva, E.; Zhuravskii, S.; Postnov, V.; Korolev, D.; Vasina, L.; Kostina, D.; Mukhametdinova, D.; Zelinskaya, I.; Skorik, Y.; et al. Biological Safety and Biodistribution of Chitosan Nanoparticles. *Nanomaterials* **2020**, *10*, 810. [CrossRef]
33. Lee, G.Y.; Kim, J.H.; Choi, K.Y.; Yoon, H.Y.; Kim, K.; Kwon, I.C.; Choi, K.; Lee, B.H.; Park, J.H.; Kim, I.S. Hyaluronic acid nanoparticles for active targeting atherosclerosis. *Biomaterials* **2015**, *53*, 341–348. [CrossRef]
34. Burdick, J.A.; Prestwich, G.D. Hyaluronic Acid Hydrogels for Biomedical Applications. *Adv. Mater.* **2011**, *23*, H41–H56. [CrossRef]
35. Diaz-Salmeron, R.; Ponchel, G.; Bouchemal, K. Hierarchically built hyaluronan nano-platelets have symmetrical hexagonal shape, flattened surfaces and controlled size. *Eur. J. Pharm. Sci.* **2019**, *133*, 251–263. [CrossRef]
36. Teder, P.; Vandivier, R.W.; Jiang, D.; Liang, J.; Cohn, L.; Puréé, E.; Henson, M.P.; Noble, P.W. Resolution of lung inflammation by CD44. *Nat. Rev. Immunol.* **2002**, *296*, 155–158. [CrossRef]
37. Sionkowska, A.; Gadomska, M.; Musial, K.; Piatek, J. Hyaluronic Acid as a Component of Natural Polymer Blends for Biomedical Applications: A Review. *Molecules* **2020**, *25*, 4035. [CrossRef]
38. Ganesh, S.; Iyer, A.K.; Morrissey, D.V.; Amiji, M.M. Hyaluronic acid based self-assembling nanosystems for CD44 target mediated siRNA delivery to solid tumors. *Biomaterials* **2013**, *34*, 3489–3502. [CrossRef] [PubMed]
39. Termeer, C.; Benedix, F.; Sleeman, J.; Fieber, C.; Voith, U.; Ahrens, T.; Miyake, K.; Freudenberg, M.; Galanos, C.; Simon, J.C. Oligosaccharides of hyaluronan activate dendritic cells via toll-like receptor 4. *J. Exp. Med.* **2002**, *195*, 99–111. [CrossRef]
40. Muthukumar, R.; Alexandar, V.; Thangam, B.; Ahmed, S. A Systems Biological Approach Reveals Multiple Crosstalk Mechanism between Gram-Positive and Negative Bacterial Infections: An Insight into Core Mechanism and Unique Molecular Signatures. *PLoS ONE* **2014**, *9*, e0089993. [CrossRef]
41. Danaei, M.; Dehghankhold, M.; Ataei, S.; Hasanzadeh Davarani, F.; Javanmard, R.; Dokhani, A.; Khorasani, S.; Mozafari, M.R. Impact of Particle Size and Polydispersity Index on the Clinical Applications of Lipidic Nanocarrier Systems. *Pharmaceutics* **2018**, *10*, 57. [CrossRef]
42. Khondker, A.; Rheinstädter, M.C. How do bacterial membranes resist polymyxin antibiotics? *Commun. Biol.* **2020**, *3*, 1–4. [CrossRef]
43. Brogden, K.A. Antimicrobial peptides: Pore formers or metabolic inhibitors in bacteria? *Nat. Rev. Microbiol.* **2005**, *3*, 238–250. [CrossRef] [PubMed]
44. Dupuy, F.G.; Pagano, I.; Andenoro, K.; Peralta, M.F.; Elhady, Y.; Heinrich, F.; Tristram-Nagle, S. Selective interaction of colistin with lipid model membranes. *Biophys. J.* **2018**, *114*, 919–928. [CrossRef]
45. Binder, W.H. Polymer-Induced Transient Pores in Lipid Membranes. *Angew. Chem. Int. Ed.* **2008**, *47*, 3092–3095. [CrossRef] [PubMed]
46. Petrova, V.A.; Chernyakov, D.D.; Poshina, D.N.; Gofman, I.V.; Romanov, D.P.; Mishanin, A.I.; Golovkin, A.S.; Skorik, Y.A. Electrospun Bilayer Chitosan/Hyaluronan Material and Its Compatibility with Mesenchymal Stem Cells. *Materials* **2019**, *12*, 2016. [CrossRef]
47. Ueno, Y.; Tanaka, Y.; Horie, K.; Tokuyasu, K. Low-angle laser light scattering measurements on highly purified sodium hyaluronate from rooster comb. *Chem. Pharm. Bull.* **1988**, *36*, 4971–4975. [CrossRef]
48. Raik, S.V.; Poshina, D.N.; Lyalina, T.A.; Polyakov, D.S.; Vasilyev, V.B.; Kritchenkov, A.S.; Skorik, Y.A. *N*-[4-(*N*,*N*,*N*-trimethylammonium)benzyl]chitosan chloride: Synthesis, interaction with DNA and evaluation of transfection efficiency. *Carbohydr. Polym.* **2018**, *181*, 693–700. [CrossRef]
49. Iudin, D.; Zashikhina, N.; Demyanova, E.; Korzhikov-Vlakh, V.; Shcherbakova, E.; Boroznjak, R.; Tarasenko, I.; Zakharova, N.; Lavrentieva, A.; Skorik, Y.; et al. Polypeptide Self-Assembled Nanoparticles as Delivery Systems for Polymyxins B and E. *Pharmaceutics* **2020**, *12*, 868. [CrossRef]
50. Horowitz, G.; Altaie, S.; Boyd, J.; Ceriotti, F.; Garg, P.; Horn, P.; Clinical and Laboratory Standards Institute. *Defining, Establishing, and Verifying Reference Intervals in the Clinical Laboratory: Approved Guideline*; Clinical and Laboratory Standards Institute: Wayne, PA, USA, 2008.

International Journal of *Molecular Sciences*

MDPI

Article

Hybrid Nanoparticles and Composite Hydrogel Systems for Delivery of Peptide Antibiotics

Dmitrii Iudin [1,2], Marina Vasilieva [1], Elena Knyazeva [3], Viktor Korzhikov-Vlakh [2], Elena Demyanova [3], Antonina Lavrentieva [4], Yury Skorik [1] and Evgenia Korzhikova-Vlakh [1,*]

1 Institute of Macromolecular Compounds, Russian Academy of Sciences, Bolshoi VO 31, 199004 St. Petersburg, Russia; dmitriy-yudin97@mail.ru (D.I.); mazarine@list.ru (M.V.); yury_skorik@mail.ru (Y.S.)
2 Institute of Chemistry, St. Petersburg State University, Universitetskii 26, Peterhof, 198504 St. Petersburg, Russia; v.korzhikov-vlakh@spbu.ru
3 State Research Institute of Highly Pure Biopreparations, Pudozhsakya 7, 197110 St. Petersburg, Russia; e@knyazeva32.ru (E.K.); lenna_22@mail.ru (E.D.)
4 Institute of Technical Chemistry, Gottfried-Wilhelm-Leibniz University of Hannover, 30167 Hannover, Germany; lavrentieva@iftc.uni-hannover.de
* Correspondence: vlakh@hq.macro.ru

Abstract: The growing number of drug-resistant pathogenic bacteria poses a global threat to human health. For this reason, the search for ways to enhance the antibacterial activity of existing antibiotics is now an urgent medical task. The aim of this study was to develop novel delivery systems for polymyxins to improve their antimicrobial properties against various infections. For this, hybrid core–shell nanoparticles, consisting of silver core and a poly(glutamic acid) shell capable of polymyxin binding, were developed and carefully investigated. Characterization of the hybrid nanoparticles revealed a hydrodynamic diameter of approximately 100 nm and a negative electrokinetic potential. The nanoparticles demonstrated a lack of cytotoxicity, a low uptake by macrophages, and their own antimicrobial activity. Drug loading and loading efficacy were determined for both polymyxin B and E, and the maximal loaded value with an appropriate size of the delivery systems was 450 μg/mg of nanoparticles. Composite materials based on agarose hydrogel were prepared, containing both the loaded hybrid systems and free antibiotics. The features of polymyxin release from the hybrid nanoparticles and the composite materials were studied, and the mechanisms of release were analyzed using different theoretical models. The antibacterial activity against *Pseudomonas aeruginosa* was evaluated for both the polymyxin hybrid and the composite delivery systems. All tested samples inhibited bacterial growth. The minimal inhibitory concentrations of the polymyxin B hybrid delivery system demonstrated a synergistic effect when compared with either the antibiotic or the silver nanoparticles alone.

Keywords: hybrid nanoparticles; core–shell structures; peptides; polymyxin; colistin; drug delivery systems; antibiotics; antimicrobial properties; composite materials

Citation: Iudin, D.; Vasilieva, M.; Knyazeva, E.; Korzhikov-Vlakh, V.; Demyanova, E.; Lavrentieva, A.; Skorik, Y.; Korzhikova-Vlakh, E. Hybrid Nanoparticles and Composite Hydrogel Systems for Delivery of Peptide Antibiotics. *Int. J. Mol. Sci.* **2022**, *23*, 2771. https://doi.org/10.3390/ijms23052771

Academic Editor: Maurizio Battino

Received: 31 January 2022
Accepted: 28 February 2022
Published: 2 March 2022

1. Introduction

The emergence of antibiotics revolutionized the treatment of infectious diseases and made a significant contribution to reducing the associated morbidity and mortality. However, the year-by-year growth in the use of antibiotics has led to the appearance of bacterial resistance to these drugs. Today, the increasing number of drug-resistant pathogenic bacteria poses a global threat to human health. Despite the fact that new antibiotics are actively being researched to overcome the resistance of microorganisms to antibacterial drugs, a steady and gradual reduction in the introduction of new drugs has been reported [1].

In the past decade, failures in the treatment of multidrug-resistant bacterial infections have led to the return of some antibiotics previously rejected by medical practice.

These include the polymyxins (PMXs), a group of cyclic peptide antibiotics consisting of encoded and non-encoded amino acids (Figure 1) and active mainly against Gram-negative pathogens [2]. Polymyxin B contains phenylalanine in the peptide structure, which is replaced by leucine in polymyxin E. Furthermore, both polymyxin B and E exist as mixtures of closely related lipopeptides differing in the structure of the *N*-terminal fatty acyl group, namely, 6-methyloctanoic acid (form B1 or E1) or 6-methylheptanoic acid (form B2 or E2). This group of antibiotics was introduced into clinical practice more than 50 years ago, but they were quickly discovered to have serious side effects, such as nephro- and neurotoxicity [3]. However, in 2011, the World Health Organization reclassified PMXs as extremely important antibiotics in practical health care for the treatment of infections for which there are now practically no alternatives [4].

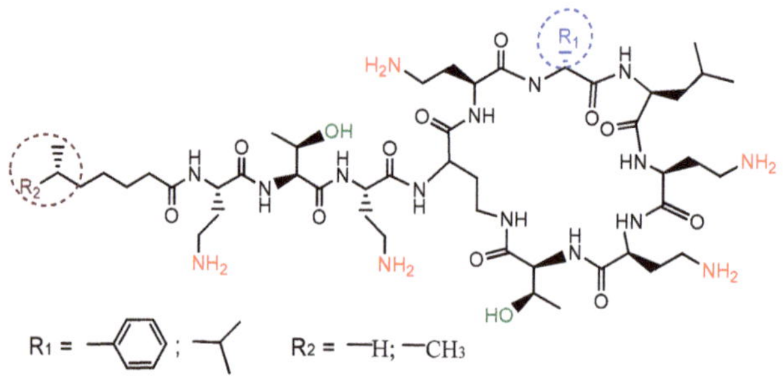

Figure 1. General structure of polymyxins.

The return of PMXs as therapeutic agents in clinical practice has prompted a search for ways to minimize their side effects. The emergence of nanomaterials has presented one strategy for developing drug delivery systems that can optimize the treatment of infectious diseases while minimizing the toxic effects of drugs like PMXs [5–7]. PMXs are positively charged, so they can interact with anionic phospholipids and anionic polymers; therefore, the modern delivery systems considered for PMXs include anionic liposomes [8–10] and anionic polymers. The anionic polymers form polyelectrolyte complexes (PECs) with PMXs due to electrostatic interactions [11,12]. For instance, Liu et al. reported the preparation of nanoparticles (NPs) by complexation of colistin (polymyxin E) with poly(glutamic acid) and further stabilization of this delivery system using 1,2-dimyristoyl-sn-glycero-3-phosphoethanolamine-N-(methoxy (polyethylene glycol)-2000) [11]. Dubashynskaya et al. evaluated the efficiency of hyaluronan/colistin PECs as PMX delivery systems [12]. The positively charged PMXs can also be successfully loaded into PECs formed by combinations of polycations and polyanions. For example, Coppi et al. developed pH-sensitive alginate/chitosan PECs for oral administration of PMX B [13,14]. Similar systems based on PECs formed by a polycation (oligochitosan) and a polyanion (anionic starch) for co-delivery of colistin and tobramycin were recently studied by Yasar et al. [15]. Hyaluronan/diethylaminoethyl chitosan PECs for improved colistin delivery have been reported by Dubashynskaya et al. [16].

The PMX structure, which includes a lipophilic tail and a hydrophobic amino acid (phenylalanine or leucine), allows PMX loading into lipid NPs and polymers, as well as into amphiphilic copolymers. Recently developed PMX delivery systems have included not only lipid NPs [17,18] and hydrophobic poly(butyl cyanoacrylate)-based NPs [19], but also poly(L-lactide)/halloysite nanotube nanofiber mats for wound treatment [20]. Therefore, in addition to systemic administration (e.g., parenteral administration), PMXs can also be used for external treatment to prevent skin infections in cases of postsurgical or burn wounds [21]. The conventional forms for prolonged external skin treatments with PMXs include patches [22] and hydrogel films [23,24]. Innovative microneedle technologies can also be a choice for effective transdermal delivery of these antibiotics [25,26]. Compared to systemic administration, local delivery of antibiotics has several advantages, including reduced systemic toxicity, high therapeutic efficacy, and low occurrence of bacterial resistance. Several biopolymers, such as hyaluronic acid, sodium alginate, chitosan, collagen, and dextran, have been widely used to prepare materials for prolonged and safe antibacterial wound treatment [24,27,28].

Like many other antibiotics PMXs can be used together with silver NPs (Ag NPs), which already have their own antimicrobial activity against different microorganisms, including *Staphylococcus aureus, Vibrio cholerae, Pseudomonas aeruginosa, Bacillus subtilis, Escherichia coli*, etc. [29,30]. One of the proposed mechanisms of the antimicrobial activity of Ag NPs is associated with a gradual release of silver ions (Ag^+), which actively interact with the cell membrane and penetrate into the microorganism. The internalized Ag^+ ions deactivate the respiratory enzymes, thereby blocking adenosine triphosphate production but retaining electron transport, which then supports the generation of reactive oxygen species that can modify DNA. The Ag^+ ions can also bind to the phosphate groups of DNA, thereby interfering with DNA replication [30]. Brown et al. compared the biological activity of Ag and Au NPs, as well as the same NPs functionalized with the beta-lactam antibiotic ampicillin [31]. The authors found no antibacterial effect with neat Au NPs, whereas the NPs modified with the antibiotic showed antimicrobial activity. By contrast, the neat Ag NPs clearly had antibacterial activity against Gram-positive and Gram-negative bacteria, even without the presence of ampicillin. Manukumar et al. reported that a preparation of Ag NPs coated with chitosan and capable of incorporating thymol showed broad bactericidal activity against both Gram-positive and Gram-negative bacteria [32]. Moreover, in some studies, antibiotics loaded into hybrid polymer–silver NPs have demonstrated enhanced antibacterial activity [33,34].

Earlier, we studied the self-assembly of amphiphilic poly(glutamic acid-*co*-polyphenylalanine) into NPs of ca. 200 nm and the use of these NPs as delivery systems for both PMXs B and E [35,36]. The developed NPs did not show cytotoxicity but demonstrated reduced uptake by macrophages. PMX B, being slightly more hydrophobic than PMX E, demonstrated a higher loading and a stronger retention than PMX E. The minimal inhibitory concentrations for the developed formulations were equal to those of the free antibiotics.

In the present study, we prepared hybrid core–shell NPs consisting of an Ag core covered by a poly(glutamic acid) (PGlu) shell, and we investigated these PGlu@Ag NPs as delivery systems for PMXs B and E. The Ag core was chosen to enhance the antibacterial effect of the PMX formulation, while PGlu was selected because of its effective PMX binding and the absence of cytotoxicity. The hybrid PMX/PGlu@Ag NPs were carefully characterized in terms of their size, polydispersity, and morphology. The loading of PMXs was evaluated to optimize the amount of loaded drug and the stability of the formulation. The PMX/PGlu@Ag NPs were used further as a filler to prepare agarose-based hydrogel composites that could be considered as systems for antibacterial treatment of skin damage. The release kinetics of PMX from both hybrid and composite delivery systems were investigated and analyzed using different mathematical models to predict the most probable mechanisms of release. The antimicrobial efficacy against *P. aeruginosa* was evaluated for both types of PMX delivery systems developed in this study.

2. Results and Discussion

2.1. Synthesis of SH-PGlu

The preparation of PGlu@Ag NPs was based on the reduction of Ag^+ in silver nitrate to Ag^0 by sodium borohydride in the presence of thiol-containing PGlu as a coating and stabilizing agent.

PGlu containing a terminal thiol group was obtained by the ring-opening polymerization of N-carboxyanhydride (NCA) of L-glutamic acid γ-benzyl ester (Glu(OBzl)) using S-acetamidomethyl-L-cysteine (Cys(Acm)) as an initiator (its amino group can initiate ring opening polymerization (ROP) via the amine mechanism). The scheme of the synthesis is illustrated in Figure 2A. At this step, the polymer had a hydrophobic nature. According to size-exclusion chromatography (SEC), the synthesized protected polymer had the following molecular weight characteristics: $M_w = 5700$, $M_n = 4300$, Đ = 1.32.

A
B

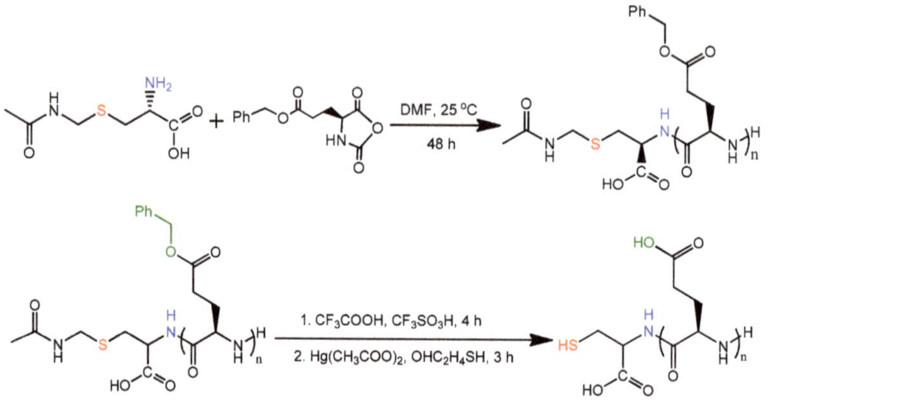

Figure 2. Scheme of synthesis of PGlu containing a terminal SH-group: (**A**) ROP and (**B**) deprotection.

In the next step, the obtained polymer was deprotected (Figure 2B). Initially, the γ-carboxyl groups of glutamic acid were deprotected using trifluoracetic/ trifluoromethanesulfonic acid (TFA/TFMSA) solution. Finally, the Acm protection of the terminal cysteine was removed to generate a free thiol group. The presence of the SH-group in PGlu was testified by Ellman's test [37]. The product of the reaction with Ellman's reagent is easily detected spectrophotometrically at a wavelength of 412 nm and can be quantified. The content of thiol groups in the polymer was 2.28 mg, which corresponded to 70 ± 5 wt% of the thiol groups in S-acetamidomethyl-L-cysteine taken as the initiator of the polymerization reaction.

2.2. Preparation of Hybrid Nanoparticles

Hybrid NPs were obtained via the reduction of the Ag^+ in silver nitrate with sodium borohydride with a simultaneous stabilization of the generated NPs by SH-PGlu due to the formation of -S-Ag bonds. The scheme of the reaction is shown in Figure 3A. During the reaction, the suspension changed in color from light beige to a rich golden brown, indicating that silver ions were reduced, and that stabilized NPs had formed. Ag NPs stabilized with a low molecular weight compound, cysteine (Cys@Ag), were also obtained for comparison (Figure 3B).

A
B

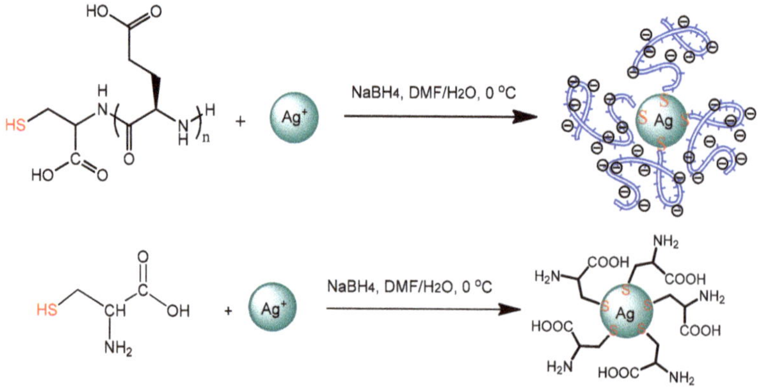

Figure 3. Scheme of preparation of PGlu@Ag (**A**) and Cys@Ag nanoparticles (**B**).

The synthesized NPs were analyzed by dynamic and electrophoretic light scattering (DLS and ELS) methods in various media to determine hydrodynamic diameter (D_H), polydispersity index (PDI), and ζ-potential. Table 1 shows that the hybrid NPs had hydrodynamic diameters close to 90–100 nm in all studied media. As expected, the PGlu@Ag NPs demonstrated a strong negative charge and provided high stability to the colloid system.

Table 1. Characteristics of PGlu@Ag nanoparticles determined by dynamic and electrophoretic light scattering.

Conditions of NPs' Redispersion	D_H (nm)	PDI	ζ-Potential (mV)
H_2O	97	0.44 ± 0.03	-57 ± 2
0.01 M PBS, pH 7.4	92	0.46 ± 0.02	-48 ± 5
0.02 M acetic buffer, pH 3.8	89	0.50 ± 0.01	-51 ± 3
0.02 M borate buffer, pH 10.5	88	0.50 ± 0.01	-55 ± 2

The morphology of the synthesized hybrid particles was investigated by TEM (Figure 4) using non-contrasted NPs and NPs stained with uranyl acetate. In the first case, the detection of only the silver core is possible, while the use of uranyl acetate allowed the detection of the size of the particle complete with its polymer shell.

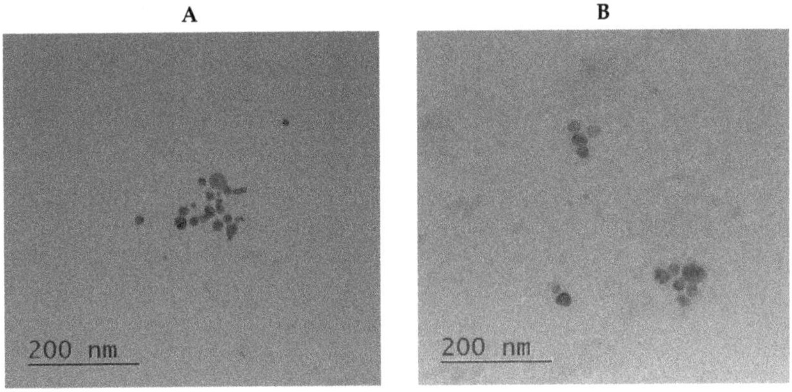

Figure 4. TEM images of PGlu@Ag nanoparticles without staining (**A**) and with uranyl acetate staining (**B**).

The DLS results for Cys@Ag NPs revealed that they were considerably larger than PGlu@Ag NPs (Table 2). Most likely, this result could be associated with less stabilization of the reducing Ag by cysteine. The Cys@Ag NPs have both amino- and carboxylic groups on their surface; therefore, they showed the ability to change the ζ-potential depending on the medium in which they were redispersed.

In addition, PGlu@Ag and Cys@Ag NPs obtained in PBS were studied by nanoparticle tracking analysis (NTA) (Figure S1 of Supplementary Materials). The hydrodynamic diameter of NPs determined by NTA is in agreement with the average D_H obtained by DLS: Ag NPs stabilized by PGlu had smaller size than those stabilized by small molecule Cys. In both cases, DLS and NTA indicate fairly wide size distribution of NPs, but, unlike PGlu@Ag NPs, a multimodal distribution was detected for Cys@Ag NPs. The appearance of several modes for Cys@Ag may be related to the poor stabilization of Ag by a small ligand as well as to the aggregation of particles due to the electrostatic interaction between the amino and carboxyl groups that are present in each cysteine molecule and ionized at pH 7.4.

Table 2. Characteristics of Cys@Ag nanoparticles determined by dynamic and electrophoretic light scattering.

Conditions of NPs' Redispersion	D_H (nm)	PDI	ζ-Potential (mV)
H_2O, 20 s	423	0.56 ± 0.01	−19 ± 1
0.01 M PBS, pH 7.4, 20 s	606	0.42 ± 0.03	−42 ± 3
0.02 M acetic buffer, pH 3.8, 45 s	505	0.32 ± 0.01	+39 ± 2
0.02 M borate buffer, pH 10.5, 30 s	780	0.36 ± 0.02	−53 ± 2

2.3. *Biological Evaluation of Hybrid Nanoparticles*

The cytotoxicity of the hybrid NPs was evaluated using HEK 293 and HepG2 cell lines by the CTB test after 72 h (Figure 5).

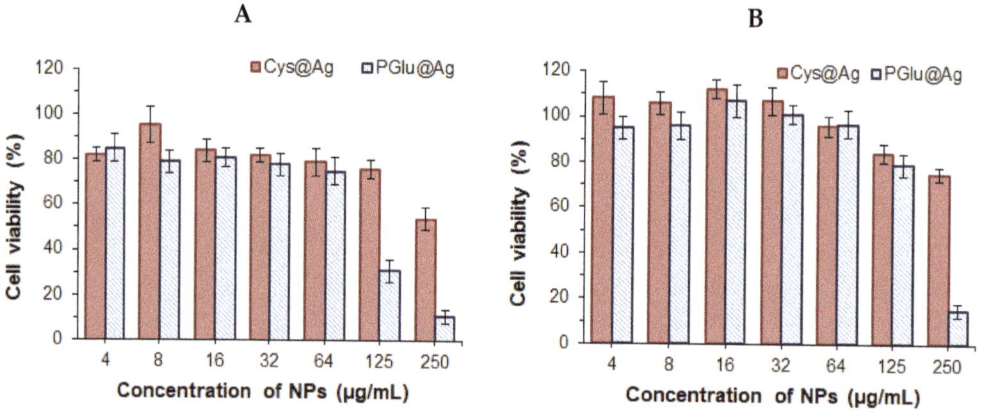

Figure 5. Viability of HEK 293 (**A**) and HepG2 (**B**) incubated with hybrid nanoparticles for 72 h.

In the case of Cys@Ag NPs, the IC_{50} values for normal (HEK 293) and cancer (HepG2) cells were 284.4 ± 9.8 and 342.1 ± 16.7 μg/mL, respectively. In turn, IC_{50} values for PGlu@Ag NPs were 84.9 ± 6.3 and 192.8 ± 9.5 μg/mL for HEK 293 and HepG2, respectively. Taking into account that PGlu is a non-toxic polymer (IC_{50} > 1000 μg/mL) [36,38], the cytotoxic effect of PGlu@Ag NPs at concentrations higher than 64 μg/mL is evidently provided by the silver core of the NPs. According to the published data, Ag NPs are quite toxic to mammalian cells. The reported cytotoxicity of non-coated Ag NPs with a size of 20–50 nm is in the range of 5–15 μg/mL for normal cells and 40 μg/mL for the cancerous cells (MCF-7) [39,40]. In turn, coating of the Ag NPs with a low molecular or polymer shell has been shown to improve their compatibility with cells [40,41]. In particular, the Janus PEG@Ag NPs 40 nm in size appeared to be non-toxic to HepG2 cells at concentrations up to 64 μg/mL [42]. Tang et al. reported IC_{50} values for Fe_3O_4@*T. spicata*/Ag NPs with a size of 50 nm of 289, 311, and 174 μg/mL against Ramos.2G6.4C10, HCT-8 [HRT-18], and HCT 116 cancer cell lines, respectively [43], while normal cells (HUVECs) retained their viability in the presence of up to 1000 μg/mL Fe_3O_4@*T. spicata*/Ag. The cytotoxicity of Ag NPs is associated with their size; smaller Ag NPs demonstrate higher cytotoxicity, presumably due to their faster penetration into the cells [40]. This may be the reason for the higher cytotoxicity of PGlu@Ag (~100 nm) compared to CyS@Ag (~420 nm) in HEK 293 and HepG2 cells.

Flow cytometry was used to study the uptake of NPs by macrophages. Besides the two hybrid NP types, the poly(L-glutamic acid-co-L-phenylalanine) (P(Glu-co-Phe)) NPs developed earlier were also applied as a benchmark. We have recently shown that uptake of P(Glu-co-Phe) about 200 nm in size by macrophages was reduced in comparison with widely used PEG-*b*-PLA NPs [36]. The capture study was carried out by incubation of mouse macrophages (J774.1A) with Cy5-labeled NPs for 0–8 h. As shown in Figure 6, the lowest uptake by macrophages was observed for the control P(Glu-co-Phe) NPs. Both kinds of Ag-based NPs demonstrated higher uptake by macrophages than was observed for the polymer NPs. Several factors, such as shape, size, surface charge, and rigidity, are known to affect the rate of phagocytosis [44,45]. Taking into account that all the tested NPs are spherical and negatively charged under the conditions of the experiment, the most probable reason for such behavior of the Ag-based NPs is their higher rigidity in comparison to the self-assembled P(Glu-co-Phe) NPs. Indeed, the published data clearly indicate that softer NPs demonstrate reduced uptake efficiency, which is associated with their tendency to undergo local deformation upon contact with macrophages [44]. Compared to PGlu@Ag, the silver NPs stabilized with cysteine (Ag@Cys) demonstrated an enhanced uptake during the first hour, while the total uptake after 8 h was slightly lower. This trend might be explained by the influence of the rigidity factor at the initial step and the size factor at a subsequent step. In particular, a larger sized NP is known to require more time for uptake, as the cell needs to build a larger phagocytic cup [44]. In general, the rate of uptake detected for PGlu@Ag is comparable to that established by us earlier for rigid PEG-b-PLA NPs of a similar size (~100 nm) [36]. Thus, it can be concluded that the coating of silver NPs with PGlu demonstrated a satisfactory result and that the hybrid NPs can be considered as viable delivery systems.

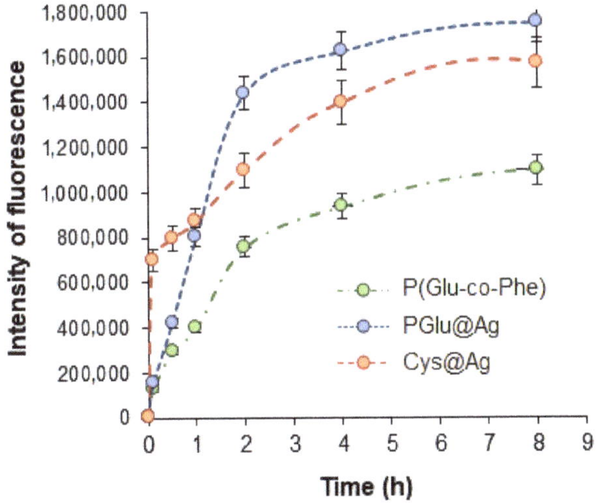

Figure 6. Uptake of different nanoparticles by macrophages (J7741.A).

As an additional experiment, the uptake of NPs by macrophages was visualized by fluorescence microscopy (Figure 7) after 3 h of co-incubation. Analysis of the images supported the results of flow cytometry regarding the capture of NPs by macrophages by nonspecific endocytosis.

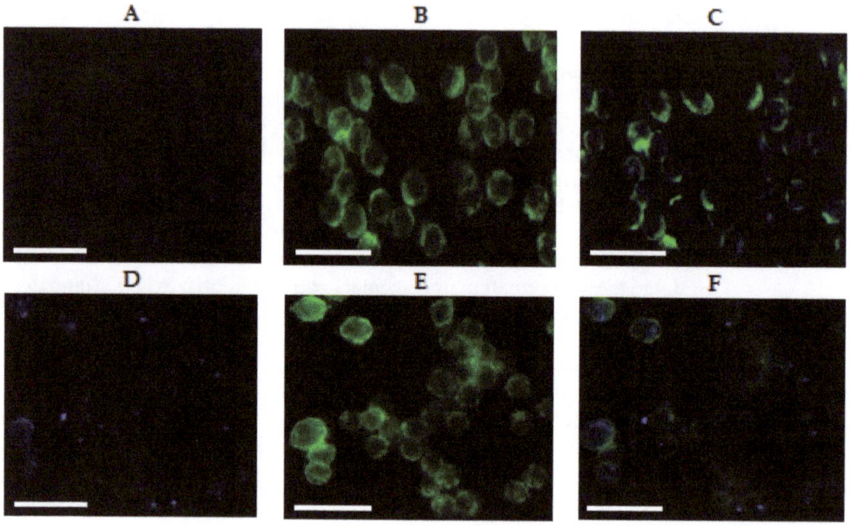

Figure 7. Fluorescence images of J7741.A cells treated for 3 h with: (**A–C**) hybrid PGlu@Ag nanoparticles labeled with Cy5; (**D–F**) Cy5-P(Glu-co-Phe) nanoparticles. The images from left to right show the fluorescence of Cy5-labeled nanoparticles in cells (blue), the stained cell membrane (green), and the overlap of two images (scale bar equal to 50 μm).

2.4. Preparation of Polymyxin Formulations Based on Hybrid Nanoparticles

PMXs were loaded by exploiting the polyelectrolyte interactions between the positively charged PMX B or E and the negatively charged PGlu. Drug loading (DL) and loading efficacy (LE) are important parameters of drug delivery systems. The dependence on DL and LE obtained for loading of PMXs B and E into PGlu@Ag is shown in Figure 8. Both antibiotics demonstrated high binding ability to PGlu, in agreement with the results previously observed for P(Glu-co-Phe) [36]. The loading of more than 1000 μg of PMX per mg of NPs occurred to be possible for both antibiotics. However, the LE was slightly higher for the more hydrophilic PMX E than for PMX B.

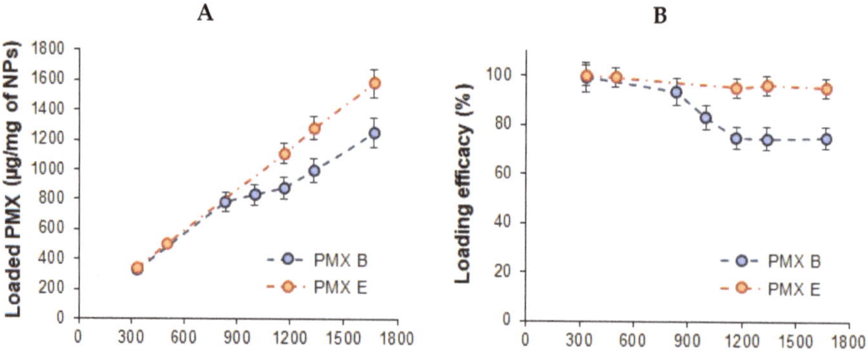

Figure 8. Drug loading (DL) (**A**) and loading efficacy (LE) (**B**) for polymyxins B and E loaded into hybrid nanoparticles.

Monitoring of the hydrodynamic diameter and ζ-potential of the delivery systems revealed a considerable change in these characteristics with an increase in PMX loading

(Table 3). The loading up to 450 μg/mg of NPs ensured the formation of stable compositions with D_H up to 210 nm. A further increase in the PMX loading led to a sharp increase in the hydrodynamic diameter (more than 1 μm) with simultaneous reduction in ζ-potential that, in turn, followed by the instability of the formulation over time. Thus, the loading of more than 450 μg of PMX per mg of NPs seems impractical.

Table 3. Changes in characteristics of hybrid PGlu@Ag NPs after loading of polymyxin B.

Loaded PMX (μg/mg of NPs)	D_H (nm)	ζ-Potential (mV)
0	93 ± 54	−48 ± 2
330	184 ± 95	−41 ± 1
450	206 ± 107	−36 ± 1
830	1215 ± 675	−28 ± 2
1250	1945 ± 874	−10 ± 1

2.5. Composite Delivery Systems

As mentioned earlier, the composite materials based on hydrogels may have potential applications as systems for wound treatment. In this work, agarose was selected to prepare the composite hydrogels for use as wound coatings. Agarose is an inexpensive natural hydrophilic polymer obtained from various algal species. The key property of agarose is its ability to form hydrogels due to the formation of hydrogen bonds between saccharide units at room temperature. Agarose-based hydrogels are biocompatible and non-toxic, making them promising for applications in medicine.

Composite films were manufactured by casting a warm agarose solution containing free antibiotics or hybrid PMX-loaded PGlu@Ag NPs onto the surface of a plastic substrate and then cooling to room temperature. The scheme for obtaining the composite material and images of hydrogels loaded with free PMX B and hybrid delivery systems are illustrated in Figure 9. The loading was 1.0–2.5 mg PMXs per 0.3 mL of hydrogel.

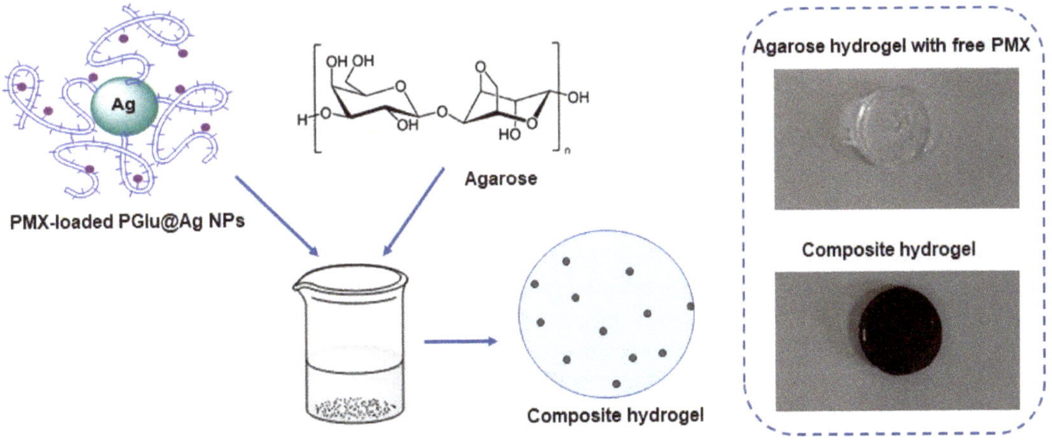

Figure 9. Scheme for the preparation of composite hydrogels with antimicrobial properties. The concentration of hybrid nanoparticles in the composite gel shown was 5 mg per specimen (0.3 mL of hydrogel).

2.6. Release of Polymyxins from Hybrid and Composite Systems

As was noted by several groups, the greater hydrophilicity of PMX E leads to less retention in formulations than is seen with PMX B; therefore, PMX E shows a faster release [8,9]. In the present study, the cumulative release of PMX E from PGlu@Ag NPs in 0.01 M PBS (pH 7.4) was two times faster than for PMX B (Figure 10A). These data are in agreement with our previous study, in which the release of both PMX E and B was studied from PGlu-containing NPs [36]. Moreover, in that study, we observed a considerable release of PMXs into human blood plasma. Here, to compare the effect of the presence of competitive macromolecules, the release of PMX B into a simulated plasma solution at 37 °C was monitored for 6 days (Figure 10A). As expected, the presence of competing protein macromolecules (human serum albumin, HSA) in solution caused a pronounced release, which reached almost 90% over the tested period.

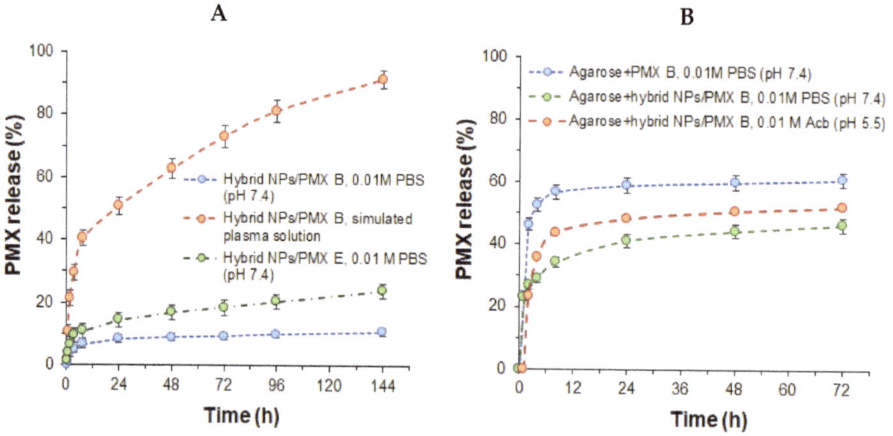

Figure 10. Release of polymyxin E and B over time from hybrid PGlu@Ag nanoparticles (**A**) and polymyxin B from composite hydrogels (**B**).

In the case of composite materials, the rate of release of PMX B was studied in 0.01 M PBS (pH 7.4) and compared with the release of a free antibiotic from the hydrogel (Figure 10B). A release of 60% of the PMX B was observed within 6 h for the antibiotic-loaded hydrogel, whereas the release was only half that value for the composite material for the same time. A comparable rate of PMX B release has also been reported recently by Shi at al., who investigated hydrogels from self-assembling peptide amphiphiles containing negatively charged carboxyl groups and loaded with PMX B [46]. In particular, the release of 60–80% PMX B after 60 h was documented by those authors.

The release from composite materials in acidic buffer (0.01 M acetic buffer, pH 5.5) was more pronounced than for the same material in 0.01 M PBS (pH 7.4) (Figure 10B). This result can be explained by better ionization of PMX amino groups in acidic buffer and, as a result, faster diffusion of peptide into the medium.

The obtained PMXs release profiles were analyzed by fitting them with several mathematical models to elucidate the mechanisms of peptide release. All regression curves and calculated data are presented in the Supplementary Materials (Table S1 and Figure S2). Here we present some conclusions from the analysis of these data [47]. First, it is obvious that release of polymyxins from both hybrid NPs and gels could hardly be fitted with zero-order and first-order models (Figure 11A,B). Thus, the release of peptides is not only a function of time, and the rate of the process is proportional not only to the amount of remaining drug in the matrix. The release process is more complex because it involves peptide–polymer interactions. It is also evident that polymyxins' release profiles poorly correlate with Hixson–Crowel and Hopfenberg models. This means that dissolutions and matrix erosion could not be considered as rate-limiting processes for release. Among standard models the best fitting was observed for Higuchi and Baker-Lonsdale models, revealing the fact that diffusion is the most important process for peptides' release from the systems under study.

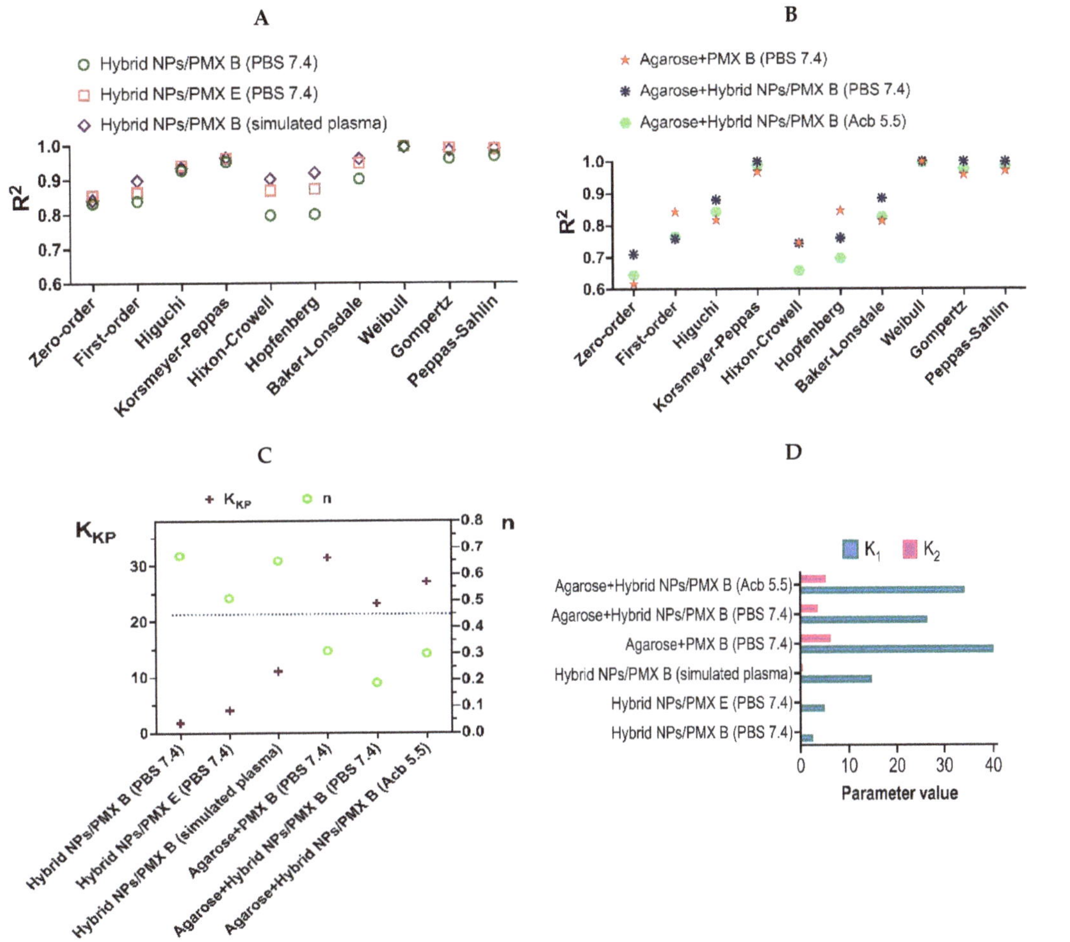

Figure 11. Results of mathematical analysis of PMX release profiles with application of standard models: (**A**) comparison of correlation coefficients of the regressions obtained with different models for release from hybrid NPs; (**B**) comparison of correlation coefficients of the regressions obtained with different models for release from agarose gel and composite gels; (**C**) results obtained by application of the Korsmeyer–Peppas model, K_{KP}—release rate constant from the Korsmeyer–Peppas equation, n—parameter from Korsmeyer–Peppas equation showing the mechanism of drug release; (**D**) results obtained by application of Korsmeyer–Peppas model, K_1—impact of diffusional mechanism, K_2—impact of relaxation on release.

Quite good correlation was obtained with Korsmeyer–Peppas model by using initial stage of release. The correlation coefficients allow to consider and analyze the rate constant (K_{KP}) and n parameters for different systems under investigation (Figure 11C). One can observe that for hybrid NPs the n parameter is above 0.45 (above the dividing line on Figure 11C), which means that mechanism of both PMX B and PMX E release is controlled by non-Fickian diffusion. In the case of gels, the n is below the dividing line, so the mechanism of release corresponds to Fickian diffusion. The release rate constants were lower for all hybrid NPs, than those for released from the agarose gel and composite gels.

The obtained results also show very good correlation with the Weibull, Gompertz, and Peppas–Sahlin models. The first model allows us to conclude that the release has some latency time (Ti) and could be described by parabolic curve (β < 1) [47]. Good correlation with the Gompertz model looks quite rational because this model describes the release of drugs possessing good solubility and an intermediate release rate [48]. The best results with this model were obtained with PMX E release from hybrid NPs (R^2 = 0.9915) and for PMX B release from composite agarose gel loaded with hybrid NPs (R^2 = 0.9993). Good fitting of the full release curves with the Peppas–Sahlin model allowed us to evaluate the impact of diffusion and relaxation on the mechanism of peptides release (Figure 11D). The values of constants from this model, namely K_1 and K_2, show the effect of diffusion and relaxation, correspondingly, on the process of release. One can observe that diffusion is the major factor affecting peptide release for all systems under study. The polymer matrix relaxation effect in the case of hybrid NPs is negligible. However, such relaxation seems to be important in the case of systems containing agarose. This looks quite rational because peptides need to diffuse through gel layer in those cases.

Thus, we can conclude that diffusion is the main mechanism of peptide release from the systems under study. The nature of PMX, the release medium and the gel layer affect the rate of release, but not the mechanism. The release is faster in simulated plasma (in the case of hybrid NPs) and in acetate buffered solution, pH 5.5, than in phosphate buffered solution, pH 7.4, but the diffusion control of the release is acting in all these systems.

2.7. Antimicrobial Activity

The antimicrobial activity of the PMX hybrid formulations was evaluated by determining the MIC against *P. aeruginosa*. The PGlu@Ag NPs loaded with PMX E or B effectively suppressed the growth of *P. aeruginosa*. For both formulations, the MIC was 1 μg/mL whereas the free PMX E and B demonstrated MICs of 1 and 4 μg/mL, respectively. For comparison, a previously developed delivery system based on P(Glu-co-Phe) NPs also demonstrated a MIC equal to 4 μg/mL for a system loaded with PMX B and 1 μg/mL for a system containing PMX E [36]. Thus, the higher antimicrobial properties of the PMX B formulation based on the hybrid NPs compared to a purely polymer delivery system or free antibiotic can be attributed to the synergistic effect of the antibiotic and the silver NPs. A synergistic effect of PMX B and Ag NPs has also been observed by Salman et al., who studied the inhibition of *P. aeruginosa* using PMX B solution and non-covered Ag NPs in a mixture [49].

Both empty PGlu@Ag and Cys@Ag NPs demonstrated an evident inhibitory effect: a MIC of 32 μg/mL was determined for both silver-containing NPs (Figure 12). Since no difference was evident if the silver core was coated with a polymer or with a low molecular weight organic shell, the detected inhibitory activity may be related only to the presence of the silver core. Our previous evaluation of the inhibitory activity of P(Glu-co-Phe) NPs did not reveal any antimicrobial effect for PGlu-containing polymer NPs against *P. aeruginosa* [36]. Thus, even the presence of a polymer shell on the surface of silver NPs did not interfere with the manifestation of their antimicrobial activity.

The suppression of *P. aeruginosa* growth by composite materials was tested by the agar disk diffusion method. First, the suspensions of PMX B/E hybrid delivery systems were placed in wells formed in an agar plate and incubated for 24 h. In this case, the inhibitory activity directly depends on the rate of antibiotic release. Free PMX E was used as a control. Figure 13A shows that the inhibition zones due to free PMX E provided at two different amounts (37.3 ± 1.1 and 74.6 ± 2.2 μg/zone) were 14 (zone 1) or 16 mm (zone 2) in size. The inhibition zones produced by PMX E/PGlu@Ag formulations, taken at the same concentrations of PMX E, were 15 (zone 1) or 17 mm (zone 2) in size. In this case, the larger inhibition zones can be a result of both the fast release of PMX E from the hybrid systems and the antimicrobial activity of the silver NPs. At the same time, in the case of PMX B, which demonstrates a slower release rate than PMX E, the inhibition zones at the same concentrations of antibiotic were 13 (zone 1) and 14 mm (zone 2) in size.

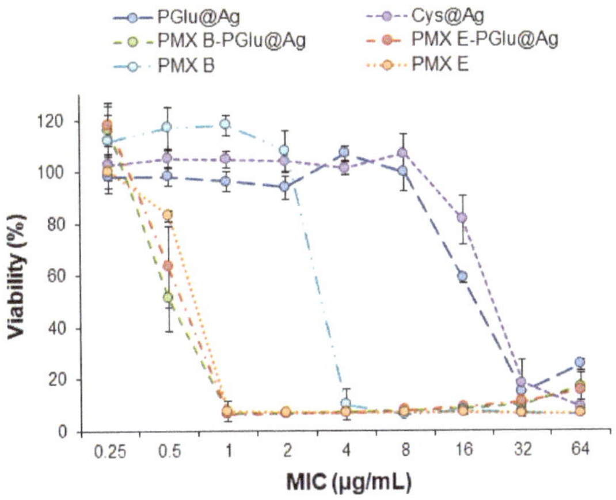

Figure 12. Antimicrobial activity of free polymyxins B and E, hybrid nanoparticles, and their nanoformulations against *Pseudomonas aeruginosa*. The data are given as mean ± SD (n = 3).

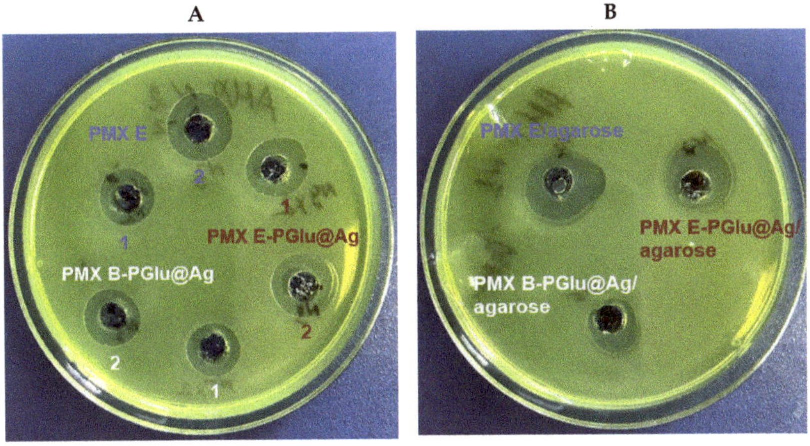

Figure 13. Suppression of *Pseudomonas aeruginosa* growth by exposure to different PMX formulations for 24 h: (**A**) dispersions of PMX hybrid delivery systems and free PMX as control: contents of the PMX in zones 1 and 2 were 37 ± 1 and 74 ± 2 μg/zone; (**B**) composite hydrogels and PMX in hydrogel as control (85.6 ± 1.7 μg of PMX per each zone).

A similar tendency was observed when testing the hydrogels. Composite hydrogels and a hydrogel containing PMX B as a control were loaded into the wells (85.6 ± 1.7 μg/zone) in the agar plate and incubated for 24 h. As in the previous case, the largest inhibition zone (19 mm) was detected for the hydrogel containing free PMX E (Figure 13B). As expected, the suppression of the bacterial growth provided by PMX E and B present in the composite hydrogels was proportional to the release rate of these antibiotics. In particular, the sizes of the inhibition zones for PMX E PGlu@Ag/agarose and PMX B PGlu@Ag/agarose were 17 and 13 mm, respectively.

Thus, all developed formulations demonstrated satisfactory antibacterial activity against *P. aeruginosa*. However, due to the faster release of PMX E, the inhibitory activity was higher for the PMX E formulations than for the PMX B formulations.

3. Materials and Methods

3.1. Chemicals and Supplements

L-glutamic acid-γ-benzyl ether (≥99.0%) (Glu(OBzl)), triphosgene (98%), α-pinene (98%), trifluoromethanesulfonic acid (TFMSA) (98%), S-acetamidomethyl-L-cysteine (Cys(Acm)) (99%), silver nitrate (≥99%), sodium tetaborate (99%), and human serum albumin (≥99%) were purchased from Sigma-Aldrich (Darmstadt, Germany) and used as received. Trifluoroacetic acid (≥99%) was purchased from Chemical Line LLC (St. Petersburg, Russia). Uranyl acetate was a product of Agar Scientific (Stansted, Essex, UK). Polymyxin B (sulfate) and polymyxin E (colistin sulfate) were acquired from Fluka (Munich, Germany) and BetaPharma (Wujiang, Shanghai, China), respectively. According to the previous HPLC-MS analysis of preparations, the commercial PMX B contained of 81.5% of PMX B1 and 18.5% of PMX B2 while the commercial PMX E contained 31.1% of PMX E1 and 68.9% of PMX E2 [36].

Dioxane, petroleum ether, ethyl acetate, and N,N-dimethylformamide (DMF) were purchased from Vecton (St. Petersburg, Russia), purified according to standard protocols, and dried before use.

Analytical-purity salts (Vecton, St. Petersburg, Russia) and deionized water were used to prepare buffer solutions. All buffer solutions were additionally filtered through a 0.45 μm Millex®® membrane microfilter (Millipore Merck, Darmstadt, Germany). Spectra/Pore®® (molecular weight cut-off (MWCO): 1000) dialysis bags were purchased from Spectra (Rancho Dominguez, CA, USA). Poly(methyl methacrylate) (PMMA) standards (Mw = 17,000–250,000; Đ ≤ 1.14) (Supelco, Bellefonte, PA, USA) were used for size exclusion chromatography (SEC) column calibration. The NPs self-assembled from poly(glutamic acid-*co*-phenylalanine) (P(Glu-co-Phe)) and used for comparisons in this work were prepared as described in our previous paper [36].

Human kidney embryonic cells (HEK 293), human liver carcinoma cells (HepG2), and mouse BALB/c monocyte macrophages (J774A.1) were acquired from Cell Lines Service GmbH (Eppelheim, Germany) and used for the evaluation of cytotoxicity and uptake of NPs by macrophages. A *P. aeruginosa* (ATCC 27,853 strain) culture was obtained from the collection of microorganisms of the State Research Institute of Highly Pure Biopreparations (St. Petersburg, Russia) and used to test the delivery systems for their minimum inhibitory concentrations (MICs).

3.2. Instruments

A VaCo 5-II Zirbus lyophilizer (Bad Grund, Germany) was used for freeze-drying. An ultrasonic homogenizer Sonopuls HD 2070 Bandelin Electronic (Berlin, Germany) was used for redispersion of the NPs. A Millipore Direct-Q 3 UV water purification system (Merck, Guyancourt, France) was used to purify water for a wide range of laboratory applications. A Zetasizer Nano ZS and Nanosight NS300, both from Malvern Instruments (Malvern, UK), were used for the determination of hydrodynamic diameter (D_H), polydispersity index (*PDI*), and electrokinetic potential (ζ-potential) of the NPs. UV absorption measurements were carried out on a UV–1800 Shimadzu spectrophotometer (Kyoto, Japan).

All obtained copolymers were analyzed by SEC using a tandem of two Agilent PLgel MIXED-D columns (7.5 mm × 300 mm, 5 μm) (USA) and a Shimadzu LC-20 Prominence system supplied with a refractometric RID 10-A detector (Kyoto, Japan). The analysis was performed in a 0.1 M solution of LiBr in dimethyl formamide (DMF) as an eluent, at a flow rate of the mobile phase of 1 mL/min and a temperature of 40 °C. Molecular weight characteristics were calculated using GPC LC Solutions software (Shimadzu, Kyoto, Japan) and the calibration curve was plotted using PMMA standards.

NMR spectroscopy was carried out using a Bruker Avance III WB 400 MHz (Karlsruhe, Germany). A Shimadzu LC-20AD Prominence HPLC system (Kyoto, Japan) equipped with a mass spectrometric detector was used for PMX analysis.

A Jeol JEM-1400 transmission electron microscope with a maximum accelerating voltage of 120 kV was used to study the morphology of NPs.

A Thermo Fischer Fluoroscan Ascent FL fluorimeter (Bradenton, FL, USA) was utilized for the cytotoxicity study. A BD Accuri C6 flow cytometer (Becton Dickinson, Franklin Lakes, NJ, USA) was used to evaluate the cellular absorption of NPs. The visualization of fluorescently labeled cells was carried out with a Cytation 5 cell imaging multi-mode reader (Bad Friedrichshall, Germany).

3.3. *Methods*

3.3.1. Synthesis and Characterization of Thiol-Containing PGlu

PGlu was obtained by the ring-opening polymerization (ROP) of L-glutamic acid-γ-benzyl ether N-carboxyanhydride (Glu(OBzl) NCA). Glu(OBzl) NCA was synthesized as described elsewhere [38,50] and polymerized in freshly distilled and anhydrous DMF. Cys(Acm) was used as an amine-type initiator (I). A Cys(Acm) solution in DMF was added to the monomer (M) solution in the same solvent to achieve a ratio of [M]/[I] = 35. The monomer concentration in the polymerization solution was 4 wt%. The polymerization was performed under stirring at 25 °C for 48 h. The obtained polymer was precipitated in 200 mL diethyl ether, and the precipitate was separated by centrifugation at 10,000 rpm for 7 min. The diethyl ether was decanted, and the precipitate was placed in a vacuum desiccator and dried for 24 h. The polymer yield was 79%. ^{1}H NMR (DMSO-d_6), δ (ppm): OBzl group: 7.33 (-C_6H_5) and 5.03 (-CH_2); glutamic acid: 4.26 (-CH).

The obtained polymer was deprotected in two steps. First, the Bzl protective group was removed using a TFA/TFMSA solution. For this, 30 mg of polymer were dissolved in 1 mL trifluoroacetic acid and left in an ice bath with stirring for 30 min to completely dissolve the polymer. TFMSA (50 μL) was added, and the reaction medium was left for 3.5 h. The PGlu was then precipitated in diethyl ether, and the precipitate was separated by centrifugation at 10,000 rpm for 10 min. The resulting precipitate was dissolved in 5 mL DMF, and the solution was transferred to a dialysis bag (MWCO 1000). Dialysis was performed against deionized water for 48 h to remove low molecular weight impurities. The content of the dialysis bag was then freeze-dried. The yield of deprotected polymer was 85 wt%.

As the second step, the Acm-protective group was removed from the terminal cysteine, which had been used as an initiator in the polymerization process. The polymer was dissolved in 10 mL 30% acetic acid. A solution of mercury (II) acetate was added to the polymer solution at a ratio of 2 eq. of mercury (II) acetate per Acm-group of cysteine. The reaction was carried out for 1 h at room temperature with intensive stirring in an argon atmosphere. A 30% solution of a 10-fold excess of β-mercaptoethanol in 30% acetic acid was then added to the reaction mixture. The resulting medium was left for an additional 2 h with stirring under the same conditions. Finally, the reaction mixture was poured into a dialysis bag (MWCO 1000), and dialysis was performed for 8 h against deionized water with frequent replacement of the water. The content of the dialysis bag was then freeze-dried. The polymer yield after this step was 96 wt%.

The completeness of removal of the Acm-group was checked by performing an Ellman's test to determine the presence of free thiol groups [51]. The test is based on the reaction of thiol with Ellman's reagent, during which the disulfide bond of the reagent breaks, and yellow-colored 2-nitro-5-thiobenzoic acid is formed. In brief, polymer solutions (10 mg/mL) and Ellman's reagent (4 mg/mL) were prepared in 0.05 M sodium phosphate buffer solution, pH 7.4, and 60 μL of Ellman's reagent and 60 μL of the polymer solution were added to 3 mL of the working buffer and thoroughly mixed. After 20 min, the absorbance of the colored product was measured at 412 nm. The concentration of the thiol groups was calculated using the extinction coefficient (ε = 14,150 M^{-1} cm^{-1}) [37].

3.3.2. Preparation of Hybrid Nanoparticles

Aqueous solutions of 2.3 equivalents (eq.) of silver (I) nitrate and 6.3 eq. of sodium borohydride, calculated relative to the cysteine content in the polymer, were added to SH-containing PGlu dispersed in DMF at a concentration of 7 mg/mL. This method requires that silver nitrate be added first to the polymer solution at 0 °C and vigorously stirred, followed by dropwise addition of sodium borohydride solution, with stirring, over 25–30 min. During the reaction, a -S-Ag bond forms between the thiol group of the terminal cysteine and the reduced silver. The obtained NPs were centrifuged at 8000 rpm, then redispersed in 0.01 M sodium borate buffer, pH 9.3, and purified by dialysis against water for 48 h. The final dispersion of particles in water was frozen and freeze-dried.

Silver NPs stabilized with cysteine (Cys@Ag) were prepared in the same way as described above for PGlu@Ag. The only difference was that all components (cysteine, silver nitrate, and sodium borohydride) were initially dissolved in a small amount of water and then mixed in DMF. After sedimentation, the NPs were dispersed in 0.05 M MES (2-(N-morpholino)ethanesulfonic acid) buffer, pH 6.5, purified by dialysis against water for 48 h.

3.3.3. Characterization of the Nanoparticles

The characteristics of the NPs, such as D_H, PDI and ζ-potential, were determined by dynamic and electrophoretic light scattering. Suspensions of NPs in different media with a volume of 1 mL and a concentration of 0.3 mg/mL were used for measurements.

Transmission electron microscopy (TEM) was used to study the morphology of the obtained hybrid NPs. Suspensions of NPs (0.3 mg/mL in water) were used for analysis. A drop (2–3 μL) of the colloids was placed on a copper grid (300 mesh) (Electron Microscopy Sciences, Hartfield, PA, USA) with the polymer (formvar) and carbon coatings and left in the air until completely dry. One part of the samples was additionally stained with 2 μL of a 2% uranyl acetate solution for 1 min, while the other part was left unstained. The grids with adhered samples at an accelerating voltage of 120 kV.

3.3.4. Cytotoxicity and Uptake by Macrophages

The Hep G2 and HEK 293 cell lines were used to study the cytotoxicity of the NPs by placing 4×10^3 cells into each well of a 96-well plate, followed by the addition of 200 μL DMEM-FBS medium containing basal medium, fetal bovine serum (FBS), penicillin, and streptomycin. The cells were cultured in a humidified atmosphere containing 5% CO_2 at 37 °C for 24 h. The medium was then replaced with a fresh medium containing different concentrations of NPs (n = 3), and the plate was incubated at 37 °C for a further 72 h. The medium was removed, and the wells were filled with 10% CTB reagent solution in culture medium and incubated for 2 h. The CTB test is based on the ability of living cells to convert resazurin into the fluorescent product resorufin ($\lambda_{ex.}$ = 545 nm, $\lambda_{em.}$ = 590 nm), with the amount of conversion proportional to the number of viable cells [52]. The obtained data were expressed as a percentage of the control.

The J7741.A cell line (mouse macrophages) was used to study the uptake of NPs by macrophages by flow cytometry. Three types of samples fluorescently labeled with the same amount of Cy5 were prepared: Cys@Ag, PGlu@Ag, and P(Glu-co-Phe). For the experiment, 0.6 mL DMEM containing 15×10^4 cells were placed into each well of a 24-well plate and cultivated in a humidified incubator with 5% CO_2 at 37 °C for 24 h. The medium was then replaced with a fresh medium containing 0.02 mg/mL of the labeled NPs. The plates were incubated in 5% CO_2 at 37 °C for 0–8 h, then the cells were detached with a cell lifter, centrifuged at 300 rpm for 5 min, and resuspended in 250 μL phosphate buffered saline (PBS; pH 7.4). The fluorescence signals were measured using a flow cytometer with a 488 nm argon-ion laser. Only viable cells (at least 30,000 events/sample) were used in the analysis.

The visualization of the uptake by macrophages was carried out in a 24-well black plate seeded with 15×10^3 cells in 0.6 mL of culture medium. The cells were incubated with 0.1 μg/mL Cy5-labeled NPs in 5% CO_2 at 37 °C for 3 h. The cell membranes were then stained with CellMask^{TM} green plasma membrane stain and visualized with a fluorescence microscope (20×).

3.3.5. Loading of Polymyxins into Hybrid Nanoparticles

Hybrid NPs loaded with PMX were prepared using a 1 mg/mL antibiotic solution in 0.01 M sodium phosphate buffer (pH 7.4). Dried NPs in the same buffer were redispersed by ultrasonication for 20 s to prepare hybrid NPs (1 mg/mL). PMX solution (100–600 μL) was added to the 300 μL of the NP dispersion, and the solution volume was brought to 1 mL. The resulting mixture was thoroughly vortexed and left at 4 °C overnight. Other manipulations, as well as the protocol for PMX HPLC analysis, were as described in our previous paper [36].

3.3.6. Preparation of Composite Materials

A weighed portion of agarose was placed into deionized water and heated to 90 °C to obtain a homogenous solution. Hybrid NPs loaded with PMX or free antibiotic were used as the fillers for the hydrogel matrix. To prepare a hydrogel, 100 μL of a PMX solution or the hybrid delivery system dispersion were placed into the wells of a 24-well plate, and 200 μL of the warm agarose solution was added, with mixing with a pipette tip. The final concentration of agarose was 1.5 wt%. The resulting mixture was rapidly cooled to form the hydrogel composites. The loading of PMX was varied from 1.0 to 2.5 mg per 0.3 mL of hydrogel. When the PMX hybrid delivery systems were used as fillers, the ratio of PMX to PGlu@Ag NPs was 1:2.

3.3.7. Release of Polymyxins

The release of PMXs from hybrid NPs and composite materials was studied using 0.01 M PBS (pH 7.4) or simulated plasma solution (8% human serum albumin solution in 0.01 M PBS, pH 7.4). In the case of the loaded hybrid NPs, the release took place in 1 mL of the selected medium. The removal of the released PMX from the system at different time intervals was carried out concomitantly using an ultrafiltration method using membrane concentrators with a MWCO of 3000. After the first round of ultrafiltration, a corresponding aliquot of the medium was added to the NPs, and the solution was filtered again. The procedure was repeated three times. At each time point, the colloid was made up to 1 mL with the working medium. In the case of composite materials, the release took place in 500 μL 0.01 M PBS (pH 7.4). At the desired time points, the medium was changed to a fresh one. The collected solutions were combined, freeze-dried, and then analyzed with an Elute UPLC chromatograph (Bruker Daltonics GmbH, Bremen, Germany) connected with a Maxis Impact Q-TOF mass spectrometer (Bruker Daltonics GmbH, Bremen, Germany) equipped with an electrospray ionization (ESI) source (Bruker Daltonics GmbH, Bremen, Germany) using a previously published procedure [12].

3.3.8. Antimicrobial Activity

The antimicrobial activity was studied by the microtiter broth dilution method described by the Clinical and Laboratory Standards Institute using *P. aeruginosa* (ATCC 27,853). The antimicrobial activity of PMX delivery systems based on the hybrid NP dispersions was assayed using a previously developed protocol [36].

For composite materials, the suppression of *P. aeruginosa* growth was assayed by the agar well diffusion method. The bacterial strain was cultured in Mueller–Hinton broth (HiMedia, Mumbai, India) at 37 °C. The OD of an overnight *P. aeruginosa* suspension in Mueller–Hinton broth was measured on a UV mini-1240 spectrophotometer (Shimadzu, Kyoto, Japan) at a wavelength of 540 nm and plated for enumeration of CFU on Luria-Bertani agar [53]. The culture suspension was then serially diluted in Mueller–Hinton broth to give approximately 1×10^7 CFU/mL. A 20 mL volume of sterile Muller–Hinton Agar (Research Center for Pharmacotherapy [RICF], St. Petersburg, Russia) was poured into sterile Petri dishes (d = 100 mm). After solidification, 1 mL of inoculum at a concentration of 1×10^7 CFU/mL was inoculated onto the Muller–Hinton Agar plates by rocking the Petri dish. Excess inoculum was removed with a pipette, and the Petri dishes were dried at room temperature (20–22 °C) for 10–15 min. Wells (6 mm diameter) were created in the agar using a sterile cylindrical template, 25 μL of colloids were added, and the wells were filled with nutrient medium. The amounts of PMXs loaded per well as a solution of free antibiotic or a dispersion of the delivery system were 37 ± 1 and 74 ± 2 μg per zone 1 and 2. In the case of hydrogels, circular pieces of hydrogel (6 mm diameter) were placed into the wells. Each hydrogel sample contained 85.6 ± 1.7 μg of PMX. The antimicrobial agents diffused to the nutrient medium and suppressed the growth of *P. aeruginosa*. The results were evaluated by measuring the diameter of the inhibition zones, including the diameter of the well.

4. Conclusions

Hybrid core–shell NPs, consisting of a silver core and poly(glutamic acid) shell, were developed as potential delivery systems for PMXs. The obtained hybrid NPs had a hydrodynamic diameter of approximately 100 nm and were capable of binding PMXs at amounts up to 450 μg per mg of NPs without aggregation. Moreover, the hybrid NPs did not show cytotoxicity at amounts up to 64 and 125 μg/mL for normal and cancer cells, respectively, and demonstrated quite low uptake by macrophages. Comparison of the developed PGlu@Ag with silver NPs stabilized by cysteine revealed that both types of NPs have similar biological properties. However, the PGlu-covered silver NPs are smaller, more stable, and allow PMX loading for combined antibacterial therapy. Approximation of the release data using several mathematical models has shown that the main release mechanism is diffusion, namely non-Fickian diffusion in the case of hybrid NPs and Fickian diffusion in the case of composite materials. The prepared PMX formulations based on the hybrid PGlu@Ag NPs demonstrated a MIC of 1 μg/mL against *P. aeruginosa*. Notably, the MIC was lower for the PMX B/PGlu@Ag delivery system than for the free antibiotic. This result can be attributed to the synergetic action of PMX and the silver NP core. The developed hybrid NPs showed antimicrobial properties when dispersed in a hydrogel. These composite hydrogels with combined antibacterial properties can be considered as potential candidates for the treatment of skin injuries, such as wounds and burns.

Supplementary Materials: The following are available online at https://www.mdpi.com/article/10.3390/ijms23052771/s1.

Author Contributions: Conceptualization, E.K.-V.; methodology, M.V., E.D. and A.L.; formal analysis, D.I. and E.K.-V.; investigation, D.I., M.V., E.D. and E.K.; data curation, E.K.-V.; mathematical approximation, V.K.-V.; supervision, E.K.-V.; resources, A.L.; writing—original draft preparation, V.K.-V. and E.K.-V.; writing—review and editing, Y.S. and E.K.-V.; visualization, V.K.-V. and E.K.-V.; project administration, Y.S.; funding acquisition, Y.S. All authors have read and agreed to the published version of the manuscript.

Funding: This research was funded by the Russian Science Foundation (project no. 19-73-20157).

Institutional Review Board Statement: Not applicable.

Informed Consent Statement: Not applicable.

Data Availability Statement: Data available on request from the authors.

Acknowledgments: The authors are grateful to Y.A. Dubrovskii (Almazov National Medical Research Centre, St. Petersburg, Russia) for HPLC-MS measurements. D. Iudin thanks the G-RISC foundation for the one-month scholarship (#L-2020b-2).

Conflicts of Interest: The authors declare no conflict of interest.

References

1. Willyard, C. The drug-resistant bacteria that pose the greatest health threats. *Nature* **2017**, *543*, 15. [CrossRef] [PubMed]
2. Dubashynskaya, N.V.; Skorik, Y.A. Polymyxin Delivery Systems: Recent Advances and Challenges. *Pharmaceuticals* **2020**, *13*, 83. [CrossRef] [PubMed]
3. Juston, J.; Bosso, J. Adverse reactions associated with systemic polymyxin therapy. *Pharmacotherapy* **2015**, *35*, 28–33. [CrossRef] [PubMed]

4. Critically Important Antimicrobials for Human Medicine. Available online: http://apps.who.int/iris/bitstream/handle/10665/77376/9789241504485_eng.pdf?sequence=1 (accessed on 3 January 2022).
5. Huh, A.; Kwon, Y. "Nanoantibiotics": A new paradigm for treating infectious diseases using nanomaterials in the antibiotics resistant era. *J. Control. Release* **2011**, *156*, 128–145. [CrossRef]
6. Zhang, L.; Pornpattananangku, D.; Hu, C.; Huang, C. Development of nanoparticles for antimicrobial drug delivery. *Curr. Med. Chem.* **2010**, *17*, 585–594. [CrossRef] [PubMed]
7. Balaure, P.; Grumezescu, A. Smart Synthetic Polymer Nanocarriers for Controlled and Site-Specific Drug Delivery. *Curr. Top. Med. Chem.* **2015**, *15*, 1424–1490. [CrossRef] [PubMed]
8. Desai, T.R.; Tyrrell, G.J.; Ng, T.; Finlay, W.H. In Vitro Evaluation of Nebulization Properties, Antimicrobial Activity, and Regional Airway Surface Liquid Concentration of Liposomal Polymyxin B Sulfate. *Pharm. Res.* **2003**, *20*, 442–447. [CrossRef]
9. Wallace, S.J.; Li, J.; Nation, R.L.; Prankerd, R.J.; Boyd, B.J. Interaction of Colistin and Colistin Methanesulfonate with Liposomes: Colloidal Aspects and Implications for Formulation. *J. Pharm. Sci.* **2012**, *101*, 3347–3359. [CrossRef] [PubMed]
10. Wang, S.; Yu, S.; Lin, Y.; Zou, P.; Chai, G.; Yu, H.H.; Wickremasinghe, H.; Shetty, N.; Ling, J.; Li, J.; et al. Co-Delivery of Ciprofloxacin and Colistin in Liposomal Formulations with Enhanced In Vitro Antimicrobial Activities against Multidrug Resistant Pseudomonas aeruginosa. *Pharm. Res.* **2018**, *35*, 187. [CrossRef]
11. Liu, Y.H.; Kuo, S.C.; Yao, B.Y.; Fang, Z.S.; Lee, Y.T.; Chang, Y.C.; Chen, T.L.; Hu, C.M.J. Colistin nanoparticle assembly by coacervate complexation with polyanionic peptides for treating drug-resistant gram-negative bacteria. *Acta Biomater.* **2018**, *82*, 133–142. [CrossRef]
12. Dubashynskaya, N.V.; Raik, S.V.; Dubrovskii, Y.A.; Shcherbakova, E.S.; Demyanova, E.V.; Shasherina, A.Y.; Anufrikov, Y.A.; Poshina, D.N.; Dobrodumov, A.V.; Skorik, Y.A. Hyaluronan/colistin polyelectrolyte complexes: Promising antiinfective drug delivery systems. *Int. J. Biol. Macromol.* **2021**, *187*, 157–165. [CrossRef] [PubMed]
13. Coppi, G.; Montanari, M.; Rossi, T.; Bondi, M.; Iannuccelli, V. Cellular uptake and toxicity of microparticles in a perspective of polymyxin B oral administration. *Int. J. Pharm.* **2010**, *385*, 42–46. [CrossRef] [PubMed]
14. Coppi, G.; Iannuccelli, V.; Sala, N.; Bondi, M. Alginate microparticles for Polymyxin B Peyer's patches uptake: Microparticles for antibiotic oral administration. *J. Microencapsul.* **2004**, *21*, 829–839. [CrossRef] [PubMed]
15. Yasar, H.; Ho, D.-K.; De Rossi, C.; Herrmann, J.; Gordon, S.; Loretz, B.; Lehr, C.-M. Starch-Chitosan Polyplexes: A Versatile Carrier System for Anti-Infectives and Gene Delivery. *Polymers* **2018**, *10*, 252. [CrossRef] [PubMed]
16. Dubashynskaya, N.V.; Raik, S.V.; Dubrovskii, Y.A.; Demyanova, E.V.; Shcherbakova, E.S.; Poshina, D.N.; Shasherina, A.Y.; Anufrikov, Y.A.; Skorik, Y.A. Hyaluronan/Diethylaminoethyl Chitosan Polyelectrolyte Complexes as Carriers for Improved Colistin Delivery. *Int. J. Mol. Sci.* **2021**, *22*, 8381. [CrossRef] [PubMed]
17. Moreno-Sastre, M.; Pastor, M.; Esquisabel, A.; Sans, E.; Viñas, M.; Bachiller, D.; Pedraz, J.L. Stability study of sodium colistimethate-loaded lipid nanoparticles. *J. Microencapsul.* **2016**, *33*, 636–645. [CrossRef]
18. Severino, P.; Silveira, E.F.; Loureiro, K.; Chaud, M.V.; Antonini, D.; Lancellotti, M.; Sarmento, V.H.; da Silva, C.F.; Santana, M.H.A.; Souto, E.B. Antimicrobial activity of polymyxin-loaded solid lipid nanoparticles (PLX-SLN): Characterization of physicochemical properties and in vitro efficacy. *Eur. J. Pharm. Sci.* **2017**, *106*, 177–184. [CrossRef]
19. Costa, J.S.R.; Medeiros, M.; Yamashiro-Kanashiro, E.H.; Rocha, M.C.; Cotrim, P.C.; Stephano, M.A.; Lancellotti, M.; Tavares, G.D.; Oliveira-Nascimento, L. Biodegradable nanocarriers coated with polymyxin b: Evaluation of leishmanicidal and antibacterial potential. *PLoS Negl. Trop. Dis.* **2019**, *13*, e0007388.
20. Zhang, X.; Guo, R.; Xu, J.; Lan, Y.; Jiao, Y.; Zhou, C.; Zhao, Y. Poly(l-lactide)/halloysite nanotube electrospun mats as dual-drug delivery systems and their therapeutic efficacy in infected full-thickness burns. *J. Biomater. Appl.* **2015**, *30*, 512–525. [CrossRef]
21. Brothers, K.M.; Stella, N.A.; Hunt, K.M.; Romanowski, E.G.; Liu, X.; Klarlund, J.K.; Shanks, R.M.Q. Putting on the brakes: Bacterial impediment of wound healing. *Sci. Rep.* **2015**, *5*, 14003. [CrossRef]
22. Alfalah, M.; Zargham, H.; Moreau, L.; Stanciu, M.; Sasseville, D. Contact Allergy to Polymyxin B Among Patients Referred for Patch Testing. *Dermatitis* **2016**, *27*, 119–122. [CrossRef]
23. Obuobi, S.; Voo, Z.X.; Low, M.W.; Czarny, B.; Selvarajan, V.; Ibrahim, N.L.; Yang, Y.Y.; Ee, P.L.R. Phenylboronic Acid Functionalized Polycarbonate Hydrogels for Controlled Release of Polymyxin B in Pseudomonas Aeruginosa Infected Burn Wounds. *Adv. Healthc. Mater.* **2018**, *7*, 1701388. [CrossRef] [PubMed]
24. Wang, L.; Li, X.; Sun, T.; Tsou, Y.-H.; Chen, H.; Xu, X. Dual-Functional Dextran-PEG Hydrogel as an Antimicrobial Biomedical Material. *Macromol. Biosci.* **2018**, *18*, 1700325. [CrossRef] [PubMed]
25. Dillon, C.; Hughes, H.; O'Reilly, N.; McLoughlin, P. Formulation and characterisation of dissolving microneedles for the transdermal delivery of therapeutic peptides. *Int. J. Pharm.* **2017**, *526*, 125–126. [CrossRef] [PubMed]
26. Zhang, X.; Chen, G.; Yu, Y.; Sun, L.; Zhao, Y. Bioinspired Adhesive and Antibacterial Microneedles for Versatile Transdermal Drug Delivery. *Research* **2020**, *2020*, 3672120. [CrossRef]
27. Kundu, R.; Payal, P. Antimicrobial Hydrogels: Promising Soft Biomaterials. *ChemistrySelect* **2020**, *5*, 14800–14810. [CrossRef]
28. Ferreira, N.N.; Perez, T.A.; Pedreiro, L.N.; Prezotti, F.G.; Boni, F.I.; Cardoso, V.M.D.O.; Venâncio, T.; Gremião, M.P.D. A novel pH-responsive hydrogel-based on calcium alginate engineered by the previous formation of polyelectrolyte complexes (PECs) intended to vaginal administration. *Drug Dev. Ind. Pharm.* **2017**, *43*, 1656–1668. [CrossRef]
29. Mühling, M.; Bradford, A.; Readman, J.; Somerfield, P.; Handy, R. An investigation into the effects of silver nanoparticles on antibiotic resistance of naturally occurring bacteria in an estuarine sediment. *Mar. Environ. Res.* **2009**, *68*, 278–283. [CrossRef]

30. Yin, I.X.; Zhang, J.; Zhao, I.S.; Mei, M.L.; Li, Q.; Chu, C.H. The Antibacterial Mechanism of Silver Nanoparticles and Its Application in Dentistry. *Int. J. Nanomed.* **2020**, *15*, 2555–2562. [CrossRef]
31. Brown, A.N.; Smith, K.; Samuels, T.A.; Lu, J.; Obare, S.O.; Scott, M.E. Nanoparticles Functionalized with Ampicillin Destroy Multiple-Antibiotic-Resistant Isolates of Pseudomonas aeruginosa and Enterobacter aerogenes and Methicillin-Resistant Staphylococcus aureus. *Appl. Environ. Microbiol.* **2012**, *78*, 2768–2774. [CrossRef]
32. Manukumar, H.M.; Umesha, S.; Kumar, H.N.N. Promising biocidal activity of thymol loaded chitosan silver nanoparticles (T-C@AgNPs) as anti-infective agents against perilous pathogens. *Int. J. Biol. Macromol.* **2017**, *102*, 1257–1265. [CrossRef] [PubMed]
33. McShan, D.; Zhang, Y.; Deng, H.; Ray, P.C.; Yu, H. Synergistic Antibacterial Effect of Silver Nanoparticles Combined with Ineffective Antibiotics on Drug Resistant Salmonella typhimurium DT104. *J. Environ. Sci. Health Part C* **2015**, *33*, 369–384. [CrossRef]
34. Zou, L.; Wang, J.; Gao, Y.; Ren, X.; Rottenberg, M.E.; Lu, J.; Holmgren, A. Synergistic antibacterial activity of silver with antibiotics correlating with the upregulation of the ROS production. *Sci. Rep.* **2018**, *8*, 11131. [CrossRef] [PubMed]
35. Zashikhina, N.N.; Yudin, D.V.; Tarasenko, I.I.; Osipova, O.M.; Korzhikova-Vlakh, E.G. Multilayered Particles Based on Biopolyelectrolytes as Potential Peptide Delivery Systems. *Polym. Sci. Ser. A* **2020**, *62*, 43–53. [CrossRef]
36. Iudin, D.; Zashikhina, N.; Demyanova, E.; Korzhikov-Vlakh, V.; Shcherbakova, E.; Boroznjak, R.; Tarasenko, I.; Zakharova, N.; Lavrentieva, A.; Skorik, Y.; et al. Polypeptide self-assembled nanoparticles as delivery systems for polymyxins B and E. *Pharmaceutics* **2020**, *12*, 868. [CrossRef] [PubMed]
37. Pilipenko, I.M.; Korzhikov-Vlakh, V.A.; Zakharova, N.V.; Urtti, A.; Tennikova, T.B. Thermo- and pH-sensitive glycosaminoglycans derivatives obtained by controlled grafting of poly(N-isopropylacrylamide). *Carbohydr. Polym.* **2020**, *248*, 116764. [CrossRef]
38. Zashikhina, N.; Sharoyko, V.; Antipchik, M.; Tarasenko, I.; Anufrikov, Y.; Lavrentieva, A.; Tennikova, T.; Korzhikova-Vlakh, E. Novel Formulations of C-Peptide with Long-Acting Therapeutic Potential for Treatment of Diabetic Complications. *Pharmaceutics* **2019**, *11*, 27. [CrossRef]
39. Liao, C.; Li, Y.; Tjong, S.C. Bactericidal and Cytotoxic Properties of Silver Nanoparticles. *Int. J. Mol. Sci.* **2019**, *20*, 449. [CrossRef]
40. Akter, M.; Sikder, M.T.; Rahman, M.M.; Ullah, A.K.M.A.; Hossain, K.F.B.; Banik, S.; Hosokawa, T.; Saito, T.; Kurasaki, M. A systematic review on silver nanoparticles-induced cytotoxicity: Physicochemical properties and perspectives. *J. Adv. Res.* **2018**, *9*, 1–16. [CrossRef]
41. Fahmy, H.M.; Mosleh, A.M.; Elghany, A.A.; Shams-Eldin, E.; Abu Serea, E.S.; Ali, S.A.; Shalan, A.E. Coated silver nanoparticles: Synthesis, cytotoxicity, and optical properties. *RSC Adv.* **2019**, *9*, 20118–20136. [CrossRef]
42. Xu, M.; Liu, J.; Xu, X.; Liu, S.; Peterka, F.; Ren, Y.; Zhu, X. Synthesis and Comparative Biological Properties of Ag-PEG Nanoparticles with Tunable Morphologies from Janus to Multi-Core Shell Structure. *Materials* **2018**, *11*, 1787. [CrossRef] [PubMed]
43. Tang, T.; Xia, Q.; Guo, J.; Chinnathambi, A.; Alrashood, S.T.; Alharbi, S.A.; Zhang, J. In situ supported of silver nanoparticles on Thymbra spicata extract coated magnetic nanoparticles under the ultrasonic condition: Its catalytic activity in the synthesis of Propargylamines and their anti-human colorectal properties in the in vitro condit. *J. Mol. Liq.* **2021**, *338*, 116451. [CrossRef]
44. Vorselen, D.; Labitigan, R.L.D.; Theriot, J.A. A mechanical perspective on phagocytic cup formation. *Curr. Opin. Cell Biol.* **2020**, *66*, 112–122. [CrossRef]
45. Vonarbourg, A.; Passirani, C.; Saulnier, P.; Benoit, J.P. Parameters influencing the stealthiness of colloidal drug delivery systems. *Biomaterials* **2006**, *27*, 4356–4373. [CrossRef] [PubMed]
46. Shi, Y.; Wareham, D.W.; Yuan, Y.; Deng, X.; Mata, A.; Azevedo, H.S. Polymyxin B-Triggered Assembly of Peptide Hydrogels for Localized and Sustained Release of Combined Antimicrobial Therapy. *Adv. Healthc. Mater.* **2021**, *10*, 2101465. [CrossRef] [PubMed]
47. Bruschi, M. Mathematical models of drug release. In *Strategies to Modify the Drug Release from Pharmaceutical Systems*; Elsevier: Amsterdam, The Netherlands, 2015; pp. 63–86.
48. Dash, S.; Murthy, P.; Nath, L.; Chowdhury, P. Kinetic modeling on drug release from controlled drug delivery systems. *Acta Pol. Pharm.* **2010**, *67*, 217–223.
49. Salman, M.; Rizwana, R.; Khan, H.; Munir, I.; Hamayun, M.; Iqbal, A.; Rehman, A.; Amin, K.; Ahmed, G.; Khan, M.; et al. Synergistic effect of silver nanoparticles and polymyxin B against biofilm produced by Pseudomonas aeruginosa isolates of pus samples in vitro. *Artif. Cells Nanomed. Biotechnol.* **2019**, *47*, 2465–2472. [CrossRef]
50. Wilder, R.; Mobashery, S. The use of triphosgene in preparation of N-carboxy alpha-amino acid anhydrides. *J. Org. Chem.* **1992**, *57*, 2755–2756. [CrossRef]
51. Ellman, G.L. Tissue sulfhydryl groups. *Arch. Biochem. Biophys.* **1959**, *82*, 70–77. [CrossRef]
52. CellTiter-Blue®Cell Viability Assay. Available online: https://www.promega.com/-/media/files/resources/protocols/technical-bulletins/101/celltiter-blue-cell-viability-assay-protocol.pdf (accessed on 3 January 2022).
53. Burman, S.; Bhattacharya, K.; Mukherjee, D.; Chandra, G. Antibacterial efficacy of leaf extracts of Combretum album Pers. against some pathogenic bacteria. *BMC Complement. Altern. Med.* **2018**, *18*, 213. [CrossRef]

International Journal of
Molecular Sciences

Article

High-Payload Buccal Delivery System of Amorphous Curcumin–Chitosan Nanoparticle Complex in Hydroxypropyl Methylcellulose and Starch Films

Li Ming Lim and Kunn Hadinoto *

School of Chemical and Biomedical Engineering, Nanyang Technological University, Singapore 637459, Singapore; particletechnology.ntu@gmail.com
* Correspondence: kunnong@ntu.edu.sg; Tel.: +65-6514-8381

Abstract: Oral delivery of curcumin (CUR) has limited effectiveness due to CUR's poor systemic bioavailability caused by its first-pass metabolism and low solubility. Buccal delivery of CUR nanoparticles can address the poor bioavailability issue by virtue of avoidance of first-pass metabolism and solubility enhancement afforded by CUR nanoparticles. Buccal film delivery of drug nanoparticles, nevertheless, has been limited to low drug payload. Herein, we evaluated the feasibilities of three mucoadhesive polysaccharides, i.e., hydroxypropyl methylcellulose (HPMC), starch, and hydroxypropyl starch as buccal films of amorphous CUR–chitosan nanoplex at high CUR payload. Both HPMC and starch films could accommodate high CUR payload without adverse effects on the films' characteristics. Starch films exhibited far superior CUR release profiles at high CUR payload as the faster disintegration time of starch films lowered the precipitation propensity of the highly supersaturated CUR concentration generated by the nanoplex. Compared to unmodified starch, hydroxypropyl starch films exhibited superior CUR release, with sustained release of nearly 100% of the CUR payload in 4 h. Hydroxypropyl starch films also exhibited good payload uniformity, minimal weight/thickness variations, high folding endurance, and good long-term storage stability. The present results established hydroxypropyl starch as the suitable mucoadhesive polysaccharide for high-payload buccal film applications.

Keywords: curcumin; nanoparticle complex; buccal drug delivery; polysaccharides

Citation: Lim, L.M.; Hadinoto, K. High-Payload Buccal Delivery System of Amorphous Curcumin–Chitosan Nanoparticle Complex in Hydroxypropyl Methylcellulose and Starch Films. *Int. J. Mol. Sci.* **2021**, *22*, 9399. https://doi.org/10.3390/ijms22179399

Academic Editor: Yury A. Skorik

Received: 30 July 2021
Accepted: 27 August 2021
Published: 30 August 2021

1. Introduction

The vast therapeutic activities of curcumin (CUR)—a polyphenol extracted from turmeric plants—have been well-established, where antioxidant, antimicrobial, anti-inflammatory, antidiabetic, and anticancer activities of CUR have been successfully demonstrated in vivo [1–3]. Numerous human clinical trials on the use of CUR in the management of several chronic diseases (e.g., cardiovascular, metabolic, neurological, cancers) have also shown promising results [4,5]. Not unlike other pharmaceuticals, the oral route represents the most commonly used delivery route of CUR owing to the convenience and cost-effectiveness of oral dosage forms (e.g., tablets, capsules, liquid suspension) for the patients [6]. The oral route, however, produces low CUR systemic bioavailability due to CUR's low solubility in the gastrointestinal fluid, and due to CUR degradation by first pass metabolism [7,8]. Consequently, high CUR dosages near its toxicity limit are often needed in clinical trials to achieve the intended therapeutic outcomes [9].

Amorphization represents one of the most effective solubility enhancement strategy of poorly soluble drugs by virtue of the ability of amorphous drugs to produce a highly supersaturated concentration of the drug upon dissolution [10,11]. The supersaturation generation is attributed to the low energy barrier for dissolution of amorphous drugs as a result of their metastable liquid-like form. The supersaturation generation results in a high kinetic drug solubility, which is multifold higher than the thermodynamic solubility exhibited by the stable crystalline form [12]. Various amorphous CUR formulations exhibiting enhanced CUR solubility have been developed in the form of amorphous solid dispersion [13–15], co-amorphous system [16–18], and amorphous nanoparticles [19–21].

Among the amorphous drug formulations, amorphous CUR–polyelectrolyte nanoparticle complex (or nanoplex in short) presented in Lim et al. [19] stands out because of its significantly simpler preparation method. The amorphous CUR nanoplex is prepared by bulk mixing of aqueous CUR solution with chitosan (CHI) solution acting as the oppositely charged polyelectrolyte (PE). Soluble CUR–CHI complexes were formed upon mixing by virtue of their electrostatic interactions. The soluble complexes subsequently form aggregates due to hydrophobic interactions among the bound CUR molecules. The aggregates precipitate upon reaching a critical mass to form the CUR–CHI nanoplex. The restricted mobility of CUR molecules due to the electrostatic binding with CHI prevent them from re-arranging to ordered crystalline structure, resulting in the formation of amorphous CUR [22]. Importantly, the bioavailability enhancement afforded by the CUR–CHI nanoplex has been demonstrated in vivo for wound healing applications [23,24].

While formulating CUR in the form of the CUR–CHI nanoplex can adequately address its low solubility issue in the gastrointestinal fluid, orally administered CUR remains vulnerable to extensive CUR metabolisms in the gut, regardless of the solubility. Thereby, low systemic bioavailability persists [25]. For this reason, parenteral delivery routes of CUR that can be easily administered by the patients themselves, such as transdermal [26–28] and buccal [29–31] routes, have been explored. Even though the transdermal and buccal delivery routes of CUR can circumvent its gut metabolism issue, the systemic bioavailability remains poor if the native form of CUR is used due to the low solubility of native CUR in the plasma fluid. Therefore, it is imperative that the solubility-enhanced CUR formulation is incorporated into whichever parenteral formulation used.

For this reason, the present work aimed to develop a buccal delivery system of the amorphous CUR–CHI nanoplex in the form of nanoplex-loaded polymer films. The buccal route for systemic drug delivery relies on transmucosal absorption of drug molecules through mucosal membranes lining the cheeks [32]. In addition to systemic CUR delivery, the buccal film of the CUR–CHI nanoplex can also have applications in local CUR administration in the oral cavity, as CUR is known to be effective in treating various periodontal diseases [33]. Unlike fast-dissolving oral thin film in which the drug payload is released almost instantaneously [34], the buccal film in the present work was prepared using mucoadhesive polysaccharides to facilitate sustained CUR release known to be ideal for its bioavailability enhancement [35].

Buccal films containing drug-loaded nanoparticles have been investigated before using a wide range of drugs and materials for both the film and nanoparticle carrier [29,36–40]. The drug payload in these studies, however, were limited to a maximum value of approximately 1 mg of drug per square centimeter of the film. While the reason was not elaborated in these studies, we rationalized that the limit on the maximum drug payload investigated could be attributed to keeping the drug payload at ≤ 1 mg/cm^2 in order to (1) minimize agglomeration among the nanoparticles and (2) to maintain the physical integrity of the buccal film. The nanoparticle agglomeration could adversely affect the drug content uniformity in the buccal film and the drug release rate [41].

A high drug payload is desired in buccal films intended for sustained drug release, as too low payload limits the amount of drug that can be delivered over time. At low drug payload, multiple dosing is needed, which makes the sustained release formulation redundant, and inevitably reduces patients' compliance to the dosing regimen. Therefore, the objective of the present work was to develop a high-payload mucoadhesive buccal film containing the amorphous CUR–CHI nanoplex (up to ≈ 4 mg/cm^2), which is capable of producing a sustained CUR release profile.

In the present work, we investigated the feasibilities of three mucoadhesive polysaccharides as the matrix former for buccal films at high drug payload. They were (1) hydroxypropylmethyl cellulose (HPMC), (2) pre-gelatinized starch, and (3) pre-gelatinized hydroxypropyl starch. The three polysaccharides exhibit well-established mucoadhesiveness, biocompatibility, and film-forming abilities [41]. HPMC represents one of the most widely used mucoadhesive polymers for buccal drug delivery owed to its hydrogel forming ability, which was ideal for producing a sustained drug release profile [42,43]. Starches, on the other hand, have only been recently investigated for buccal drug delivery, despite their well-established film-forming ability [44–46]. Starches, nevertheless, have many applications in pharmaceutical formulations, for example, as fillers/binders/disintegrants in oral tablets [47].

The abilities of the HPMC and starch films to accommodate the high drug payload were individually examined using the following criteria: (1) high CUR entrapment efficiency, (2) minimal variations in the film's weight and thickness, (3) good CUR payload uniformity, and (4) high folding endurance signifying physical robustness. In addition, the buccal films ought to be able to produce sustained CUR release profiles at high drug payload. The ideal sustained CUR release profile would be one that follows the zero-order kinetics, where a constant amount of CUR was released from the film as a function of time, resulting in uniform CUR concentrations over time.

The effects of the plasticizer's type (i.e., glycerol and propylene glycol) and inclusion of adjuvants (i.e., sodium alginate, polyvinyl alcohol) on the HPMC film's characteristics were investigated. From the results of the above evaluations, the optimal high-payload nanoplex-loaded buccal film formulation was determined. Subsequently, the physical stability of the amorphous CUR–CHI nanoplex in the optimal buccal film formulation was characterized after long-term storage to examine the crystallization tendency of the embedded nanoplex during its shelf life. The amorphous form stability of the nanoplex during storage is crucial for the nanoplex to maintain its solubility enhancement capability.

2. Results and Discussion

2.1. Physical Characteristics of Amorphous CUR–CHI Nanoplex

From DLS, the CUR–CHI nanoplex was found to exhibit size and zeta potential of 244 $\pm$ 24 nm and 32 $\pm$ 2 mV, respectively. The positive zeta potential indicated the predominant presence of cationic CHI on the nanoplex surface. The CUR content of the nanoplex was 61 $\pm$ 2% (w/w) with CHI making up the remaining mass. The CUR encapsulation efficiency into the nanoplex was 83 $\pm$ 3% (w/w). The FESEM image of the CUR–CHI nanoplex after lyophilization showed the appearance of agglomerates of the nanoplex exhibiting roughly spherical shapes with individual sizes in the range of 100–150 nm (Figure 1). The nanoplex agglomerates shown in the FESEM image readily dissociated into the individual nanoplex upon their reconstitution in deionized water, resulting in similar size and zeta potential (i.e., 258 $\pm$ 13 nm and 34 $\pm$ 1 mV, respectively).

Figure 1. FESEM image of amorphous CUR–CHI nanoplex prior to its incorporation into buccal film.

2.2. HPMC-Based Films

2.2.1. Effects of Plasticizer

Physical characteristics of HPMC films prepared using either Gly or PG as the plasticizer are presented in Table 1. The HPMC films were prepared at two theoretical CUR payloads of 1 and 5 mg/cm^2. As mentioned before, the former represented the typical drug payload used in the previous studies on buccal films of drug nanoparticles, whereas the latter represented the higher drug payload pursued in the present work.

For HPMC films prepared using Gly as the plasticizer, the effects of increasing the CUR payload on the CUR entrapment efficiency were found to be minimal, with both payloads exhibiting entrapment efficiency of around 81–82% (*w*/*w*). Hence, approximately 20% of the supplied CUR–CHI nanoplex was not successfully embedded into the HPMC matrix during the HPMC film formation. As these free nanoplexes were present on the film's surface, they were easily removed during the convective drying step. This resulted in experimental CUR payloads of 0.8 ± 0.03 and 4.1 ± 0.3 mg/cm^2 for the theoretical CUR payloads of 1 and 5 mg/cm^2, respectively, for HPMC (Gly) films.

For HPMC films prepared using PG as the plasticizer, the CUR entrapment efficiencies at both payloads were similar in magnitude to the values in the HPMC (Gly) films at around 80% (*w*/*w*). As a result, HPMC (Gly) and HPMC (PG) films exhibited similar experimental CUR payloads. In this regard, as the CUR:CHI ratio in the nanoplex was roughly equal to 60:40 (*w*/*w*), the CHI payloads in the films were equal to approximately two-thirds of the CUR payloads. Therefore, the CHI payloads in the films were equal to roughly 0.5 and 2.6 mg/cm^2 at theoretical CUR payloads of 1 and 5 mg/cm^2, respectively.

Table 1. Physical characteristics of HPMC (Gly) and HPMC (PG) films (n = 10, mean ± stdev).

Type of Film	Theoretical CUR Payload (mg/cm^2)	% CUR Entrapment (*w*/*w*)	CUR Payload (mg/cm^2)	Weight (mg/cm^2)	Thickness (µm)
HPMC (Gly)	1	81 ± 3	0.8 ± 0.03	13 ± 1	110 ± 3
HPMC (Gly)	5	82 ± 5	4 ± 0.3	30 ± 1	231 ± 7
HPMC (PG)	1	83 ± 3	0.8 ± 0.03	17 ± 2	138 ± 6
HPMC (PG)	5	79 ± 8	4 ± 0.4	31 ± 1	254 ± 5

The similar CUR entrapment efficiency between the HPMC (Gly) and HPMC (PG) films were not unexpected as Gly and PG are highly similar chemically in terms of the alcohol functional groups (3 vs. 2 OH groups), molecular weight (92 vs. 76 g/mol), and density (1.2 vs. 1.0 g/cm^3). While both Gly and PG are miscible in water, Gly exhibits a significantly higher viscosity (1.41 vs. 0.04 Pa·s). The multifold difference in their viscosities was nevertheless not found to have any impact on the CUR entrapment efficiency. Noticeably, the minimal effect of increased CUR payload on the CUR entrapment efficiency was observed in both the HPMC (Gly) and HPMC (PG) films. These results signified that both HPMC (Gly) and HPMC (PG) films had the capacity to accommodate the larger amount of the CUR–CHI nanoplex at high CUR payload without any adverse effect on the CUR entrapment efficiency.

While the CUR entrapment efficiency was not affected by the increased CUR payload in the films, the same was not observed for the films' weight and thickness. The weight and thickness of the HPMC (Gly) and HPMC (PG) films increased from roughly 13–17 mg/cm^2 and 110–138 µm, respectively, at the low CUR payload to roughly 30–31 mg/cm^2 and 231–254 µm, respectively, at the high CUR payload (Table 1). The increased thickness at higher CUR payload suggested that the nanoplex accumulated in the films as vertical layers. Importantly, both HPMC films passed the weight and thickness variation tests. All ten independent samples exhibited weight and thickness that were within ± 10% of the average weight and thickness signifying batch-to-batch consistency of their preparation.

Before the HPMC films were characterized further in terms of their CUR payload uniformity and folding endurance, the CUR dissolution from the films was characterized first to verify that their CUR payload could be released at the desired rate. Being a hydrogel, HPMC films swelled upon dissolution, resulting in the formation of viscous gel layers acting as physical barriers for the CUR release. The CUR release rate from the film was influenced by diffusion of CUR molecules across the gel layers, as well as the CUR dissolution from the CUR–CHI nanoplex. As the CUR–CHI nanoplex was designed as a supersaturating delivery system of CUR, the nanoplex dissolved rapidly in the swollen HPMC matrix to produce a highly supersaturated CUR concentration in the film. Thus, the CUR release rate from the film was essentially governed by the molecular diffusion across the gel layers. Nevertheless, the supersaturated CUR concentration in the film was thermodynamically unstable, thereby, CUR precipitation in the film might take place in the absence of sufficient crystallization inhibition mechanisms [48].

For the HPMC (Gly) film prepared at theoretical CUR payload of 1 mg/cm^2, the results of the dissolution tests in the SSF showed that roughly only 25% (*w*/*w*) of the CUR payload was released from the HPMC film (Gly) after 0.5 h. This signified the absence of a burst CUR release profile as expected from HPMC-based drug delivery systems (Figure 2). In this regard, the burst release is generally defined as rapid dissolution in which more than 85% (*w*/*w*) of the drug payload is released after 0.5 h [49]. The % CUR release then increased to reach approximately 60% after 1 h. However, the % CUR release plateaued afterwards to remain at around 60% after 4 h. As HPMC is a highly hydrophilic hydrogel, the incomplete CUR release was not likely to be caused by the lack of water uptake in certain segments of the HPMC (Gly) film, which would have suppressed the nanoplex dissolution in those segments.

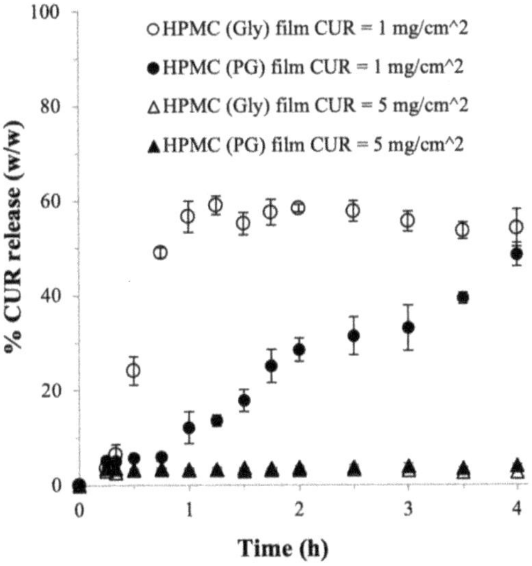

Figure 2. CUR release from HPMC (Gly) and HPMC (PG) films (*n* = 6, error bars represent the standard deviations).

Therefore, we postulated that the incomplete CUR release was caused by precipitation of the supersaturated CUR concentration in the HPMC (Gly) film due to slow outward diffusion of the CUR molecules across the viscous gel layers. In a separate experiment, the HPMC (Gly) films were found to disintegrate very slowly in the SSF with disintegration time of 550 ± 17 min until its complete disintegration. The slow disintegration of the HPMC (Gly) film resulted in prolonged presence of intact gel layers that in turn slowed down the outward diffusion of CUR molecules. The slow diffusion rate caused the CUR supersaturation in the film to be at an unsustainably high level, resulting in high CUR precipitation propensity. Theoretically, the likelihood of the supersaturated CUR concentration in the film to precipitate increased with increasing CUR payload. This was evidenced experimentally by the severe inhibition of the CUR release from the HPMC (Gly) film prepared at theoretical CUR payload of 5 mg/cm^2, where less than 5% of the CUR payload was released after 4 h (Figure 2).

Compared to the HPMC (Gly) film, the HPMC (PG) film prepared at theoretical CUR payload of 1 mg/cm^2 exhibited a significantly slower CUR release rate in the beginning with less than 15% (*w*/*w*) of the CUR payload was released after 1 h (Figure 2). Nevertheless, the CUR release picked up afterwards to reach around 50% (*w*/*w*) after 4 h. The slower CUR release in the HPMC (PG) film could be attributed to its thicker structure than the HPMC (Gly) film, resulting in longer diffusion pathways for the CUR molecules. Significantly, not unlike the finding in the HPMC (Gly) film, the HPMC (PG) film prepared at theoretical CUR payload of 5 mg/cm^2 also exhibited severely inhibited CUR release (Figure 2). This was not unexpected, as the HPMC (PG) film also exhibited slow disintegration time at 520 ± 25 min.

Importantly, the failures of both the HPMC (Gly) and HPMC (PG) films in releasing the CUR at high CUR payload led us to conclude that bare formulations of HPMC films were inadequate for high drug payload applications. Therefore, the effects of adding adjuvants to the HPMC films were investigated next. Gly was used as the plasticizer in the subsequent studies, as the results in Table 1 and Figure 2 showed the minimal role of the plasticizer's type in the resultant characteristics of the HPMC films. Moreover, Gly was preferred over PG owing to its better cytotoxicity profile.

2.2.2. Effects of Adjuvants

Besides precipitation of the supersaturated CUR concentration, another possible reason for the suppressed CUR release at high CUR payload was the aggregation of the nanoplex in the increasingly confined space of the film as the film solution was dried up. The aggregation limited the nanoplex's surface areas exposed to the aqueous surrounding in the swollen polymer matrix, resulting in inhibited CUR release. Therefore, the effects of adding PVA—a widely used amphiphilic polymeric surfactant for colloids stabilization—were investigated to prevent the aggregation of the nanoplex in HPMC films. Moreover, PVA is also known for its good film-forming ability and mucoadhesive properties, rendering it a suitable adjuvant for buccal film applications [5].

In a separate study, the effects of adding AGN—another mucoadhesive polymer with good film-forming ability [50]—were investigated as the addition of anionic AGN to nonionic HPMC had been shown to improve the swelling properties of HPMC films [51]. We postulated that improved swelling properties of the HPMC film would lead to an improved CUR release profile. Physical characteristics of the HPMC–PVA and HPMC–AGN films prepared using Gly as the plasticizer are presented in Table 2. Both PVA and AGN were added at HPMC:PVA and HPMC:AGN mass ratios of 10:1. On this note, higher HPMC:PVA and HPMC:AGN ratios had also been investigated, but they were found to have significantly adverse effects on the film's physical integrity and CUR entrapment efficiency; hence, they are not presented here.

Table 2. Physical characteristics of the HPMC–PVA (Gly) and HPMC–AGN (Gly) films (n = 10, mean ± stdev).

Type of Film	Theoretical CUR Payload (mg/cm^2)	% CUR Entrapment (*w*/*w*)	CUR Payload (mg/cm^2)	Weight (mg/cm^2)	Thickness (μm)
HPMC–PVA (Gly)	1	59 ± 1	0.6 ± 0.01	16 ± 1	130 ± 5
HPMC–PVA (Gly)	5	57 ± 9	3 ± 0.4	24 ± 1	194 ± 5
HPMC–AGN (Gly)	1	58 ± 1	0.6 ± 0.01	20 ± 2	159 ± 15
HPMC–AGN (Gly)	5	63 ± 2	3 ± 0.1	27 ± 3	230 ± 5

The results showed that the addition of PVA and AGN resulted in lower CUR entrapment efficiency from roughly 81–82% in their absence to 57–63% in their presence. This trend was observed at both theoretical CUR payloads of 1 and 5 mg/cm^2. The lower CUR entrapment efficiency in the HPMC–PVA (Gly) and HPMC–AGN (Gly) films resulted in their lower experimental CUR payloads of around 0.6 and 3 mg/cm^2 for theoretical CUR payloads of 1 and 5 mg/cm^2, respectively. The impacts of adding PVA and AGN on the experimental CUR payloads of HPMC films were found to be similar.

The lower CUR entrapment efficiency in the presence of PVA and AGN suggested that fewer CUR–CHI nanoplexes were embedded into the HPMC films in their presence. We postulated that the increased colloidal stability of the CUR–CHI nanoplex in the presence of PVA caused a larger proportion of the nanoplex to remain in the bulk fluid during the HPMC film formation, resulting in fewer nanoplexes incorporated into the film. For AGN, we postulated that AGN, being polyanions, interacted with the positively charged CUR–CHI nanoplex, resulting in destabilization of the nanoplex, where CUR was released from the nanoplex prematurely and in turn precipitated due to its low aqueous solubility. This resulted in less CUR available for incorporation into the HPMC film.

In terms of their weight and thickness, the HPMC–PVA (Gly) and HPMC–AGN (Gly) films were also comparable with each other, albeit the HPMC–AGN (Gly) film was slightly denser and thicker (Table 2). Both films also passed the weight and thickness variations at high CUR payload denoting the batch-to-batch consistency of their preparation. Compared to the bare HPMC (Gly) film, the weight and thickness of the HPMC–PVA (Gly) and HPMC–AGN (Gly) films were increased.

The lower CUR payloads obtained upon the addition of PVA and AGN would be acceptable if the CUR release from the films was enhanced compared to the CUR release from the bare HPMC film. Indeed, for the films prepared at theoretical CUR payload of 1 mg/cm^2, the results showed that the CUR release from the HPMC–PVA (Gly) film was better than the CUR release from the bare HPMC (Gly) film (Figure 3). Specifically, the CUR release from the HPMC–PVA (Gly) film reached > 80% (*w*/*w*) after 4 h, in contrast to the plateau observed at ≈ 60% after 1 h for the HPMC (Gly) film (as shown earlier in Figure 2). The CUR release from the HPMC–PVA (Gly) film was slower in the beginning, where only roughly 20% (*w*/*w*) of the CUR payload was released after 1 h, resulting in a more sustained CUR release profile. Hence, the inclusion of long-chain PVA in the HPMC film was found to slow down the CUR release likely due to the increased physical barrier for outward molecular diffusion.

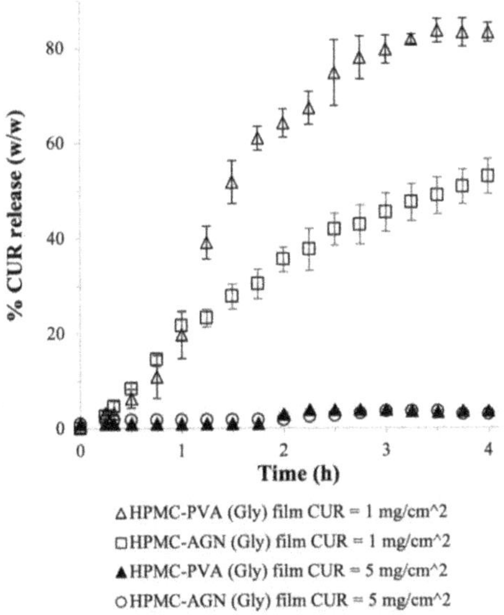

Figure 3. CUR release from HPMC–PVA (Gly) and HPMC–AGN (Gly) films (n = 6, error bars represent the standard deviations).

A more sustained CUR release profile without a plateau after 1 h was also observed in the HPMC–AGN (Gly) film prepared at theoretical CUR payload of 1 mg/cm^2 (Figure 3). In fact, the HPMC–AGN (Gly) film exhibited the ideal CUR release profile for sustained release with nearly zero-order kinetics. This could be attributed to the abovementioned improvement in the swelling properties of HPMC films with the addition of AGN. Importantly, the improvements in the CUR release profiles exhibited by the HPMC–PVA (Gly) and HPMC–AGN (Gly) films could compensate for their lower CUR entrapment efficiency.

Unfortunately, despite the improved CUR release profiles at theoretical CUR payload of 1 mg/cm^2, both HPMC–PVA (Gly) and HPMC–AGN (Gly) films remained unsuccessful in producing uninhibited CUR release at high theoretical CUR payload of 5 mg/cm^2 (Figure 3). In this regard, even though our postulate that the presence of PVA could reduce the aggregation tendency of the nanoplex at high CUR payload might be true, the slower CUR release in the presence of PVA would keep the supersaturated CUR concentration in the HPMC film at an unsustainably high level, which increased the precipitation propensity of CUR in the film. The same phenomenon was believed to occur upon inclusion of AGN, which also slowed down the CUR release from the HPMC film.

2.3. *Starch-Based Films*

2.3.1. Physical Characteristics

The inability of the HPMC-based films to effectively release their CUR payloads upon an increase in the payload above 1 mg/cm^2 required us to explore the use of an alternative mucoadhesive polymer, such as starch. Two types of starch were investigated, i.e., unmodified starch and HP starch. The latter is a modified starch in which some of the hydroxyl groups of the starch's amylose and amylopectin molecules are substituted with hydroxypropyl groups at varying degree of substitution. HP starch is known to exhibit superior swelling and aqueous solubility than unmodified starch owing to its higher hydrophilicity [52]. The physical characteristics of the starch films prepared using Gly as the plasticizer are presented in Table 3. Like the HPMC films, both starch films were also prepared at two theoretical CUR payloads of 1 and 5 mg/cm^2.

Table 3. Physical characteristics of the unmodified starch and HP starch films (n = 10, mean ± stdev).

Type of Film	Theoretical CUR Payload (mg/cm^2)	% CUR Entrapment (*w*/*w*)	CUR Payload (mg/cm^2)	Weight (mg/cm^2)	Thickness (µm)
Starch	1	74 ± 5	0.7 ± 0.05	65 ± 2	404 ± 7
Starch	5	79 ± 5	4 ± 0.2	65 ± 4	453 ± 23
HP starch	1	74 ± 3	0.7 ± 0.03	44 ± 1	307 ± 5
HP starch	5	77 ± 6	4 ± 0.3	51 ± 2	352 ± 10

The unmodified starch and HP starch films exhibited similar CUR entrapment efficiencies in the range of 70 to 80%. The CUR entrapment efficiency in the starch-based films was comparable, albeit slightly lower, to that of the bare HPMC films. Between the starch-based films prepared at theoretical CUR payloads of 1 and 5 mg/cm^2, the difference in their CUR entrapment efficiencies were found to be statistically insignificant (Student's t-test, $p \leq 0.05$). Therefore, similar to the trend observed in the HPMC films, the effects of increasing the CUR payload had little effect on the CUR entrapment efficiency. This signified the ability of the starch-based films to accommodate the high CUR payload.

With this CUR entrapment efficiency, the experimental CUR payloads of the starch-based films were determined to be approximately equal to 0.7 and 4 mg/cm^2 at theoretical CUR payloads of 1 and 5 mg/cm^2, respectively. The CUR payloads in the starch-based films were slightly lower than the CUR payloads in the bare HPMC films. The CHI contents in the starch-based films were approximately equal to 0.5 and 2.6 mg/cm^2 at theoretical CUR payloads of 1 and 5 mg/cm^2, respectively.

In terms of the weight and thickness, the unmodified starch films were denser and thicker than the HP starch films due to the higher density and viscosity of the unmodified starch. Specifically, the weight and thickness of the HP starch films were 44–50 mg/cm^2 and 300–350 µm, respectively, compared to 65 mg/cm^2 and 400–450 µm for the unmodified starch films. The thickness of the starch-based films was also shown to increase with increasing CUR payload indicating the accumulation of nanoplex as vertical layers in the starch films. Both starch films also passed the weight and thickness variation tests at high CUR payload, denoting their consistent preparation.

In comparison to the HPMC films, the starch-based films were denser and thicker due to the higher concentration of the precursor solution required in the starch film to produce films with good physical integrity, i.e., 15% (*w*/*v*) for starch and 5% (*w*/*v*) for HPMC. On the one hand, the increased thickness exhibited by the starch-based films could affect patients' experience upon administration of the buccal film either positively or negatively. This merits its own investigation in the future. On the other hand, the increased film thickness could bode well for the present goal of producing sustained CUR release profile at high CUR payload as we investigated in the next section.

2.3.2. CUR Release Profile

The CUR release profile from the unmodified starch film prepared at theoretical CUR payload of 1 mg/cm^2 in Figure 4a was found to closely resemble that of the HPMC (Gly) film shown earlier in Figure 2. More specifically, the CUR release was relatively fast in the beginning, when approximately 52% (*w*/*w*) of the CUR payload was released from the unmodified starch film after 1 h. Afterwards, the CUR release slowed down greatly and reached a plateau at around 60% after 4 h. However, unlike the HPMC (Gly) film, the CUR release from the unmodified starch film was not severely suppressed at the high CUR payload. In fact, the percentage CUR release from the unmodified starch film prepared at theoretical CUR payload of 5 mg/cm^2 was slightly higher after 4 h (Figure 4a). Nevertheless, the plateau in the CUR release at around 65% (*w*/*w*) remained evident after 2 h.

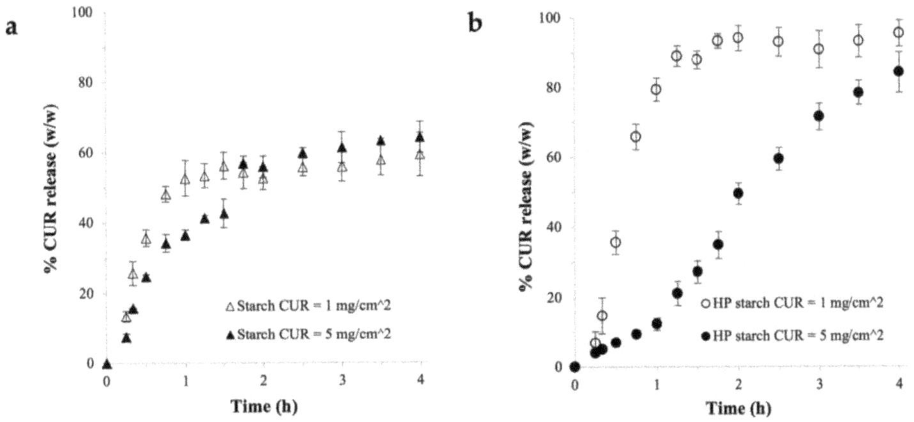

Figure 4. CUR release from (**a**) unmodified starch and (**b**) HP starch films (n = 6, error bars represent the standard deviations).

The CUR release was greatly improved in the HP starch films, as evidenced by the absence of a plateau in the CUR release profile (Figure 4b). For the HP starch film prepared at theoretical CUR payload of 1 mg/cm^2, the initial CUR release was fast, with nearly 80% (*w*/*w*) of the CUR payload released after 1 h. The CUR release slowed down significantly afterwards to reach around 95% after 4 h. The CUR release from the HP starch film at this CUR payload thus did not exhibit the desired sustained CUR release profile.

In contrast, at theoretical CUR payload of 5 mg/cm^2, the CUR release from the HP starch film was much slower, with roughly only 15% (*w*/*w*) of the CUR payload released after 1 h (Figure 4b). The CUR release rate picked up greatly afterwards with around 50% and 85% released after 2 h and 4 h, respectively. The HP starch film thus was able to produce the desired sustained release profile over 4 h at high CUR payload. The sustained CUR release profile fitted the zero-order kinetics closely, as shown in Figure A1 in Appendix A. Fitting the CUR release profiles to first-order kinetics and the Higuchi model led to poorer fitting, hence indicating that the CUR release from the starch-based film was independent of the CUR concentration in the film. Compared to the unmodified starch films, the superior CUR release observed in the HP starch films could be attributed to the aforementioned superior swelling and solubility of the HP starch, as well as the difference in the weight and thickness between the two films.

Significantly, the results in Figure 4 established that starches were more suitable than HPMC for use as the buccal film matrix at high CUR payload. The ability of the starch-based films to produce uninhibited CUR release at high CUR payload suggested that the precipitation propensity of the supersaturated CUR concentration in the film was minimized. This occurred when the supersaturated CUR concentration generated by the nanoplex could diffuse out of the film in a timely manner. Compared to the HPMC films, both starch films, despite being thicker and denser, exhibited much shorter disintegration time of 360 ± 50 and 330 ± 25 min for the unmodified starch and HP starch, respectively. The shorter disintegration time enabled the entrapped CUR molecules to navigate the gel layers more quickly, resulting in faster outward diffusion of CUR.

2.4. *Further Characterizations of HP Starch Films*

2.4.1. FESEM and FTIR

Having established the HP starch films as the optimal buccal film formulation at high CUR payload, further characterizations of the HP starch films were carried out. The macroscopic image of the HP starch film was presented in Figure A2 of Appendix A. The CUR–CHI nanoplex embedded in the HP starch films was visible in the FESEM image shown in Figure 5a using the HP starch film prepared at theoretical CUR payload of 5 mg/cm^2 as the representative sample. The nanoplexes in the film were shown to be well dispersed as individual nanoparticles in the size range of 150–300 nm with minimal agglomeration among them. The FESEM image showed that the CUR–CHI nanoplex was well preserved upon its incorporation into the HP starch films.

a b

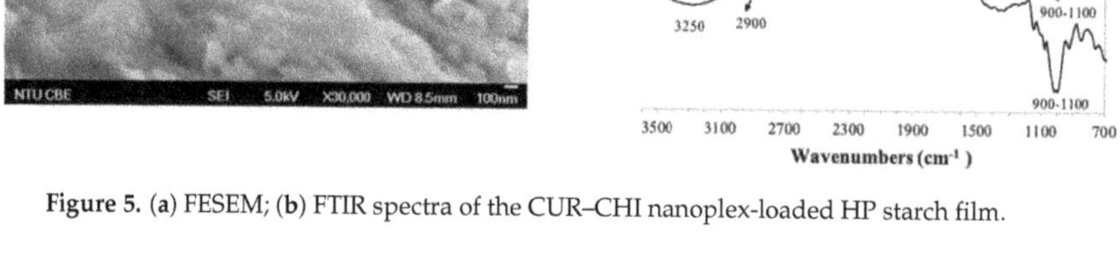

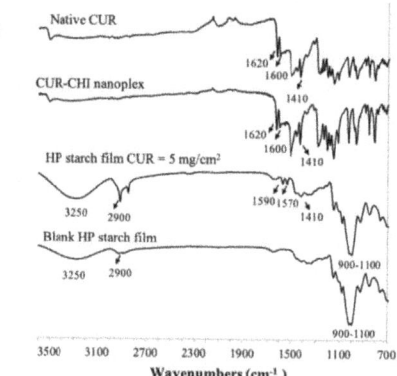

Figure 5. (**a**) FESEM; (**b**) FTIR spectra of the CUR–CHI nanoplex-loaded HP starch film.

The presence of CUR in the HP starch film was verified by FTIR analysis via the appearance of the characteristic peaks of CUR at 1590, 1570, and 1410 cm^{-1} in the FTIR spectrum of the nanoplex-loaded HP starch film (Figure 5b). These three peaks were attributed to the (C=C) and (C=O) vibrations, C=C aromatic ring stretching vibration, and

OH bending of the phenol group of CUR, respectively [19]. The three peaks in the HP starch films were shifted from higher wavenumbers of 1620, 1600, and 1410 cm^{-1} in the FTIR spectra of the native CUR and CUR–CHI nanoplex. These peaks were not visible in the FTIR spectrum of the blank HP starch film. The peaks at 900–1100, 2900, and 3250 cm^{-1} were attributed to the C-O-H bending, C-H stretching, and OH vibration of the amylose groups of the HP starch, respectively [53].

2.4.2. CUR Payload Uniformity and Folding Endurance

The CUR payload uniformity among the independent samples of HP starch films (n = 10) prepared at theoretical CUR payloads of 1 and 5 mg/cm^2 was examined (Table 4). The results showed that both HP starch films exhibited the CUR payload's acceptance values (AV) equal to lower or slightly higher than the 15% maximum threshold value set by USP for AV. Hence, the HP starch films met the USP's requirement for uniformity of a dosage unit [54]. Nevertheless, we recognized that the AV value at high CUR payload barely met the acceptance limit; thus, improvements in the HP starch film formulation will be needed in the future. In terms of their physical robustness, both HP starch films exhibited good folding endurance with values around of 2.5 to 2.7 denoting no film breakage was observed after ≥300 double folds (Table 4).

Table 4. CUR payload uniformity (n = 10) and folding endurance of HP starch films (n = 3).

CUR Payload (mg/cm^2)	AV for CUR Payload (%)	Folding Endurance
1	13.0	2.5 ± 0.07
5	15.2	2.7 ± 0.04

2.4.3. Amorphous form Stability and Thermal Stability

PXRD analysis of the nanoplex-loaded HP starch film performed after the accelerated storage did not show the appearance of strong intensity peaks, which were present in the PXRD pattern of the native CUR crystals (Figure 6a). The HP starch film prepared at theoretical CUR payload of 5 mg/cm^2 was used as the representative sample for PXRD. Thus, the CUR–CHI nanoplex in the HP starch film maintained its amorphous form after the accelerated storage equivalent to twelve-month storage at ambient condition. Nevertheless, the amorphous halo at $2\theta \approx 15$–$25°$ visible in the PXRD pattern of the HP starch film before storage became less pronounced after storage. The amorphous halo was replaced by low-intensity peaks, indicating decreased amorphous contents as crystallization of some of the nanoplex took place during storage.

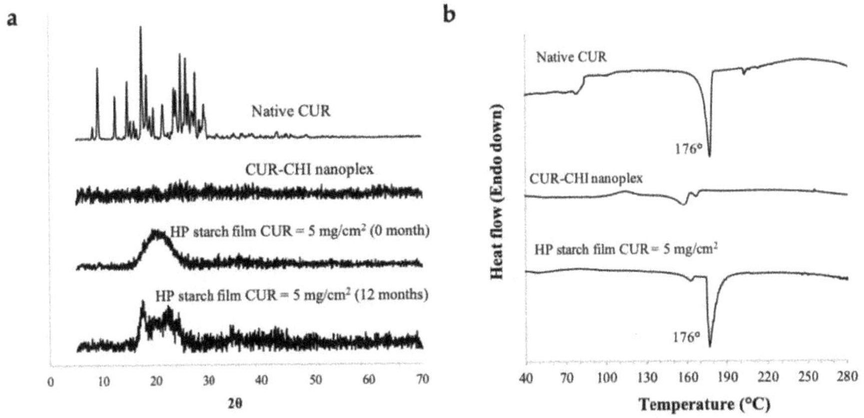

Figure 6. (**a**) PXRD; (**b**) DSC of the CUR–CHI nanoplex-loaded HP starch film.

The TGA results showed that the native CUR and HP starch film started to decompose at temperatures above 280 °C (Figure A3 in Appendix A); thus, the DSC thermograph for thermal stability was analyzed at temperatures below the decomposition temperature. DSC thermograph of the HP starch film prepared at theoretical CUR payload of 5 mg/cm^2 showed the appearance of a sharp endothermic peak at around 176 °C, which was typical of the melting point of crystalline CUR (Figure 6b). Not unexpectedly, the same peak at 176 °C appeared in the DSC thermograph of the native CUR crystals. The DSC results indicated that the CUR–CHI nanoplex in the HP starch film experienced amorphous to crystalline transition upon heating above 170 °C. In contrast, the melting point peak was not evident in the DSC thermograph of the free CUR–CHI nanoplex, where only solid transition events at around 160–170 °C were recorded. This signified the higher thermal stability of the free CUR–CHI nanoplex compared to the nanoplex embedded in the film. Nevertheless, the CUR–CHI nanoplex embedded in the HP starch film remained stable upon heating up to 150 °C, hence, the drying step in the film preparation should not adversely affect the amorphous form stability of the CUR–CHI nanoplex.

3. Materials and Methods

3.1. Materials

Materials for CUR–CHI nanoplex's preparation and characterization: curcumin (CUR) from turmeric rhizome (>95% curcuminoid content) was purchased from Alfa Aesar (Singapore). Chitosan (CHI) (190–310 kDa, 75–85% deacetylation), potassium hydroxide (KOH), disodium phosphate ($Na_2HPO_4.7H_2O$), ethanol, potassium dihydrogen phosphate (KH_2PO_4), sodium chloride (NaCl), hydrogen chloride (HCl), phosphoric acid (H_3PO_4), and glacial acetic acid (AA) were purchased from Sigma Aldrich (Singapore). *Materials for buccal film's preparation*: hydroxypropyl methylcellulose (HPMC) (MW = 26 kDa), polyvinyl alcohol (PVA) (90 kDa, 99% hydrolyzed), sodium alginate (AGN), glycerol (Gly), and propylene glycol (PG) were purchased from Sigma Aldrich (Singapore). Hydroxypropyl (HP) starch (LYCOAT® NG720) and pre-gelatinized starch (LYCATAB®) were generously provided by Roquette (Singapore).

3.2. Methods

3.2.1. Preparation and Characterization of CUR–CHI Nanoplex

The amorphous CUR–CHI nanoplex was prepared following the protocols presented in Lim et al. [19]. Briefly, CUR was dissolved at 5 mg/mL in 0.1M KOH (pH 13) and separately, CHI was dissolved at 5.9 mg/mL in 1.2% (*w*/*v*) AA (pH 2.7), both at room temperature. Equal volumes of the CUR and CHI solutions (10 mL each) were then mixed immediately after their preparation under gentle stirring to minimize alkaline degradation of CUR. The resultant CUR–CHI suspension was ultrasonicated for 15 s at 20 kHz (VC 505, Sonics, New Town, CT, USA) to break up large agglomerates (if any). The nanoplex suspension then underwent two cycles of ultracentrifugation (14,000× *g*, 10 min) and washing with deionized water to remove free CUR and free CHI that did not form the nanoplex. The CUR encapsulation efficiency into the nanoplex was characterized by measuring the free CUR concentration in the supernatant after the first centrifugation step using high performance liquid chromatography (HPLC) as described below. Afterwards, the washed CUR–CHI nanoplex suspension was lyophilized for 24 h at −52 °C and 0.05 mbar in Alpha 1–2 LD Plus freeze dryer (Martin Christ, Osterode am Harz, Germany) for characterization purposes.

The size and zeta potential of the CUR–CHI nanoplex suspension were characterized in triplicates by dynamic light scattering (DLS) after 100× dilution, using Brookhaven 90 Plus Nanoparticle Size Analyzer (Brookhaven Instruments Corporation, Holtsville, NY, USA). The CUR content in the nanoplex, which was defined as the amount of CUR per unit mass of the CUR–CHI nanoplex, was determined in triplicates by dissolving 1 mg of the lyophilized nanoplex powder in 10 mL 80% (*v*/*v*) ethanol. The amount of CUR in the ethanol was subsequently determined by HPLC (Agilent 1100, Agilent Technologies,

Santa Clara, CA, USA) at CUR detection wavelength of 423 nm. The HPLC was performed using ZORBAX Eclipse Plus C18 column (250 × 4.6 mm, 5 µm particle size) and 75% (*v*/*v*) acetonitrile solution as the mobile phase at flow rate of 1 mL/min, resulting in CUR retention time of approximately 2.8 min. The HPLC chromatogram of the CUR detection was presented in Figure A4 of Appendix A. The physical appearances of the CUR–CHI nanoplex before and after incorporation into the buccal film were examined by field emission scanning electron microscope (FESEM) (JSM 6700F, JEOL, Peabody, MA, USA).

3.2.2. Preparation of CUR–CHI Nanoplex-Loaded Buccal Film

The precursor solution for the HPMC film was prepared by overnight dissolution of HPMC in deionized water at 5% (*w*/*v*) under gentle stirring. The plasticizer, Gly or PG, was added to the HPMC solution at 5% (*w*/*w*). In the study on the effects of adjuvants inclusion, aqueous AGN or PVA solution was added to the HPMC + plasticizer solution at 0.5% (*w*/*v*). Freshly prepared CUR–CHI nanoplex was added last at theoretical CUR payloads of 1 or 5 mg CUR per cm^2 of the film. The precursor solution was then vortexed for 1 min to ensure its homogeneity. Afterwards, the precursor solution was casted onto a 9 cm diameter glass petri dish at a liquid height of 4 mm. Next, the petri dish was transferred to a convective laminar flow oven for drying at 60 °C for 3 h. The resultant dried HPMC film was peeled off the petri dish and stored in a sealed plastic bag for characterizations. The starch films were prepared by the same procedures at starch concentrations of 15% (*w*/*v*) for both unmodified and HP starches using 5% (*w*/*w*) Gly as the plasticizer for both. The HP starch was pre-gelatinized in deionized water at 80 °C prior to the film preparation. The CUR–CHI nanoplex was also added at theoretical CUR payloads of 1 and 5 mg CUR per cm^2 of the starch films.

3.2.3. Experimental CUR Payload, CUR Payload Uniformity, CUR Entrapment Efficiency

The experimental CUR payload in the buccal film (in mg of CUR per cm^2 of film) was determined by dissolving a 2 × 2 cm^2 square film samples in 25 mL 80% (*w*/*v*) ethanol solution for 1 h. Afterwards, the amount of CUR dissolved in the ethanol was determined by HPLC as previously described. The CUR payloads of ten independently prepared films were determined from which the CUR payload uniformity was characterized. According to the United States Pharmacopeia's (USP) criteria, the payload uniformity of a drug dosage was deemed acceptable when the acceptance value (AV) was less than 15%. The definition for AV is elaborately explained in USP monograph "<905> Uniformity of Dosage Units" [54] and not repeated here for brevity. After the experimental CUR payload was determined, the CUR entrapment efficiency (n = 10) was calculated from the ratio of the experimental CUR payload to the theoretical CUR payload.

3.2.4. Weight, Thickness, Folding Endurance of the Buccal Film

The weight and thickness of the buccal film were characterized from 9 cm diameter circular films using analytical balance and digital caliper, respectively. The variations in the buccal film's weight and thickness were characterized using ten independently prepared films. According to the USP's criteria, the variations in the weight and thickness among the independent samples were determined to be acceptable if each sample exhibited weight (or thickness) within ±10% of the arithmetic average of the weight (or thickness) [54]. The folding endurance was defined as the logarithmic (log10) of the number of double folds that resulted in the breakage of the buccal film. Briefly, triplicates of 2 × 2 cm^2 square film samples were manually double folded and the number of folds at which the film started showing signs of breakage was noted.

3.2.5. CUR Dissolution from Buccal Film and Film Disintegration

CUR dissolution from the buccal film was characterized in six replicates in simulated saliva fluid (SSF) under a sink condition at $\frac{1}{4}$ of the thermodynamic saturation solubility of CUR in the SSF (C_{Sat}). The SSF was prepared following the formulation of Peh and Wong [43] in which 4.49 g of $Na_2HPO_4.7H_2O$, 0.19 g KH_2PO_4, and 8.00 g NaCl were dissolved in one 1 L of deionized water. The pH of the SSF was adjusted to pH 6.75 by the addition of H_3PO_4. C_{Sat} of CUR in the SSF was experimentally determined by incubating excess CUR in 100 mL SSF maintained at 37 °C in a shaking incubator. After 24-h incubation, the CUR concentration in the SSF was determined by HPLC as previously described, resulting in C_{Sat} equal to 6.2 μg/mL.

Briefly, 2 × 2 cm^2 square film samples were completely immersed in 100 mL SSF maintained at 37 °C in a shaking incubator. At specific timepoints over a 4-h period, 1 mL of aliquot was withdrawn from the dissolution flask and the same volume of fresh SSF was added back to the flask as replenishment. The aliquot was syringe filtered (0.22 pore size), followed by 3-min ultracentrifugation at 14,000× *g*. Afterwards, the CUR concentration in the supernatant was determined by HPLC as previously described.

In a separate experiment, the disintegration time of the buccal film was characterized in triplicates by immersing 2 × 2 cm^2 square film samples in 10 mL SSF in a petri dish maintained at 37 °C in a shaking incubator. The smaller volume of SSF used in the disintegration test was to simulate the aqueous environment of oral cavity more closely. Herein, the disintegration time was defined as the time at which the buccal film had been completely disintegrated to aqueous suspension of the nanoplex and polymers, where no visible film fragments were present.

3.2.6. PXRD, DSC, and FTIR

The long-term stability of the amorphous form of the CUR–CHI nanoplex in the buccal film was examined by storing the nanoplex-loaded film in an open container placed inside a desiccator for three months under accelerated storage condition of 40 °C and 75% relative humidity. The accelerated storage condition was approximately equivalent to twelve-month storage under ambient condition (i.e., 25 °C and 60% relative humidity). The 75% relative humidity was generated inside the desiccator by placing an open container of saturated NaCl solution at 40 °C. At the end of the storage, the amorphous form of the nanoplex in the buccal film was examined by powder X-ray diffraction (PXRD) using D8 Advance X-ray Diffractometer (Bruker, Berlin, Germany) performed between 10° and 70° (2θ) with a step size of 0.02°/s. For comparison, the PXRD analysis was also carried out for the native CUR, the free CUR–CHI nanoplex, and the nanoplex-loaded film before storage.

The thermal stability of the amorphous form of the CUR–CHI nanoplex in the buccal film was characterized by thermal gravimetric analysis (TGA) (Pyris Diamond TGA, PerkinElmer, Waltham, MA, USA) and differential scanning calorimetry (DSC) (DSC 822E, Mettler Toledo, Columbus, OH, USA). The TGA analysis was performed at heating rate of 10 °C/min between 30 °C and 400 °C. The DSC analysis was performed at heating rate of 2 °C/min between 30 °C to 300 °C. Lastly, the presence of CUR in the buccal film was verified by Fourier transform infrared spectroscopy (FTIR) from 450 and 4000 cm^{-1} at 4 cm^{-1} spectral resolution in Spectrum One (Perkin–Elmer, Waltham, MA, USA). The DSC/TGA and FTIR analyses were also performed for the native CUR and the free CUR–CHI nanoplex.

3.2.7. Statistical Analysis

All experiments were performed with a minimum of three replicates and the results are presented as mean ± standard deviation. The statistical significance was analyzed using Student's *t*-test in GraphPad Prism software (GraphPad Software, San Diego, CA, USA). All *p*-values were two-sided and considered significant at $p \leq 0.05$, unless stated otherwise.

4. Conclusions

The feasibilities of HPMC and pre-gelatinized starch-based films as buccal sustained release delivery systems of amorphous CUR–CHI nanoplex at high CUR payload were investigated. HPMC and starch films were found to exhibit similar CUR entrapment efficiencies ($\approx$ 80% *w*/*w*), resulting in their similar CUR payloads in the range of 0.6 to 4 mg/cm^2. Both HPMC and starch films were able to accommodate higher CUR payloads without any adverse effect on the CUR entrapment efficiency. The starch films were denser and thicker than HPMC films, as higher film precursor concentrations were needed to produce starch films with good physical integrity. Despite being denser and thicker, starch films disintegrated faster than HPMC films. The faster disintegration time of the starch films resulted in its significantly superior CUR release profiles at high CUR payload (4 mg/cm^2) compared to the HPMC films. This was because the faster disintegration time enabled faster outward CUR molecular diffusion across the swollen polymer matrix, which in turn reduced the precipitation propensity of the highly supersaturated CUR concentration generated in the film by the amorphous nanoplex dissolution. The superior CUR release exhibited by the starch films, nevertheless, were not as evident at lower CUR payloads ($\leq$1 mg/cm^2) due to the lower CUR supersaturation level generated in the film. Between the unmodified starch and HP starch films, the HP starch films exhibited superior CUR release profiles in which at least 85% (*w*/*w*) of the CUR payload was released within 4 h, whereas the CUR release from the unmodified starch plateaued at around 65% (*w*/*w*) after the same period. At the high CUR payload, HP starch films exhibited the ideal sustained CUR release profile following the zero-order kinetics. The films prepared from different batches (n = 10) exhibited good CUR payload uniformity and minimal weight/thickness variations. The HP starch films were physically robust with high folding endurance and the embedded nanoplex was thermally stable with good long-term storage stability.

Author Contributions: Conceptualization, L.M.L. and K.H.; methodology, L.M.L.; validation, K.H.; formal analysis, L.M.L. and K.H.; investigation, L.M.L. and K.H.; resources, K.H.; data curation, L.M.L.; writing—original draft preparation, K.H.; writing—review and editing, K.H.; visualization, L.M.L.; supervision, K.H.; project administration, K.H.; funding acquisition, K.H. All authors have read and agreed to the published version of the manuscript.

Funding: This work was funded by the Nanyang Research Programme (NRP SCBE01jr 2017) of Nanyang Technological University, Singapore.

Data Availability Statement: Not applicable.

Acknowledgments: The authors would like to thank Roquette Singapore Pte Ltd. for gratuitously providing the starch samples.

Conflicts of Interest: The authors declare no conflict of interest.

Abbreviations

AA	Acetic acid
AGN	Alginate
C_{Sat}	Thermodynamic saturation solubility
CHI	Chitosan
CUR	Curcumin
DLS	Dynamic light scattering
DSC	Differential scanning calorimetry
FESEM	Field emission scanning electron microscope
FTIR	Fourier transform infrared spectroscopy
Gly	Glycerol
HPMC	Hydroxypropyl methyl cellulose
HP	Hydroxypropyl

PG	Propylene glycol
PXRD	Powder X-ray diffraction
PVA	Polyvinyl alcohol
SSF	Simulated saliva fluid
TGA	Thermal gravimetry analysis
USP	United States Pharmacopeia

Appendix A

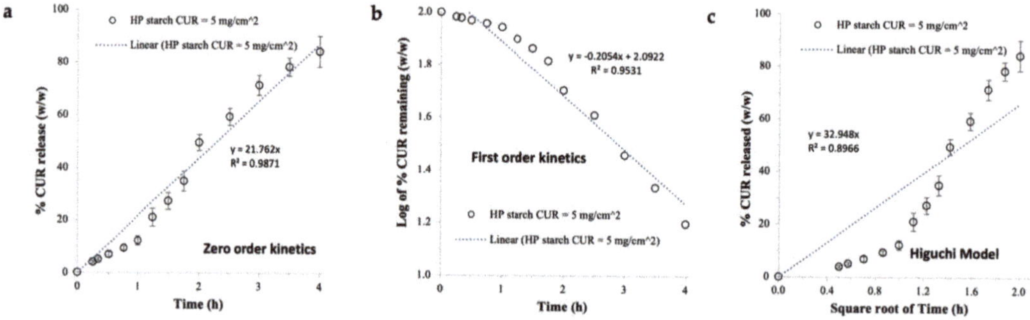

Figure A1. Dissolution kinetics modelling of the CUR release from HP starch film prepared at 5 mg/cm^2 (**a**) zero order kinetics, (**b**) first order kinetics, and (**c**) Higuchi model.

Figure A2. Macroscopic image of the CUR–CHI nanoplex-loaded HP starch film.

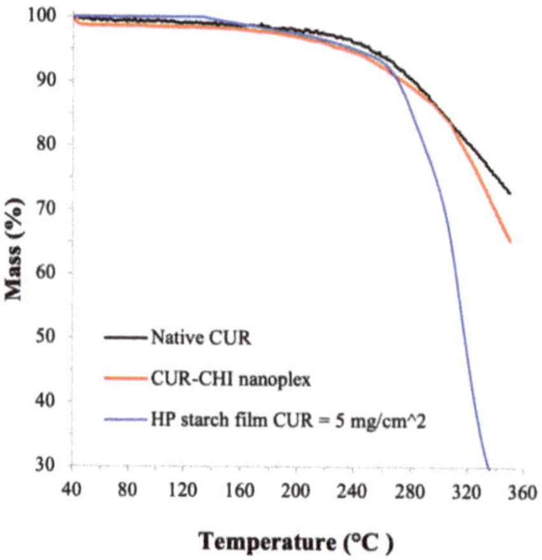

Figure A3. TGA of the HP starch film prepared at theoretical CUR payload of 5 mg/cm^2.

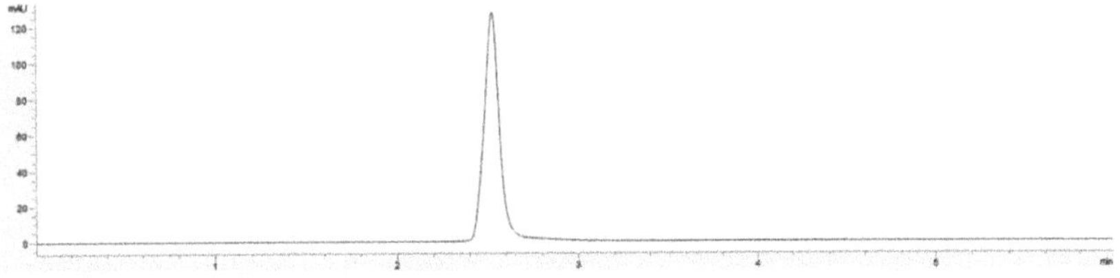

Figure A4. HPLC chromatogram of CUR detection.

References

1. Hay, E.; Lucariello, A.; Contieri, M.; Esposito, T.; De Luca, A.; Guerra, G.; Perna, A. Therapeutic effects of turmeric in several diseases: An overview. *Chem.-Biol. Interact.* **2019**, *310*, 108729. [CrossRef]
2. Tomeh, M.A.; Hadianamrei, R.; Zhao, X.B. A Review of Curcumin and Its Derivatives as Anticancer Agents. *Int. J. Mol. Sci.* **2019**, *20*, 1033. [CrossRef]
3. Barchitta, M.; Maugeri, A.; Favara, G.; San Lio, R.M.; Evola, G.; Agodi, A.; Basile, G. Nutrition and Wound Healing: An Overview Focusing on the Beneficial Effects of Curcumin. *Int. J. Mol. Sci.* **2019**, *20*, 1119. [CrossRef]
4. Kunnumakkara, A.B.; Harsha, C.; Banik, K.; Vikkurthi, R.; Sailo, B.L.; Bordoloi, D.; Gupta, S.C.; Aggarwal, B.B. Is curcumin bioavailability a problem in humans: Lessons from clinical trials. *Expert Opin. Drug Metab. Toxicol.* **2019**, *15*, 705–733. [CrossRef]
5. Salehi, B.; Stojanović-Radić, Z.; Matejić, J.; Sharifi-Rad, M.; Anil Kumar, N.V.; Martins, N.; Sharifi-Rad, J. The therapeutic potential of curcumin: A review of clinical trials. *Eur. J. Med. Chem.* **2019**, *163*, 527–545. [CrossRef]
6. Bhutani, U.; Basu, T.; Majumdar, S. Oral Drug Delivery: Conventional to Long Acting New-Age Designs. *Eur. J. Pharm. Biopharm.* **2021**, *162*, 23–42. [CrossRef]
7. Sharma, R.A.; Steward, W.P.; Gescher, A.J. Pharmacokinetics and pharmacodynamics of curcumin. *Adv. Exp. Med. Biol.* **2007**, *595*, 453–470. [PubMed]
8. Iurciuc, C.-E.; Atanase, L.I.; Jérôme, C.; Sol, V.; Martin, P.; Popa, M.; Ochiuz, L. Polysaccharides-Based Complex Particles' Protective Role on the Stability and Bioactivity of Immobilized Curcumin. *Int. J. Mol. Sci.* **2021**, *22*, 3075. [CrossRef] [PubMed]
9. Hassanzadeh, K.; Buccarello, L.; Dragotto, J.; Mohammadi, A.; Corbo, M.; Feligioni, M. Obstacles against the Marketing of Curcumin as a Drug. *Int. J. Mol. Sci.* **2020**, *21*, 6619. [CrossRef] [PubMed]
10. Arioglu-Tuncil, S.; Voelker, A.L.; Taylor, L.S.; Mauer, L.J. Amorphization of Thiamine Mononitrate: A Study of Crystallization Inhibition and Chemical Stability of Thiamine in Thiamine Mononitrate Amorphous Solid Dispersions. *Int. J. Mol. Sci.* **2020**, *21*, 9370. [CrossRef] [PubMed]
11. Rho, K.; Hadinoto, K. Dry powder inhaler delivery of amorphous drug nanoparticles: Effects of the lactose carrier particle shape and size. *Powder Technol.* **2013**, *233*, 303–311.
12. Cheow, W.S.; Hadinoto, K. Green Amorphous Nanoplex as a New Supersaturating Drug Delivery System. *Langmuir* **2012**, *28*, 6265–6275. [CrossRef]
13. Wegiel, L.A.; Zhao, Y.; Mauer, L.J.; Edgar, K.J.; Taylor, L.S. Curcumin amorphous solid dispersions: The influence of intra and intermolecular bonding on physical stability. *Pharm. Dev. Technol.* **2014**, *19*, 976–986. [CrossRef]
14. Sunagawa, Y.; Miyazaki, Y.; Funamoto, M.; Shimizu, K.; Shimizu, S.; Nurmila, S.; Katanasaka, Y.; Ito, M.; Ogawa, T.; Ozawa-Umeta, H.; et al. A novel amorphous preparation improved curcumin bioavailability in healthy volunteers: A single-dose, double-blind, two-way crossover study. *J. Funct. Foods* **2021**, *81*, 104443. [CrossRef]
15. Sekitoh, T.; Okamoto, T.; Fujioka, A.; Tramis, O.; Takeda, K.; Matsuura, T.; Imanaka, H.; Ishida, N.; Imamura, K. Sole-amorphous-sugar-based solid dispersion of curcumin and the influence of formulation composition and heat treatment on the dissolution of curcumin. *Dry. Technol.* **2020**, 1–10. [CrossRef]
16. Skieneh, J.M.; Sathisaran, I.; Dalvi, S.V.; Rohani, S. Co-Amorphous Form of Curcumin–Folic Acid Dihydrate with Increased Dissolution Rate. *Cryst. Growth Des.* **2017**, *17*, 6273–6280. [CrossRef]
17. Suresh, K.; Mannava, M.K.C.; Nangia, A. A novel curcumin–artemisinin coamorphous solid: Physical properties and pharmacokinetic profile. *RSC Adv.* **2014**, *4*, 58357–58361. [CrossRef]
18. Bhagwat, A.; Pathan, I.B.; Chishti, N.A.H. Design and optimization of pellets formulation containing curcumin ascorbic acid co-amorphous mixture for ulcerative colitis management. *Part. Sci. Technol.* **2020**, 1–9. [CrossRef]
19. Lim, L.M.; Tran, T.T.; Long Wong, J.J.; Wang, D.; Cheow, W.S.; Hadinoto, K. Amorphous ternary nanoparticle complex of curcumin-chitosan-hypromellose exhibiting built-in solubility enhancement and physical stability of curcumin. *Colloids Surf. B Biointerfaces* **2018**, *167*, 483–491. [CrossRef]
20. Araki, K.; Yoshizumi, M.; Kimura, S.; Tanaka, A.; Inoue, D.; Furubayashi, T.; Sakane, T.; Enomura, M. Application of a Microreactor to Pharmaceutical Manufacturing: Preparation of Amorphous Curcumin Nanoparticles and Controlling the Crystallinity of Curcumin Nanoparticles by Ultrasonic Treatment. *AAPS PharmSciTech* **2019**, *21*, 17. [CrossRef]
21. Ubeyitogullari, A.; Ciftci, O.N. A novel and green nanoparticle formation approach to forming low-crystallinity curcumin nanoparticles to improve curcumin's bioaccessibility. *Sci. Rep.* **2019**, *9*, 19112. [CrossRef] [PubMed]

22. Cheow, W.S.; Kiew, T.Y.; Hadinoto, K. Amorphous nanodrugs prepared by complexation with polysaccharides: Carrageenan versus dextran sulfate. *Carbohydr. Polym.* **2015**, *117*, 549–558. [CrossRef]
23. Nguyen, M.H.; Lee, S.E.; Tran, T.T.; Bui, C.B.; Nguyen, T.H.; Vu, N.B.; Tran, T.T.; Nguyen, T.H.; Nguyen, T.T.; Hadinoto, K. A simple strategy to enhance the in vivo wound-healing activity of curcumin in the form of self-assembled nanoparticle complex of curcumin and oligochitosan. *Mater. Sci. Eng. C Mater. Biol. Appl.* **2019**, *98*, 54–64. [CrossRef] [PubMed]
24. Nguyen, M.H.; Vu, N.B.; Nguyen, T.H.; Le, H.S.; Le, H.T.; Tran, T.T.; Le, X.C.; Le, V.T.; Nguyen, T.T.; Bui, C.B.; et al. In vivo comparison of wound healing and scar treatment effect between curcumin-oligochitosan nanoparticle complex and oligochitosan-coated curcumin-loaded-liposome. *J. Microencapsul.* **2019**, *36*, 156–168. [CrossRef]
25. Adiwidjaja, J.; McLachlan, A.J.; Boddy, A.V. Curcumin as a clinically-promising anti-cancer agent: Pharmacokinetics and drug interactions. *Expert Opin. Drug Metab. Toxicol.* **2017**, *13*, 953–972. [CrossRef] [PubMed]
26. Sintov, A.C. Transdermal delivery of curcumin via microemulsion. *Int. J. Pharm.* **2015**, *481*, 97–103. [CrossRef]
27. Sun, Y.; Du, L.; Liu, Y.; Li, X.; Li, M.; Jin, Y.; Qian, X. Transdermal delivery of the in situ hydrogels of curcumin and its inclusion complexes of hydroxypropyl-β-cyclodextrin for melanoma treatment. *Int. J. Pharm.* **2014**, *469*, 31–39. [CrossRef]
28. Eckert, R.W.; Wiemann, S.; Keck, C.M. Improved Dermal and Transdermal Delivery of Curcumin with SmartFilms and Nanocrystals. *Molecules* **2021**, *26*, 1633. [CrossRef]
29. Mazzarino, L.; Borsali, R.; Lemos-Senna, E. Mucoadhesive films containing chitosan-coated nanoparticles: A new strategy for buccal curcumin release. *J. Pharm. Sci.* **2014**, *103*, 3764–3771. [CrossRef]
30. Hazzah, H.A.; Farid, R.M.; Nasra, M.M.; El-Massik, M.A.; Abdallah, O.Y. Lyophilized sponges loaded with curcumin solid lipid nanoparticles for buccal delivery: Development and characterization. *Int. J. Pharm.* **2015**, *492*, 248–257. [CrossRef]
31. Gowthamarajan, K.; Jawahar, N.; Wake, P.; Jain, K.; Sood, S. Development of buccal tablets for curcumin using Anacardium occidentale gum. *Carbohydr. Polym.* **2012**, *88*, 1177–1183. [CrossRef]
32. Gilhotra, R.M.; Ikram, M.; Srivastava, S.; Gilhotra, N. A clinical perspective on mucoadhesive buccal drug delivery systems. *J. Biomed Res.* **2014**, *28*, 81–97.
33. Forouzanfar, F.; Forouzanfar, A.; Sathyapalan, T.; Orafai, H.M.; Sahebkar, A. Curcumin for the Management of Periodontal Diseases: A Review. *Curr. Pharm. Des.* **2020**, *26*, 4277–4284. [CrossRef] [PubMed]
34. Karki, S.; Kim, H.; Na, S.-J.; Shin, D.; Jo, K.; Lee, J. Thin films as an emerging platform for drug delivery. *Asian J. Pharm. Sci.* **2016**, *11*, 559–574. [CrossRef]
35. Madhavi, D.; Kagan, D. Bioavailability of a Sustained Release Formulation of Curcumin. *Integr. Med.* **2014**, *13*, 24–30.
36. Krull, S.M.; Susarla, R.; Afolabi, A.; Li, M.; Ying, Y.; Iqbal, Z.; Bilgili, E.; Davé, R.N. Polymer strip films as a robust, surfactant-free platform for delivery of BCS Class II drug nanoparticles. *Int. J. Pharm.* **2015**, *489*, 45–57. [CrossRef]
37. Sneha, R.; Vedha Hari, B.N.; Ramya Devi, D. Design of antiretroviral drug-polymeric nanoparticles laden buccal films for chronic HIV therapy in paediatrics. *Colloid Interface Sci. Commun.* **2018**, *27*, 49–59. [CrossRef]
38. Dos Santos, T.C.; Rescignano, N.; Boff, L.; Reginatto, F.H.; Simões, C.M.O.; de Campos, A.M.; Mijangos, C.U. Manufacture and characterization of chitosan/PLGA nanoparticles nanocomposite buccal films. *Carbohydr. Polym.* **2017**, *173*, 638–644. [CrossRef]
39. Rana, P.; Murthy, R.S.R. Formulation and evaluation of mucoadhesive buccal films impregnated with carvedilol nanosuspension: A potential approach for delivery of drugs having high first-pass metabolism. *Drug Deliv.* **2013**, *20*, 224–235. [CrossRef]
40. Chonkar, A.D.; Rao, J.V.; Managuli, R.S.; Mutalik, S.; Dengale, S.; Jain, P.; Udupa, N. Development of fast dissolving oral films containing lercanidipine HCl nanoparticles in semicrystalline polymeric matrix for enhanced dissolution and ex vivo permeation. *Eur. J. Pharm. Biopharm.* **2016**, *103*, 179–191. [CrossRef]
41. Morales, J.O.; Brayden, D.J. Buccal delivery of small molecules and biologics: Of mucoadhesive polymers, films, and nanoparticles. *Curr. Opin. Pharm.* **2017**, *36*, 22–28. [CrossRef]
42. Kraisit, P.; Limmatvapirat, S.; Nunthanid, J.; Sriamornsak, P.; Luangtana-Anan, M. Preparation and Characterization of Hydroxypropyl Methylcellulose/Polycarbophil Mucoadhesive Blend Films Using a Mixture Design Approach. *Chem. Pharm. Bull.* **2017**, *65*, 284–294. [CrossRef] [PubMed]
43. Peh, K.K.; Wong, C.F. Polymeric films as vehicle for buccal delivery: Swelling, mechanical, and bioadhesive properties. *J. Pharm. Pharm. Sci.* **1999**, *2*, 53–61. [PubMed]
44. Chan, S.Y.; Goh, C.F.; Lau, J.Y.; Tiew, Y.C.; Balakrishnan, T. Rice starch thin films as a potential buccal delivery system: Effect of plasticiser and drug loading on drug release profile. *Int. J. Pharm.* **2019**, *562*, 203–211. [CrossRef] [PubMed]
45. Soe, M.T.; Pongjanyakul, T.; Limpongsa, E.; Jaipakdee, N. Modified glutinous rice starch-chitosan composite films for buccal delivery of hydrophilic drug. *Carbohydr. Polym.* **2020**, *245*, 116556. [CrossRef]
46. Cheow, W.S.; Kiew, T.Y.; Hadinoto, K. Combining inkjet printing and amorphous nanonization to prepare personalized dosage forms of poorly-soluble drugs. *Eur. J. Pharm. Biopharm.* **2015**, *96*, 314–321. [CrossRef] [PubMed]
47. Niazi, S. *Handbook of Pharmaceutical Manufacturing Formulations*, 3rd ed.; CRC Press: Boca Raton, FL, USA, 2020.
48. Fan, N.; He, Z.; Ma, P.; Wang, X.; Li, C.; Sun, J.; Sun, Y.; Li, J. Impact of HPMC on inhibiting crystallization and improving permeability of curcumin amorphous solid dispersions. *Carbohydr. Polym.* **2018**, *181*, 543–550. [CrossRef]
49. Diaz, D.A.; Colgan, S.T.; Langer, C.S.; Bandi, N.T.; Likar, M.D.; Van Alstine, L. Dissolution Similarity Requirements: How Similar or Dissimilar Are the Global Regulatory Expectations? *AAPS J.* **2016**, *18*, 15–22. [CrossRef]
50. Pamlényi, K.; Kristó, K.; Jójárt-Laczkovich, O.; Regdon, G., Jr. Formulation and Optimization of Sodium Alginate Polymer Film as a Buccal Mucoadhesive Drug Delivery System Containing Cetirizine Dihydrochloride. *Pharmaceutics* **2021**, *13*, 619. [CrossRef]

51. Okeke, O.C.; Boateng, J.S. Composite HPMC and sodium alginate based buccal formulations for nicotine replacement therapy. *Int. J. Biol. Macromol.* **2016**, *91*, 31–44. [CrossRef]
52. Fu, Z.; Zhang, L.; Ren, M.-H.; BeMiller, J.N. Developments in Hydroxypropylation of Starch: A Review. *Starch-Stärke* **2019**, *71*, 1800167. [CrossRef]
53. Warren, F.J.; Gidley, M.J.; Flanagan, B.M. Infrared spectroscopy as a tool to characterise starch ordered structure—A joint FTIR-ATR, NMR, XRD and DSC study. *Carbohydr. Polym.* **2016**, *139*, 35–42. [CrossRef] [PubMed]
54. USP-34. <905> "Uniformity of Dosage Unit". In *United States Pharmacopeia (USP 34-NF29)*; The United States Pharmacopeial Convention: Rockville, MD, USA, 2011.

International Journal of
Molecular Sciences

MDPI

Article

Development and Characterization of Gentamicin-Loaded Arabinoxylan-Sodium Alginate Films as Antibacterial Wound Dressing

Abdulaziz I. Alzarea [1], Nabil K. Alruwaili [2], Muhammad Masood Ahmad [2], Muhammad Usman Munir [3], Adeel Masood Butt [4], Ziyad A. Alrowaili [5], Muhammad Syafiq Bin Shahari [6], Ziyad S. Almalki [7], Saad S. Alqahtani [8], Anton V. Dolzhenko [6] and Naveed Ahmad [2,*]

[1] Department of Clinical Pharmacy, College of Pharmacy, Jouf University, Sakaka 72388, Saudi Arabia; aizarea@ju.edu.sa
[2] Department of Pharmaceutics, College of Pharmacy, Jouf University, Sakaka 72388, Saudi Arabia; nkalruwaili@ju.edu.sa (N.K.A.); mmahmad@ju.edu.sa (M.M.A.)
[3] Department of Pharmaceutical Chemistry, College of Pharmacy, Jouf University, Sakaka 72388, Saudi Arabia; mumunir@ju.edu.sa
[4] Institute of Pharmaceutical Sciences, University of Veterinary & Animal Sciences, Lahore 54000, Pakistan; adeel.masood@uvas.edu.pk
[5] Department of Physics, College of Sciences, Jouf University, Sakaka 72388, Saudi Arabia; zalrowaili@ju.edu.sa
[6] School of Pharmacy, Monash University Malaysia, Jalan Lagoon Selatan, Bandar Sunway 47500, Malaysia; muhammad.binshahari@monash.edu (M.S.B.S.); anton.dolzhenko@monash.edu (A.V.D.)
[7] Department of Clinical Pharmacy, College of Pharmacy, Prince Sattam Bin Abdulaziz University, Riyadh 11942, Saudi Arabia; z.almalki@psau.edu.sa
[8] Department of Clinical Pharmacy, College of Pharmacy, Jazan University, Jazan 45142, Saudi Arabia; ssalqahtani@jazanu.edu.sa
* Correspondence: naveedpharmacist@yahoo.com or nakahmad@ju.edu.sa

Citation: Alzarea, A.I.; Alruwaili, N.K.; Ahmad, M.M.; Munir, M.U.; Butt, A.M.; Alrowaili, Z.A.; Shahari, M.S.B.; Almalki, Z.S.; Alqahtani, S.S.; Dolzhenko, A.V.; et al. Development and Characterization of Gentamicin-Loaded Arabinoxylan-Sodium Alginate Films as Antibacterial Wound Dressing. *Int. J. Mol. Sci.* **2022**, 23, 2899. https://doi.org/10.3390/ijms23052899

Academic Editor: Yury A. Skorik

Received: 25 January 2022
Accepted: 4 March 2022
Published: 7 March 2022

Abstract: Biopolymer-based antibacterial films are attractive materials for wound dressing application because they possess chemical, mechanical, exudate absorption, drug delivery, antibacterial, and biocompatible properties required to support wound healing. Herein, we fabricated and characterized films composed of arabinoxylan (AX) and sodium alginate (SA) loaded with gentamicin sulfate (GS) for application as a wound dressing. The FTIR, XRD, and thermal analyses show that AX, SA, and GS interacted through hydrogen bonding and were thermally stable. The AXSA film displays desirable wound dressing characteristics: transparency, uniform thickness, smooth surface morphology, tensile strength similar to human skin, mild water/exudate uptake capacity, water transmission rate suitable for wound dressing, and excellent cytocompatibility. In Franz diffusion release studies, >80% GS was released from AXSA films in two phases in 24 h following the Fickian diffusion mechanism. In disk diffusion assay, the AXSA films demonstrated excellent antibacterial effect against *E.coli, S. aureus*, and *P. aeruginosa.* Overall, the findings suggest that GS-loaded AXSA films hold potential for further development as antibacterial wound dressing material.

Keywords: arabinoxylan; sodium alginate; wound healing; antibacterial dressing; drug delivery

1. Introduction

A wound can be defined as a breakage in intact tissue (particularly skin), caused by physical, thermal, chemical, or mechanical trauma, or may result from complicated pathological conditions [1,2]. Immediately after injury, a programmed and complex wound healing process takes place, which can be categorized into four major interlinked healing stages: the hemostasis (clot formation), inflammatory (influx of inflammatory mediators and cells), proliferative (dermal cells proliferation, epithelialization, angiogenesis, and formation of granular tissues), and maturation (remolding and scar formation) stages [3–6]. The complete duration of healing may vary depending upon the nature of the wound [7,8].

Depending on the time required for complete healing, wounds can be classified into acute (1–12 weeks) and chronic (>12 weeks) wounds [8,9]. Wound infections, comorbidities (cancer, diabetes mellitus, obesity, etc.), malnutrition, and poor wound care are the main factors leading to chronic wounds that put a physical, psychological, social, and financial burden on patients [2,9,10]. Wound infections caused by bacteria (mainly *Escherichia coli*, *Staphylococcus aureus*, *Pseudomonas aeruginosa*, *Bacteroides fragilis*) are one of the most predominant reasons for the prolongation of wound healing (in the inflammatory stage) [1,11–13]. These bacteria form colonies at the wound site and may overcome the patient's immunity, resulting in tissue damage and life-threatening consequences [14,15]. The development of an active dressing may prevent wound infections and augment the healing process.

Numerous kinds of dressing materials have been in human use since ancient times, with the primary aim being to cover the wound to provide protection against the environment and prevent bleeding and further injury [5,12,16]. Wound dressing in current practice can be broadly classified into dry/traditional (cotton, gauze, bandage) and moist/modern (films, foams, hydrogels, sponges, hydro actives, hydrofibers) dressings [8,17]. Dry dressings are still in clinical practice despite possessing the main limitations of wound dehydration and risk of infection [12,18,19]. Conversely, modern dressings augment wound healing by preventing dehydration and facilitating cell proliferation [12,17]. Additionally, antimicrobial agents can be incorporated into the polymeric dressings for delivery into the wound bed to prevent wound infections [8,20]. In addition to providing a moist environment and promoting healing, an ideal dressing should be biocompatible, mechanically durable, and flexible, allow water and gaseous exchange, and possess antimicrobial properties [5,8,16,21]. Among the modern dressings, polymeric films offer the advantage of gas permeability, moisture transmission, wound inspection (due to transparency), impermeability to microbes, ease of application to joints (due to flexibility), and drug incorporation and delivery to wound site [8,22]. However, difficulty in removing them from the wound, limited inherent antimicrobial effect, and lesser absorption of the exudate are the major limitations of the films [8]. Therefore, researchers are keen to develop films with novel combinations of polymers and antibacterial agents to provide optimal healing conditions [23,24].

Several natural, semi-synthetic, and synthetic polymers have been explored to develop film dressings, as reviewed recently by Savencu et al. [8]. Among the natural polymers, alginate (sodium and calcium salts), a linear unbranched polysaccharide (containing mannuronic and guluronic acid) found in brown algae, is commonly employed in wound dressing formulations owing to its excellent physicochemical, biocompatible, hemostatic, water sorption, and film-forming properties [8,17,25–30]. Moreover, SA may also trigger macrophages and the release of inflammatory mediators that augment the healing process [29]. Numerous alginate-containing dressing materials are commercially available; recent research on antibiotics-loaded alginate dressings has demonstrated their potential to prevent infections [26]. Most of these dressings were prepared by a combination of sodium alginate (SA) and other natural or synthetic polymers to improve various features required in the dressings [25,27–30].

Arabinoxylan (AX), isolated from the husk of *Plantago ovata* (psyllium) seed husk (PSH), is a branched polysaccharide that mainly contains xylose backbone and arabinose side chains with a minor percentage of rhamnose and uronic acids [31]. AX is an attractive material for biomedical applications owing to its biocompatible, film-forming, water absorption, and drug delivery characteristics [32,33]. In our previous studies, AX and AX-gelatin (AX-GL) films exhibited potential for film dressing application [32,33]. However, there is a need to improve the features of these AX-based films, especially their mechanical stability in the wound exudate. Previously, it was found that AX and AX-GL films rapidly uptake water, expand in the simulated wound environment, and start deteriorating. Preparing composite films is a simple and efficient approach to improving polymeric films' physicochemical characteristics [29]. Therefore, in this work, AX was combined with SA to improve the physicochemical and wound healing properties of AX-based films. The combination of AX with SA is expected to improve the mechanical stability of the AX-based films in a moist wound environment owing to the hydrogen bonding interaction between AX and SA. Moreover, adding the SA in AX-based dressings formulation will also augment the healing process by triggering the release of inflammatory mediators (as discussed above). However, both AX and SA lack antibacterial activity; therefore, gentamicin sulfate (GS), a broad-spectrum aminoglycoside, is incorporated into AXSA film dressings due to its efficacy again bacteria causing wound infection [34,35]. To the best of our knowledge, GS-loaded films composed of AX and SA have not been previously reported.

In this study, novel GS-loaded AXSA films were fabricated for application as an antibacterial wound dressing. The thickness, mechanical properties, morphology, water transmission, water uptake, chemical nature, thermal degradation behavior, drug release, antimicrobial, and cell viability were investigated to assess the suitability of AXSA films for antibacterial wound dressing application.

2. Materials and Methods

2.1. Materials

Sodium alginate (SA) (from brown algae) (Mol. Wt. 120,000–190,000, viscosity: 5 to 25 cps, M/G ratio: 1.56), glycerol (Gly). PBS tablets, $CaCl_2$ granules (anhydrous), pen-strep solution (containing 10,000 penicillin units and 10 mg of streptomycin per mL), cellulose acetate membranes (0.45 μm, 25 mm), MTT (3-(4,5-dimethylthiazol-2-yl)-2,5-diphenyl-tetrazolium bromide), polystyrene petri dishes (10 cm × 20 mm), and gentamicin sulfate were procured from MilliporeSigma (St. Louis, MO, USA). MRC-5 cells (ATCC CCL-171) were cultured in RPMI-1640 culture medium supplemented with FBS (Nacalai Tesque, Kyoto, Japan). Three bacterial strains *S. aureus* (ATCC 25923), *E. coli* (ATCC 8739), and *P. aeruginosa* (ATCC 9027) were all cultured in Mueller-Hinton broth (MHB) and Agar (Sigma-Aldrich, MO, USA). Arabinoxylan (AX), containing Xyl*p* (72.5%), Ara*f* (21.8%), Rha*p* (2.3%), and uronic acids (1.4%) was isolated from *Plantago ovata* (PO) seeds husk (Mol. Wt. 364,470) as reported previously [31,32].

2.2. Fabrication of AXSA Films

The AXSA films were fabricated by solvent casting technique [33]. Briefly, uniform dispersions of SA (3.5, 3, 2.5, and 2% *w/w*) were prepared in distilled water by high-speed stirring (1000 rpm) at 40 °C for 12 h. Then, AX (1.5, 2, 2.5, and 3% *w/w*) and glycerol (2.5%) were added to SA dispersion and stirred until homogenized gel was obtained. The resultant AXSA film casting gels were sonicated to eliminate air. After that, to cast blank AXSA films, 25 g of AXSA gels were poured into polystyrene plates and kept for drying in an oven at 40 °C for 48 h. While for GS-loaded AXSA films, GS (0.1% *w/w*) was added in homogenized AXSA gels and films were cast as described for blank films. After drying, films were peeled from plates, and their physical appearances (transparency, flexibility/foldability, smoothness) were visually observed. The films were kept in a desiccator until further analysis. The composition and codes of AXSA film formulations are displayed in Table 1.

Table 1. Formulation codes and composition of arabinoxylan-sodium alginate (AXSA) films.

Composition (% *w/w*)	F1	F2	F3	F4	GF1	GF2	GF3	GF4
Arabinoxylan (AX)	1.5	2	2.5	3	1.5	2	2.5	3
Sodium alginate (SA)	3.5	3	2.5	2	3.5	3	2.5	2
Glycerol (Gly)	2.5	2.5	2.5	2.5	2.5	2.5	2.5	2.5
Gentamicin (GS)	–	–	–	–	0.1	0.1	0.1	0.1
Distilled water	92.5	92.5	92.5	92.5	92.4	92.4	92.4	92.4

2.3. *Characterization of AXSA Films*

2.3.1. Film Thickness

The cross-sectional thicknesses of all dried films were measured at five randomly selected points. The thickness measurements were performed using a micrometer with 0.01 mm accuracy (APT measuring instruments, Omaha, NE, USA).

2.3.2. Water Vapor Transmission Rate (WVTR)

The WVTR of AXSA films was measured following the previously reported procedure with some modifications [32]. Briefly, AXSA films were cut into ~2.5 cm diameter and placed on openings of glass vials containing 2 g of $CaCl_2$ (desiccant). Then, caps with a 1 cm opening were fitted on the film containing vials. Then, these vials were weighed and kept in a desiccator at 85 $\pm$ 3% RH and 25 $\pm$ 2 °C. The change in the weights of the vials was noted at constant time intervals up to 24 h. WVTR was calculated according to the formula described earlier [32].

2.3.3. Mechanical Properties

Mechanical tests were performed using a 5 kN load cell (LS5) tensile testing machine (Lloyd instruments, West Sussex, UK). For this purpose, dumbbell-shaped specimens (0.5 cm width and 3 cm length) of AXSA films were prepared. To protect the film from damage due to grip, the film's area in contact with the grip was covered with double adhesive tape. The films were stretched at a crosshead speed of 10 mm·min^{-1} until breakage. Finally, tensile strengths (TS) and percent elongations at break (% EAB) were calculated according to the formula described previously [36].

2.3.4. Surface Morphology

The surface morphology of gold sputter-coated AXSA films (with and without GS) was observed using a Quanta 250-FEG scanning electron microscope (SEM) (FEI, OR, USA) at 5 kV operating voltage and 1000$\times$ and 2000$\times$ magnifications.

2.3.5. FTIR Spectroscopy

ATR-FTIR spectroscopic analyses were performed with an FTIR-7600 spectrophotometer (Lambda-Scientific, Marion, South Australia, Australia). Transmittance spectra were recorded at 4 cm^{-1} resolution in the wavenumber range of 4000 to 550 cm^{-1}.

2.3.6. X-ray Diffraction

X-ray diffraction (XRD) analyses were performed with Shimadzu X-ray diffractometer (MAXima_X XRD-7000, Kyoto, Japan) using CuKα radiations. The X-ray diffractograms were recorded in the 2θ range of 10 to 80° by operating at 30 mA, 40 kV, and 2° min^{-1} scanning speed.

2.3.7. Thermogravimetric Analyses (TGA)

The TGA analyses were performed with Shimadzu TGA-50 analyzer (Kyoto, Japan). TG curves were obtained by heating weighed AXSA films from 30 to 600 °C at 20 °C. min^{-1} heating rate and 20 mL. min^{-1} nitrogen gas flow.

2.3.8. Differential Scanning Calorimetry (DSC)

The DSC curves were acquired on DSC3 STAR instrument (Mettler Toledo, Columbus, OH, USA), operated at 20 °C. min^{-1} heating rate. The changes in the heat flow of AXSA films were recorded from 30 to 450 °C under 20 mL.min^{-1} nitrogen gas flow.

2.3.9. Expansion Profile

The exudate absorbing profile of the AXSA films was investigated by measuring the expansion of films over a gelatin solution (4 g. 100 mL^{-1}). Briefly, to create an environment similar to exudating wounds, 25 mL of gelatin solution was allowed to solidify in the polystyrene plates for 18 h at 25 °C. Then, pieces of AXSA films (~25 mm diameter) were placed over solidified gelatin, and expansion in the size of the films was measured using a ruler at predefined intervals of time. The percent expansion was calculated according to the previously described formula [29].

2.4. Gentamicin Release Profile and Kinetics

The release profile of GS from GF2 and GF3 films was studied with DHC-6T Franz diffusion cell (Logan-Instruments, Somerset, NJ, USA) using 10 mL release media (PBS, pH 7.4) in the receiver compartment. Cellulose acetate membranes were placed over the receiver compartment, and then weighed GS-loaded AXSA films (~1.5 cm diameter) were mounted over these membranes. GS release experiments were performed at 37 °C, with constant stirring of the release media. Aliquots were withdrawn from the receiver compartment at pre-programmed time points. The concentration of GS in the aliquots was determined according to the *o*-phtaldialdehyde derivatization method using a Genesys 10s Uv-Vis spectrophotometer (Thermo Fisher Scientific, Waltham, MA, USA). The GS release profile was plotted as percent cumulative release versus time [37].

To determine GS release kinetics and mechanism of release from AXSA films, the following kinetic models were fitted to the release data:

Zero order $$Q_t = Q_0 + K_0 t \quad (1)$$

First order $$InQ_t = InQ_0 + K_1 t \quad (2)$$

Higuchi $$Q_t = K_H t^{1/2} \quad (3)$$

Korsmeyer–Pappas $$Q_t = K_k\, t^{\,n} \quad (4)$$

where Q_t and Q_0 are the amounts of GS released from AXSA films at time t and 0, while K_0, K_1, K_H, and K_k represent kinetic constants and n is the release exponent.

2.5. Antibacterial Effect of AXSA Films

Antibacterial activities of the GS-loaded (GF2 and GF3) films were estimated using the agar disk (filter paper) diffusion method. For this purpose, AXSA films were cut into ~6.8 mm with filter paper backing and sterilized by UV exposure for 60 min. Autoclaved MHA (40 mL) was poured into glass petri plates and allowed to solidify. Then, standardized inoculums (20 µL) of bacterial strains (*S. aureus, E.coli, and P. aeruginosa*) were uniformly spread over MHA using a glass spreader. After that, GS-loaded AXSA films (GF2 and GF3), GS standard, and blank AXSA film (F2) were placed over the bacterial cultures, and plates were incubated for 24 h at 37 °C. Clear zones of inhibition formed around the samples were measured.

2.6. Indirect Cell Viability Assay

The effect of AXSA films on the viability of MRC-5 cells was investigated by indirect MTT assay. For this purpose, extracts of the blank (F2) and GS-loaded (GF2) AXSA films were prepared by incubating weighed films in culture media for 24 h under aseptic conditions, as reported previously [38]. MRC-5 cells were cultured in RPMI-1640 with 10% v/v FBS, 1% v/v pen-strep at 37 °C in humidified 5% CO_2 incubator. Then, cells were harvested and seeded in 96-well plates (~7500 cells per well) and incubated for another 48 h. After that, culture media in the 96-well plates was replaced with AXSA films extract, except the control (untreated) well, and incubated for 48 h. Finally, according to the previously reported method, an MTT assay was performed to determine the percent cell viability (% CV) [39].

2.7. Statistical Analysis

Statistical analysis of the data was performed using GraphPad Prism software. The differences were considered statistically significant when p was found to be less than 0.05 by one-way ANOVA. The results are expressed as mean ± standard deviation (SD).

3. Results and Discussion

3.1. Fabrication of AXSA Films

The AXSA gels containing 5% (*w*/*w*) AX and SA, prepared by high-speed stirring, were easy to pour, and free from air bubbles and polymer aggregates. This 5% (*w*/*w*) gel concentration for casting AXSA films was selected based on previous studies on alginate composite films [29]. However, when AX concentration increased above 3% in AXSA gels, the gel became highly viscous and difficult to pour [33]. Therefore, F1, F2, F3, and F4 gels were used to cast blank AXSA films. The resultant AXSA blank films were visually observed for their appearance and suitability for wound dressing application. The pictures of the blank and GS-loaded AXSA films are shown in Figure S1. Overall, all the AXSA films were transparent, smooth, flexible (foldable), and free from air bubbles. There was no significant difference in GS-loaded films' physical appearance compared to blank AXSA films of the same concentration. However, F1 films were thin and difficult to peel from the plates. Therefore, F2, F3, and F4 films were selected for further studies. The transparency and foldability of the dressing are important parameters as transparency allows observation of the healing wound while foldability facilitates dressing of wounds in bending on parts of the body [29]. To study the effect of the plasticizer on the handling of the films, AX (3% *w*/*v*), SA (3% *w*/*v*), and AXSA (AX 2% and SA 3%) films without glycerol were prepared and compared with F2 film (with 2.5% glycerol), as shown in Figure S2. These results demonstrated that the dried films (without glycerol) were transparent but not flexible (fractured upon folding). Therefore, the films without plasticizers were not considered for further characterization for wound dressing application.

Gentamicin sulfate (0.1%) was incorporated into AXSA film formulation to fabricate GS-loaded films. The selection of GS dose was based on the previous reports where 0.1% GS was reported to be safe and effective for wound healing applications [14,40,41]. Moreover, topical formulations (creams and ointments) containing 0.1% GS are commercially available for topical application and considered non-toxic [40]. The total GS-loaded in 10 cm (diameter) AXSA films is 25 mg.

3.2. Characterization of AXSA Films

3.2.1. Thickness

The mean thicknesses of F2, GF2, F3, GF3, F4, and GF4 AXSA films were 227 ± 6, 253 ± 6, 263 ± 11, 273 ± 6, 276 ± 14, and 276 ± 9 μm, respectively, as shown in Figure 1a. These results indicate that there was no significant difference in the thickness of the AXSA films. However, a slight increase in film thickness was found with increasing AX concentration and the addition of the GS in the film formulations. According to previous reports, the major factors that control the thickness of polymeric films include composition, the weight of the gel used for film casting, the interactions between the polymers, drying conditions, and the flatness of the drying surface [25,29]. Since the total amount of AX and SA (5%) was constant in AXSA films, the slight increase in the thickness of the AXSA film formulations was due to AX, which has previously been reported to form thick films [33]. Despite containing higher total polymers contents (AX + SA), the thickness of the AXSA films was lower than previously reported AX and AX-GL films, which may be due to lower casting volume and hydrogen bonding between the AX and SA. Nevertheless, the thickness of the AXSA films was in the range of the average thickness of the human skin (200 to 500 μm). Therefore, it can be suggested that these AXSA films will be suitable for application to skin wounds, as thin films are more comfortable for patients [25].

3.2.2. Water Vapor Transmission Rate (WVTR)

The results of the WVTR of the AXSA films (blank and GS-loaded) are presented in Figure 1b. The average WVTR values for the F2, GF2, F3, GF3, F4, and GF4 films were 1687 ± 41, 1537 ± 69, 1604 ± 39, 1498 ± 65, 1403 ± 59, and 1326 ± 57 $g \cdot m^{-2}.day^{-1}$, respectively. In blank AXSA films, WVTR of the F2 and F3 films was significantly higher than in F3 films, while in GS-loaded films, WVTR of the GF2 and GF3 films was significantly higher than that of GF4 films. Moreover, the WVTR of GS-loaded films was less than that of the blank AXSA films, but the difference was statistically insignificant. The WVTR of the film dressings usually depends on the chemistry of the polymers, thickness, composition, and porosity [29]. The WVTR results of the AXSA films indicate that this parameter decreases in tandem with decreasing SA concertation in films. Since the thickness of the AXSA films was not significantly different, the decrease in WVTR with smaller quantities of SA in film can be explained by the decrease in the number of carboxylic groups interacting with water molecules [42]. Moreover, an increase in AX in the film formulations might result in a denser hydrogen bonding network between the -OH groups of AX and the -COOH groups of SA, thus forming films with a lower WVTR [43]. Similar trends in the decreasing WVTR of the SA-containing blended films have been reported previously [42,43]. The WVTR of the AXSA films was higher than previously reported values for AX-GL films; this can be attributed to the lower thickness of the AXSA films [32]. The WVTR is vital in developing a wound dressing aimed for drug delivery. The WVTR of dressing for clinical application ranges between 139 to 10,973 $g{\cdot}m^{-2}{\cdot}day^{-1}$, and the WVTR of injured skin may be up to 4274 $g{\cdot}m^{-2}{\cdot}day^{-1}$ [44]. Hence, the WVTR of our AXSA films reported here is adequate for wound dressing application.

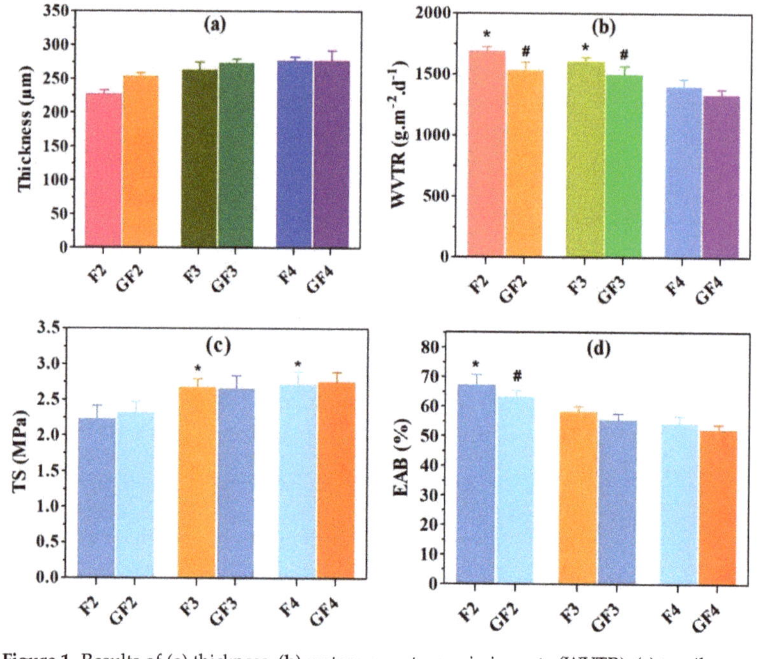

Figure 1. Results of (**a**) thickness, (**b**) water vapor transmission rate (WVTR), (**c**) tensile strength (TS), and (**d**) elongation at break (EAB) of the arabinoxylan (AX)-sodium alginate (SA) films (mean ± SD, $n = 3$). * and # show statistically significant differences ($p < 0.05$) from parallel blank and gentamicin (GS)-loaded films.

3.2.3. Mechanical Properties

The mechanical properties (TS and EAB) of the AXSA films are presented in Figure 1c,d. The TS of the AXSA films ranged between 2.31 MPa and 2.75 MPa, while the EAB varied from 54% to 67%. Overall, the TS of the AXSA films increased, and EAB decreased with increasing AX concentration in the films. Moreover, the TS of the GS-loaded AXSA films were slightly more than that of blank AXSA films. The TS of F3 and F4 films were significantly higher than the TS of F2 films, while the EAB of the F2 was significantly higher than that of F3 and F4 films. These differences in the mechanical properties suggest that intermolecular interactions between AX and SA resulted in the increased rigidness of the films and decreased flexibility [29]. The mechanical properties of the film dressings were also reported to be influenced by the concentration of plasticizer (glycerol) and the total amount of solids in the films [29]. However, the amount of plasticizer and the total amount of polysaccharides (AX + SA) was the same in all films; therefore, their influence on the mechanical properties of the AXSA films was not prominent. The F2 and GF2 films (with 2% AX) were not damaged during mechanical testing as compared to the previously repowered AX films, where films prepared by 2% AX were damaged during mechanical testing, suggesting that the addition of SA improved the mechanical stability of the films [33]. The TS and EAB of the AXSA films were lesser than earlier reported AX-GL films that may be due to lesser thickness and lower amount of plasticizer in of AXSA films [32]. The TS and the flexibility of the wound dressing are essential because an ideal material for wound dressing should be strong enough to endure the pressure and stretch during application and flexible for sustaining possible stretch during body movements [21,36]. According to the literature, the TS of the human skin varies from 2.5 to 16 MPa [21]. Hence, the TS of the prepared AXSA films is adequate for wound dressing applications.

3.2.4. Surface Morphology

SEM was used to investigate the surface microstructure and smoothness of the blank and GS-loaded AXSA films, as presented in Figure 2. The surface morphology of the blank films (F3) was compact and smooth, without any cracks, aggregation of the AX or SA, or particles (Figure 2). This compactness and smoothness of the surface structure indicates the physical compatibility and high miscibility between the AX and SA [43]. Conversely, small particles were observed in the GS-loaded (GF3) film (Figure 2), which were uniformly distributed throughout the surface of the films. Since the major difference between the composition of the GF3 and F3 films is the presence of GS in GF3 film. Therefore, particles observed in SEM images of GF3 film might be GS particles located at the film's surface. Similar particles were previously found on the surface of AX and AX-GL films after loading the GS in films [32,33]. The uniform distribution of drugs in films formulations is essential from the drug delivery perspective as it ensures the uniformity of the drug dosage [32,33].

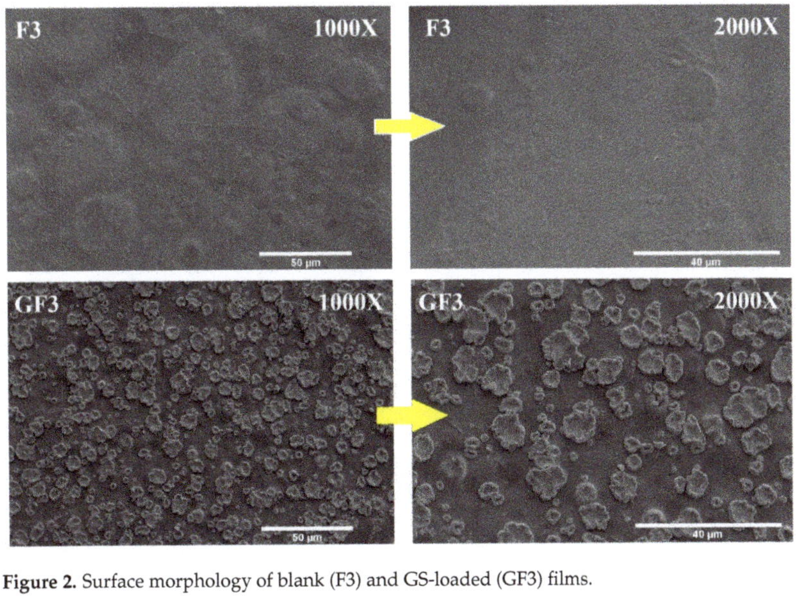

Figure 2. Surface morphology of blank (F3) and GS-loaded (GF3) films.

3.2.5. FTIR Spectroscopy

The ATR-FTIR spectra of AX, SA, GS, and AXSA films are presented in Figure 3. Characteristic bands of the polysaccharide functional groups were present in the spectra of AX and SA (Figure 3a). In spectra of both AX and SA, peaks attributed to the –OH, –CH_2, –COOH, C–O–C (glycosidic linkage), and C–O–H (primary alcohol) are present at around 3650–3300 cm^{-1}, 2930 cm^{-1}, 1660 cm^{-1}, 1450 cm^{-1}, 1160 cm^{-1}, and 1055 cm^{-1}, respectively [30,39]. The spectrum of GS exhibited major transmittance bands of –OH and Amide A (3650–2550 cm^{-1}), Amide I (1645 cm^{-1}), Amide II (1540 cm^{-1}), and C–O–C (1050 cm^{-1}) [31]. On the other hand, all characteristic peaks of both polysaccharides (AX and SA) were present in the spectra of both blank and GS-loaded AXSA films (Figure 3b). However, the peaks attributed to carboxylic groups were shifted from 1660 cm^{-1} and 1450 cm^{-1} to 1620 and 1425 cm^{-1}, indicating the hydrogen bonding interaction between polysaccharides. The shift in the vibrational frequency of hydroxyl groups from 3423 cm^{-1} (SA) and 3360 cm^{-1} (AX) to 3350 cm^{-1} (in films) confirms that hydrogen bonding is present in the films due to the interaction between hydroxyl, carboxylic, and amide groups of AX, SA and GS [39]. The hydrogen-bonding interactions contribute to increasing the tensile strength of the AXSA films (Section 3.2.3). Moreover, in the spectra of AXSA films, some functional group peaks were broad due to the overlapping of the polysaccharide peaks with glycerol and GS peaks (in GS-loaded films). These findings of the FTIR spectra of AXSA films suggest that the film components (AX, SA, GS, and glycerol) were finely blended, chemically compatible, and no chemical reaction occurred during the fabrication of the AXSA films.

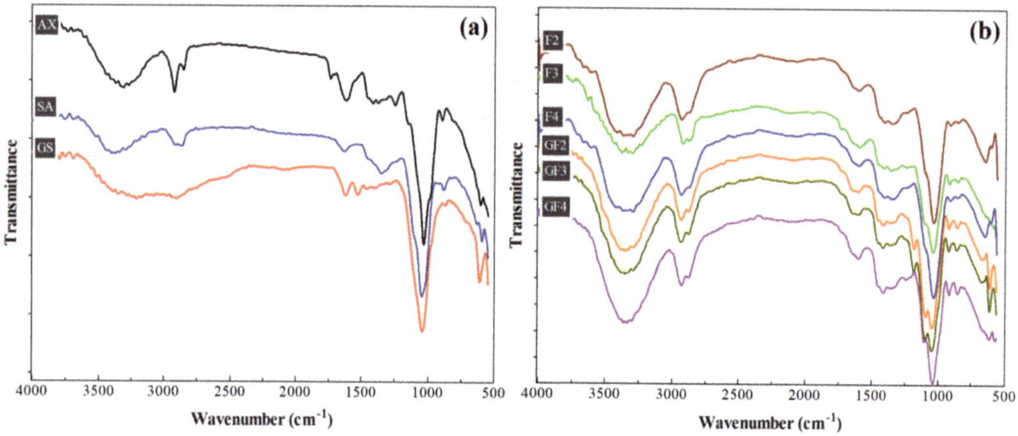

Figure 3. ATR-FTIR spectra of (**a**) pure components and (**b**) AXSA films.

3.2.6. X-ray Diffraction

The results of the XRD analyses of AX, SA, F3 (blank film), and GF3 (GS-loaded film) are presented in Figure 4. SA is semi-crystalline and exhibits characteristic peaks at around 2θ 13°, 56°, 20.6°, 20.1°, 29° and 36.4° [45]. The purpose of performing XRD analyses was to investigate the effect of combining film components on the semi-crystalline nature of the SA in AXSA films. The XRD pattern of SA had peaks between 2θ 13° to 38°, whereas AX did not show any sharp peaks attributed to its amorphous nature. The intensities of the semi-crystalline peaks of the SA were reduced in the AXSA films (F3 and F4). This reduction in the peak intensities may originate from the hydrogen bonding interaction of the SA with other amorphous components of the films (AX, Gly, and GS). Previously, Ramakrishnan et al. 2021 suggested a similar decrease in the SA crystallinity due to hydrogen bonding for SA/gum kondagogu blended films [43].

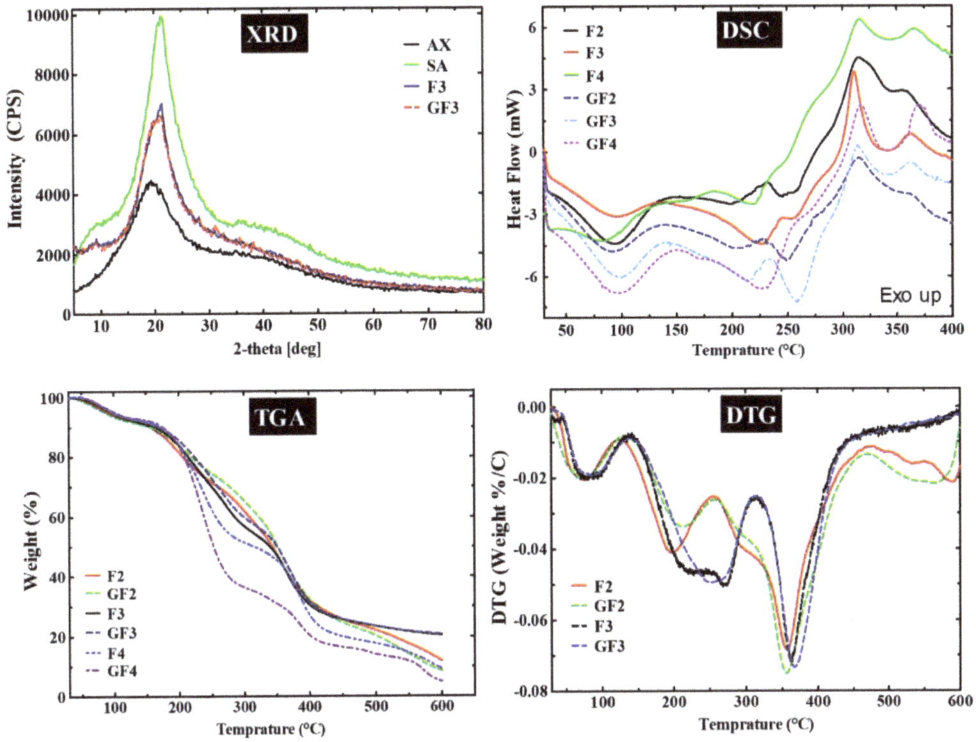

Figure 4. XRD, DSC, TGA and DTG analyses of the AXSA films.

3.2.7. Thermogravimetric Analyses (TGA)

The thermal degradation and stability of the AXSA films were investigated by TGA. The TG and DTG curves of the blank (F2, F3 and F3) and GS-loaded (GF2, GF3 and GF4) films are presented in Figure 3, while Figure S2 shows the results of the TGA of film components (AX, SA, Gly, and GS). In the first step of thermal degradation, all tested films and pure components (except Gly) exhibited weight loss up to ~120 °C, which was attributable to the loss of free water molecules and water interacting with hydroxyl and carboxylic groups [27]. Glycerol underwent characteristic single-step degradation at temperature range (T_{range}) ~125–300 °C (weight loss (ΔW) = 99.3%) with maximum weight loss temperature (T_{max}) at 256.2 °C. The TG and DTG curves show that AX exhibited degradation at T_{range} < 120 °C (ΔW = 5.1%), 260–320 °C (ΔW = 15.8%, T_{max} = 291), 325–395 °C (ΔW = 28.9%, T_{max} = 354 °C), and > 395 °C (ΔW = 21.7%) attributed to the loss of water, AX backbone and complete pyrolysis [32]. On the other hand, SA showed thermal decomposition at T_{range} < 105 °C (ΔW = 5.7%), 265–415 °C (ΔW = 71.3%, T_{max} = 330 °C), and >415 °C (ΔW = 12.5%) due to water loss, fragmentation of polysaccharide backbone of SA and complete degradation [27]. Similarly, GS underwent thermal degradation at T_{range} < 120 °C (ΔW = 13.7%), 245–295 °C (ΔW = 13.4%, T_{max} = 274 °C), 295–370 °C (ΔW = 25.9%, T_{max} = 330 °C), and >370 °C (ΔW = 35.5%), which shows that thermal decomposition of GS starts at 245 °C [46].

The F2 and GF2 films exhibited similar thermal degradation events at Trange <120 °C (ΔW up to 8%), 125–245 °C (ΔW up to 19.8%), 250–430 °C (ΔW up to 47.1%), and >430 °C (ΔW up to 18.8%). Similarly, F3 and GF3 films show similar degradation events at Trange <120 °C (ΔW up to 7.2%), 140–310 °C (ΔW up to 36.6%), 310–410 °C (ΔW up to 28.8%), and >410 °C (ΔW up to 20.6%). These results suggest that AXSA films undergo four main events of thermal degradation corresponding to the degradation events of the individual components of these films. However, the effect of the composition of the AXSA films was observed in the TGA of the films. In F3 and GF3 films, the second degradation event was broader than F2 and GF2 due to the higher AX contents in the F3 and GF3 films, resulting in the merging of the degradation step for Gly and the first step for AX. Similarly, the second degradation step was slightly broader in GS-loaded films than in the blank films due to the merging of degradations steps for Gly and the first degradation step of GS in GS-loaded films. The TG curves of F4 and GF4 also exhibited similar four-step thermal degradation behavior. It was observed that due to the plasticizer's effect, the onset temperature for polysaccharide (AX and SA) degradation shifted to a lower temperature in films vs. pure polysaccharides. These shifts in the TG and DTG curves for the AXSA films indicate the miscibility of the components [47]. These findings for the thermogravimetric analysis of the AXSA films are consistent with previous studies on AX and SA-based films [27,32,33].

3.2.8. Differential Scanning Calorimetry (DSC)

The results of the DSC analyses of the AXSA films (F2, F3, F4 GF2, GF3, and GF4) are depicted in Figure 4, while DSC curves of AX, SA, and GS are given in Figure S4. The DSC curves of the tested samples exhibited endotherms from 60 °C up to ~125 °C, indicating the removal of the water molecules. SA exhibited prominent exotherm at 297 °C, corresponding to the thermal degradation of the SA polysaccharide backbone as discussed above [27]. The DSC thermogram of AX showed characteristic exotherms at 290 and 312 °C (due to polysaccharide degradation), while a melting endotherm was found in the DSC of the GS at ~250 °C, which are in close agreement with the previous report [32]. On the other hand, two prominent endotherms were found in the DSC curves of the tested AXSA films at ~95 °C (water loss) and ~210 °C (due to glycerol). However, in DSC GS-loaded GF2 and GF3 films, a third endotherm was present at ~245 and 250 °C, respectively, attributed to the melting of GS. The second endotherm was broader in the DSC curve of the GF4, which might result from merging the endotherms due to glycerol and melting of GS. The exothermic peaks due to degradation of SA and AX were merged in the AXSA films and are present at ~310 °C. The DSC curves of the AXSA films did not show any significant shift in the DSC peaks of the components of the film. These findings of DSC analysis are consistent with previous studies of AX- and SA-based films and suggest that components of the AXSA films were highly miscible, and their stability was not affected during the casting process [27,32].

3.2.9. Expansion Profile

Water uptake (or hydration) is a vital characteristic of polymeric materials designed for wound dressing application. After application to the wounds, the dressings should absorb water (exudate) and expand in size while maintaining their integrity and shape [48]. Hydration also plays a key role in the dissolution of the drugs entrapped in the films [38]. Therefore, the expansion profiles of the AXSA films were determined using the gelatin model [32], and the results are shown in Figure 5a,b.

The expansion profile of AXSA films shows that films hydrated (uptake water) and expanded rapidly during the first hour of contact with solidified gelatin. After that, the expansion of the AXSA films slowed down. Among the blank AXSA films, F2 exhibited the highest expansion, followed by F3 and F4, respectively. Similarly, the expansion of GF2 films was highest among the GS-loaded AXSA films. Moreover, all AXSA films maintained their shape after 24 h. Higher expansion of the F2 films can be attributed to greater SA content in these films. The carboxylic group of SA converts to carboxylate ions in water, resulting in electrostatic repulsion among the polymeric chains, leading to the film expansion [49]. Therefore, higher expansion was observed in films containing more SA. As discussed above, an increase in the AX concentration in the films results in hydrogen bonding between the polymers, leading to the formation of a denser network and a decrease in free carboxylate ions available for interaction with water [50]. These factors lead to a further reduction in the expansion of the F3 and F4 films. The expansion behavior of AXSA films was in close agreement with previously reported simvastatin-loaded SA-pectin films [29]. However, the expansion percentage of the AXSA films was less than that previously reported for pure AX and AX-GL films, which can be explained by the hydrogen bonding interaction in the AXSA films [32,33]. Moreover, the AXSA films were intact (maintained their shape) throughout the expansion study compared to previously reported AX and AX-GL films, which started degrading after 5 to 6 in the expansion study [32,33]. Nevertheless, the expansion profile of AXSA films suggests that they hold the capacity to absorb the wound exudate while maintaining their shape and are suitable for wound dressing application [30].

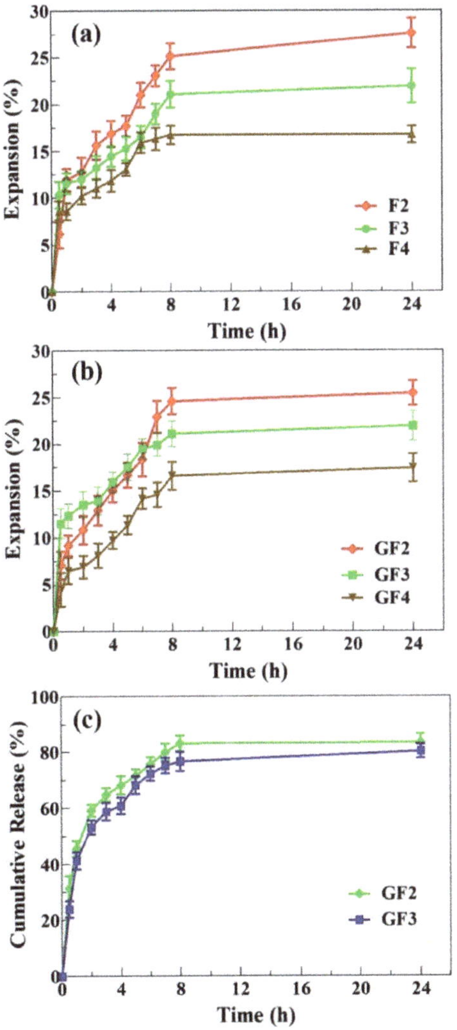

Figure 5. Expansion profiles of blank (**a**) and GS-loaded (**b**) AXSA films; and (**c**) drug release profile (mean ± SD, $n = 3$).

3.3. Gentamicin Release Profile and Kinetics

The GS release profile from AXSA (GF2 and GF3) films, investigated by Franz diffusion cell, is presented in Figure 5c. The GF2 and GF3 films were selected for the antibiotic release experiment owing to their better characteristics (EAB%, WVTR and expansion profile) for wound dressing application. The release study results show that GF2 and GF3 released 45% and 41% of the incorporated GS during the first hour of the release study. The GS release became slower after that, and reached equilibrium by 24 h. The maximum cumulative release after 24 h was 83.4% and 81.2% for GF2 and GF4, respectively. This initial release from the AXSA films is less than that of previously reported pure AX films and higher than that of AX-GL films [32,33]. These results suggest that 10 cm (diameter) GF2 and GF3 films will release up to 21 mg GS in 24 h. Previous studies on GS-based wound dressing materials suggest that 0.1% GS is effective for wound infections [14]. Moreover, topical delivery of 0.1% GS is considered safe for avoiding the toxic effects (ototoxicity and nephrotoxicity) associated with systematic GS administration [41,51].

The drug release from the dressing material depends on various factors, such as the solubility of the drug in the release media, relaxation, swelling and erosion of the polymeric network, and diffusion rate of the drug [33,38,52]. Therefore, mathematical models were applied to investigate the GS release kinetics and mechanism of release from the GS-loaded AXSA films. The results are shown in Table 2. The higher regression coefficient "R^2" values indicate the best fitting model. The results (Table 2) indicate that "R^2" values, obtained by fitting zero-order (0.4498 and 0.4243), first-order (0.543 and 0.5856), and Higuchi (0.6804 and 0.7015) equations, were less than those obtained with the Korsmeyer–Peppas equation (0.9788 and 0.9451). These results suggest that the Korsmeyer–Peppas model is the best-fitted model ("R^2" close to 1) for GS release from GF2 and GF3 films. The values of n (release exponent) for the Korsmeyer–Peppas model were 0.4603 and 0.4982 for GF2 and GF3 films, respectively. According to Korsmeyer–Peppas, for thin films, n values below 0.5 indicate that Fickian diffusion is the mechanism governing drug release [53]. Therefore, the n values of GS release from the AXSA films indicate that Fickian diffusion was the predominant mechanism, with the rate of GS release being limited by solvent penetration rate. A similar release mechanism was reported by Pires et al. for extract release from chitosan and alginate membranes intended for wound dressing [54]. These findings for the GS release profile suggest that the AXSA films hold potential for the treatment of infected wounds by delivering an initial high concentration of GS, followed by slower release up to 24 h.

Table 2. Mathematical modeling of GS release from AXSA films.

Model	Parameters	GF2	GF3
Zero order	R^2	0.4498	0.4243
	K_0	1.7521	1.6294
First order	R^2	0.543	0.5856
	K_1	−0.0225	−0.0211
Higuchi	R^2	0.6804	0.7015
	K_H	11.884	12.618
Korsmeyer–Peppas	R^2	0.9788	0.9451
	K_K	1.6417	1.5627
	n	0.4603	0.4982

3.4. *Antibacterial Effect of the Films*

The results of the antibacterial effect of GS solution, F3, GF2, and GF3 films are depicted in Figure 6a,b. In this experiment, blank AXSA film (F3) was used as a negative control, and filter paper disks dipped in 0.1% (w/w) GS solution were used as a positive control. The F3 (blank) films formed no clear inhibition zones, suggesting an insignificant antibacterial effect of AXSA films against the tested bacterial strains. However, opaque regions were observed around F2 films due to bacterial growth in the expanded blank AXSA films. In contrast, positive control (GS std.) and GS-loaded AXSA films (GF2 and GF3) exhibited significant antibacterial effects against both Gram-positive (*S. aureus*) and Gram-negative (*E. coli and P. aeruginosa*) bacteria. The values of inhibition zones for GF2 and GF3 were slightly higher than for the positive control. The antibacterial effect of the films was slightly higher against *E. coli* compared to the other two test strains, but this difference was statistically insignificant. The antibacterial effect of the AXSA films was better than the effect of recently reported chitosan-SA/GS nanofibrous wound dressing [35]. These results indicate GS-loaded AXSA films' potential to protect wounds against infections caused by Gram-positive and Gram-negative bacteria.

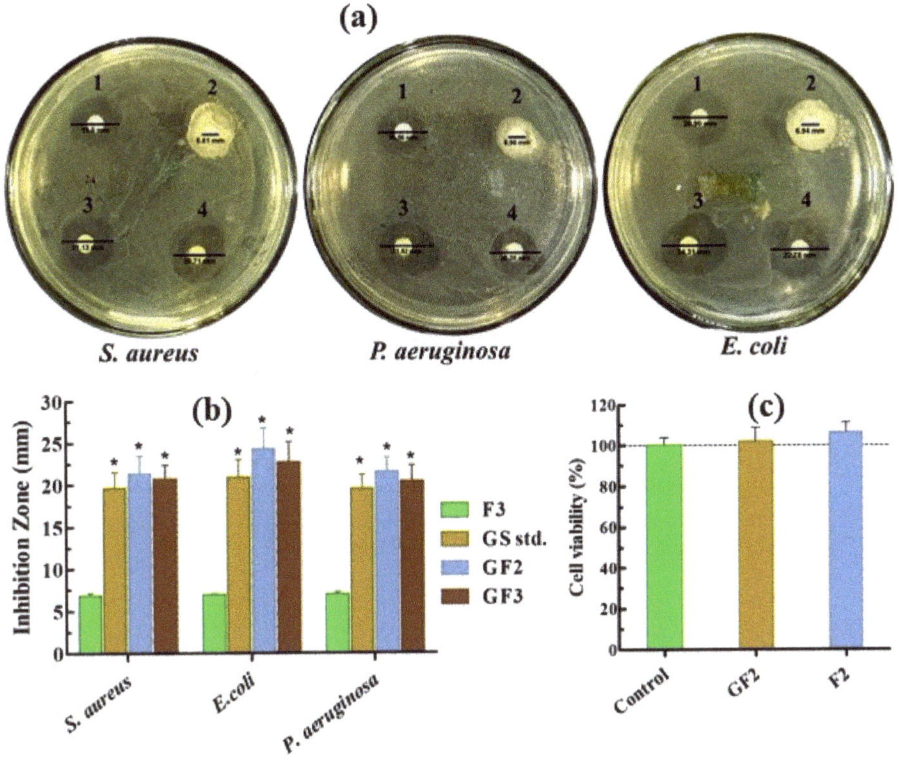

Figure 6. (**a**) Antibacterial effect of F3 (1), gentamicin sulphate (GS) (2), GF2 (3), GF3 (4); (**b**) inhibition zones and (**c**) cell viability of arabinoxylan (AX)-sodium alginate (SA) films (mean ± SD, n = 3). Asterisk (*) indicate statistically significant difference ($p < 0.05$) as compared to blank film (F3).

3.5. Indirect Cell Viability Assay

The indirect cell viability assay results of AXSA films against MRC-5 are depicted in Figure 6c. The results display that the viability of the cells treated with the extract of F2 and GF2 films exhibited no significant difference from that of untreated (control) cells. Cell viability assays are typically performed to predict films' safety for application on living tissues [7]. The extracts of F2 and GF2 films demonstrated no cytotoxic effects on normal cells and did not decrease cell viability. The cell viability (%) of the tested AXSA films was higher than the previously reported AX films indicating that the biocompatibility of the films increased by combining AX with SA in film formulation [33]. Therefore, it can be suggested that AXSA films are safe for in vivo applications.

4. Conclusions

The findings of this study demonstrate that the optimized AXSA films possess various desirable characteristics for antibacterial wound dressings. FTIR, XRD, and TGA analyses suggested that AX, SA, Gly, and GS were finely blended and formed hydrogen bonds. The prepared AXSA films were smooth, flexible, and transparent, which can aid in inspecting wounds without removing the dressing. The skin-like mechanical strength and flexibility of AXSA films make them suitable for their application on wounds in problematic body areas like joints. The moderate water uptake (expansion) and transmission (WVTR) of these films could be helpful for protecting moderately exudating wounds from dehydration and to maintain a moist environment for healing. The AXSA films released more than 80% of the loaded GS in 24 h with the initial rapid release (in the first hour), which can be used to control infections. The inhibition zones formed by the GS-loaded AXSA films in Gram-positive and Gram-negative bacteria were similar to those of GS solution, suggesting that these films can be used to control bacterial growth in infected wounds. The AXSA films also demonstrated excellent cytocompatibility in MTT assay, suggesting their safety for in vivo applications. Overall, we demonstrated that the AXSA films are a potential candidate for antibacterial wound dressing applications.

Supplementary Materials: The following supporting information can be downloaded at: https://www.mdpi.com/article/10.3390/ijms23052899/s1.

Author Contributions: Conceptualization, A.I.A., N.K.A. and N.A.; methodology, N.A.; formal analysis, A.I.A., M.M.A. and N.A.; investigation, A.I.A., M.M.A., A.M.B., M.U.M., M.S.B.S., Z.A.A. and N.A.; resources, A.I.A., A.V.D., Z.S.A. and S.S.A.; data curation, A.I.A.; writing—original draft preparation, A.I.A., M.M.A. and N.A.; writing—review and editing, A.M.B. and A.V.D.; supervision, A.I.A. and N.A.; project administration, A.I.A.; funding acquisition, A.I.A. All authors have read and agreed to the published version of the manuscript.

Funding: This research was funded by Deanship of Scientific Research at Jouf University through grant number 40/267. The APC was funded by Jouf University, Saudi Arabia.

Data Availability Statement: Data sharing is not applicable to this article.

Acknowledgments: The authors extend their appreciation to the Central laboratory at Jouf University for technical support.

Conflicts of Interest: The authors declare no conflict of interest.

References

1. Ambrogi, V.; Pietrella, D.; Donnadio, A.; Latterini, L.; Di Michele, A.; Luffarelli, I.; Ricci, M. Biocompatible alginate silica supported silver nanoparticles composite films for wound dressing with antibiofilm activity. *Mater. Sci. Eng. C* **2020**, *112*, 110863. [CrossRef] [PubMed]
2. Wang, M.; Huang, X.; Zheng, H.; Tang, Y.; Zeng, K.; Shao, L.; Li, L. Nanomaterials applied in wound healing: Mechanisms, limitations and perspectives. *J. Control. Release* **2021**, *337*, 236–247. [CrossRef] [PubMed]
3. Saghazadeh, S.; Rinoldi, C.; Schot, M.; Kashaf, S.S.; Sharifi, F.; Jalilian, E.; Nuutila, K.; Giatsidis, G.; Mostafalu, P.; Derakhshandeh, H.; et al. Drug delivery systems and materials for wound healing applications. *Adv. Drug Deliv. Rev.* **2018**, *127*, 138–166. [CrossRef] [PubMed]
4. Kim, H.S.; Sun, X.; Lee, J.-H.; Kim, H.-W.; Fu, X.; Leong, K.W. Advanced drug delivery systems and artificial skin grafts for skin wound healing. *Adv. Drug Deliv. Rev.* **2019**, *146*, 209–239. [CrossRef]
5. Arif, M.M.; Khan, S.M.; Gull, N.; Tabish, T.A.; Zia, S.; Khan, R.U.; Awais, S.M.; Butt, M.A. Polymer-based biomaterials for chronic wound management: Promises and challenges. *Int. J. Pharm.* **2021**, *598*, 120270. [CrossRef]
6. Tamer, T.M.; Alsehli, M.H.; Omer, A.M.; Afifi, T.H.; Sabet, M.M.; Mohy-Eldin, M.S.; Hassan, M.A. Development of Polyvinyl Alcohol/Kaolin Sponges Stimulated by Marjoram as Hemostatic, Antibacterial, and Antioxidant Dressings for Wound Healing Promotion. *Int. J. Mol. Sci.* **2021**, *22*, 13050. [CrossRef]
7. Colobatiu, L.; Gavan, A.; Potarniche, A.-V.; Rus, V.; Diaconeasa, Z.; Mocan, A.; Tomuta, I.; Mirel, S.; Mihaiu, M. Evaluation of bioactive compounds-loaded chitosan films as a novel and potential diabetic wound dressing material. *React. Funct. Polym.* **2019**, *145*, 104369. [CrossRef]
8. Savencu, I.; Iurian, S.; Porfire, A.; Bogdan, C.; Tomuță, I. Review of advances in polymeric wound dressing films. *React. Funct. Polym.* **2021**, *168*, 105059. [CrossRef]
9. Lou, P.; Liu, S.; Xu, X.; Pan, C.; Lu, Y.; Liu, J. Extracellular vesicle-based therapeutics for the regeneration of chronic wounds: Current knowledge and future perspectives. *Acta Biomater.* **2021**, *119*, 42–56. [CrossRef]
10. Rathod, L.; Bhowmick, S.; Patel, P.; Sawant, K. Calendula flower extract loaded PVA hydrogel sheet for wound management: Optimization, characterization and in-vivo study. *J. Drug Deliv. Sci. Technol.* **2021**, 103035. [CrossRef]
11. Boateng, J.S.; Pawar, H.V.; Tetteh, J. Polyox and carrageenan based composite film dressing containing anti-microbial and anti-inflammatory drugs for effective wound healing. *Int. J. Pharm.* **2013**, *441*, 181–191. [CrossRef] [PubMed]

12. Mayer, S.; Tallawi, M.; De Luca, I.; Calarco, A.; Reinhardt, N.; Gray, L.A.; Drechsler, K.; Moeini, A.; Germann, N. Antimicrobial and physicochemical characterization of 2, 3-dialdehyde cellulose-based wound dressings systems. *Carbohydr. Polym.* **2021**, *272*, 118506. [CrossRef] [PubMed]
13. Reczyńska-Kolman, K.; Hartman, K.; Kwiecień, K.; Brzychczy-Włoch, M.; Pamuła, E. Composites Based on Gellan Gum, Alginate and Nisin-Enriched Lipid Nanoparticles for the Treatment of Infected Wounds. *Int. J. Mol. Sci.* **2022**, *23*, 321. [CrossRef] [PubMed]
14. Ng, S.-F.; Leow, H.-L. Development of biofilm-targeted antimicrobial wound dressing for the treatment of chronic wound infections. *Drug Dev. Ind. Pharm.* **2015**, *41*, 1902–1909. [CrossRef] [PubMed]
15. Kirketerp-Møller, K.; Zulkowski, K.; James, G. Chronic wound colonization, infection, and biofilms. In *Biofilm Infections*; Springer: Berlin/Heidelberg, Germany, 2011; pp. 11–24.
16. Dong, R.; Guo, B. Smart wound dressings for wound healing. *Nano Today* **2021**, *41*, 101290. [CrossRef]
17. Varaprasad, K.; Jayaramudu, T.; Kanikireddy, V.; Toro, C.; Sadiku, E.R. Alginate-based composite materials for wound dressing application: A mini review. *Carbohydr. Polym.* **2020**, *236*, 116025. [CrossRef]
18. De Luca, I.; Pedram, P.; Moeini, A.; Cerruti, P.; Peluso, G.; Di Salle, A.; Germann, N. Nanotechnology Development for Formulating Essential Oils in Wound Dressing Materials to Promote the Wound-Healing Process: A Review. *Appl. Sci.* **2021**, *11*, 1713. [CrossRef]
19. Moeini, A.; Pedram, P.; Makvandi, P.; Malinconico, M.; Gomez d'Ayala, G. Wound healing and antimicrobial effect of active secondary metabolites in chitosan-based wound dressings: A review. *Carbohydr. Polym.* **2020**, *233*, 115839. [CrossRef]
20. Teoh, J.H.; Tay, S.M.; Fuh, J.; Wang, C.-H. Fabricating scalable, personalized wound dressings with customizable drug loadings via 3D printing. *J. Control. Release* **2022**, *341*, 80–94. [CrossRef]
21. Thomas, D.; Nath, M.S.; Mathew, N.; Reshmy, R.; Philip, E.; Latha, M. Alginate film modified with aloevera gel and cellulose nanocrystals for wound dressing application: Preparation, characterization and in vitro evaluation. *J. Drug Deliv. Sci. Technol.* **2020**, *59*, 101894. [CrossRef]
22. Hafezi, F.; Scoutaris, N.; Douroumis, D.; Boateng, J. 3D printed chitosan dressing crosslinked with genipin for potential healing of chronic wounds. *Int. J. Pharm.* **2019**, *560*, 406–415. [CrossRef] [PubMed]
23. Garms, B.C.; Borges, F.A.; de Barros, N.R.; Marcelino, M.Y.; Leite, M.N.; Del Arco, M.C.; de Souza Salvador, S.L.; Pegorin, G.S.A.; Oliveira, K.S.M.; Frade, M.A.C.; et al. Novel polymeric dressing to the treatment of infected chronic wound. *Appl. Microbiol. Biotechnol.* **2019**, *103*, 4767–4778. [CrossRef] [PubMed]
24. Teleky, B.-E.; Vodnar, D.C. Recent Advances in Biotechnological Itaconic Acid Production, and Application for a Sustainable Approach. *Polymers* **2021**, *13*, 3574. [CrossRef]
25. de Espíndola Sobczyk, A.; Luchese, C.L.; Faccin, D.J.L.; Tessaro, I.C. Influence of replacing oregano essential oil by ground oregano leaves on chitosan/alginate-based dressings properties. *Int. J. Biol. Macromol.* **2021**, *181*, 51–59. [CrossRef] [PubMed]
26. Ahmed, A.; Getti, G.; Boateng, J. Medicated multi-targeted alginate-based dressings for potential treatment of mixed bacterial-fungal infections in diabetic foot ulcers. *Int. J. Pharm.* **2021**, *606*, 120903. [CrossRef] [PubMed]
27. Bialik-Wąs, K.; Pluta, K.; Malina, D.; Barczewski, M.; Malarz, K.; Mrozek-Wilczkiewicz, A. Advanced SA/PVA-based hydrogel matrices with prolonged release of Aloe vera as promising wound dressings. *Mater. Sci. Eng. C* **2021**, *120*, 111667. [CrossRef] [PubMed]
28. Kanikireddy, V.; Varaprasad, K.; Jayaramudu, T.; Karthikeyan, C.; Sadiku, R. Carboxymethyl cellulose-based materials for infection control and wound healing: A review. *Int. J. Biol. Macromol.* **2020**, *164*, 963–975. [CrossRef] [PubMed]
29. Rezvanian, M.; Amin, M.C.I.M.; Ng, S.-F. Development and physicochemical characterization of alginate composite film loaded with simvastatin as a potential wound dressing. *Carbohydr. Polym.* **2016**, *137*, 295–304. [CrossRef]
30. Rezvanian, M.; Ahmad, N.; Mohd Amin, M.C.I.; Ng, S.-F. Optimization, characterization, and in vitro assessment of alginate-pectin ionic cross-linked hydrogel film for wound dressing applications. *Int. J. Biol. Macromol.* **2017**, *97*, 131–140. [CrossRef]
31. Saghir, S.; Iqbal, M.S.; Koschella, A.; Heinze, T. Ethylation of arabinoxylan from Ispaghula (*Plantago ovata*) seed husk. *Carbohydr. Polym.* **2009**, *77*, 125–130. [CrossRef]
32. Ahmad, N.; Ahmad, M.M.; Alruwaili, N.K.; Alrowaili, Z.A.; Alomar, F.A.; Akhtar, S.; Alsaidan, O.A.; Alhakamy, N.A.; Zafar, A.; Elmowafy, M.; et al. Antibiotic-Loaded Psyllium Husk Hemicellulose and Gelatin-Based Polymeric Films for Wound Dressing Application. *Pharmaceutics* **2021**, *13*, 236. [CrossRef] [PubMed]
33. Ahmad, N.; Tayyeb, D.; Ali, I.; KAlruwaili, N.; Ahmad, W.; Khan, A.H.; Iqbal, M.S. Development and Characterization of Hemicellulose-Based Films for Antibacterial Wound-Dressing Application. *Polymers* **2020**, *12*, 548. [CrossRef] [PubMed]
34. Phaechamud, T.; Issarayungyuen, P.; Pichayakorn, W. Gentamicin sulfate-loaded porous natural rubber films for wound dressing. *Int. J. Biol. Macromol.* **2016**, *85*, 634–644. [CrossRef] [PubMed]
35. Bakhsheshi-Rad, H.; Hadisi, Z.; Ismail, A.; Aziz, M.; Akbari, M.; Berto, F.; Chen, X. In vitro and in vivo evaluation of chitosan-alginate/gentamicin wound dressing nanofibrous with high antibacterial performance. *Polym. Test.* **2020**, *82*, 106298. [CrossRef]
36. Akkaya, N.E.; Ergun, C.; Saygun, A.; Yesilcubuk, N.; Akel-Sadoglu, N.; Kavakli, I.H.; Turkmen, H.S.; Catalgil-Giz, H. New biocompatible antibacterial wound dressing candidates; agar-locust bean gum and agar-salep films. *Int. J. Biol. Macromol.* **2020**, *155*, 430–438. [CrossRef] [PubMed]
37. Chang, H.I.; Lau, Y.C.; Yan, C.; Coombes, A. Controlled release of an antibiotic, gentamicin sulphate, from gravity spun polycaprolactone fibers. *J. Biomed. Mater. Res. Part A Off. J. Soc. Biomater. Jpn. Soc. Biomater. Aust. Soc. Biomater. Korean Soc. Biomater.* **2008**, *84*, 230–237. [CrossRef]

38. Alavi, T.; Rezvanian, M.; Ahmad, N.; Mohamad, N.; Ng, S.-F. Pluronic-F127 composite film loaded with erythromycin for wound application: Formulation, physicomechanical and in vitro evaluations. *Drug Deliv. Transl. Res.* **2019**, *9*, 508–519. [CrossRef]
39. Trevisol, T.; Fritz, A.; de Souza, S.; Bierhalz, A.; Valle, J. Alginate and carboxymethyl cellulose in monolayer and bilayer films as wound dressings: Effect of the polymer ratio. *J. Appl. Polym. Sci.* **2019**, *136*, 46941. [CrossRef]
40. El-Gendy, N.; Abdelbary, G.; El-Komy, M.; Saafan, A. Design and evaluation of a bioadhesive patch for topical delivery of gentamicin sulphate. *Curr. Drug Deliv.* **2009**, *6*, 50–57. [CrossRef]
41. Griffis, C.D.; Metcalfe, S.; Bowling, F.L.; Boulton, A.J.; Armstrong, D.G. The use of gentamicin-impregnated foam in the management of diabetic foot infections: A promising delivery system? *Expert Opin. Drug Deliv.* **2009**, *6*, 639–642. [CrossRef]
42. Phisalaphong, M.; Suwanmajo, T.; Tammarate, P. Synthesis and characterization of bacterial cellulose/alginate blend membranes. *J. Appl. Polym. Sci.* **2008**, *107*, 3419–3424. [CrossRef]
43. Ramakrishnan, R.K.; Wacławek, S.; Černík, M.; Padil, V.V. Biomacromolecule assembly based on gum kondagogu-sodium alginate composites and their expediency in flexible packaging films. *Int. J. Biol. Macromol.* **2021**, *177*, 526–534. [CrossRef] [PubMed]
44. Costa, N.N.; de Faria Lopes, L.; Ferreira, D.F.; de Prado, E.M.L.; Severi, J.A.; Resende, J.A.; de Paula Careta, F.; Ferreira, M.C.P.; Carreira, L.G.; de Souza, S.O.L. Polymeric films containing pomegranate peel extract based on PVA/starch/PAA blends for use as wound dressing: In vitro analysis and physicochemical evaluation. *Mater. Sci. Eng. C* **2020**, *109*, 110643. [CrossRef]
45. Patra, T.; Gupta, M.K. Evaluation of sodium alginate for encapsulation-vitrification of testicular Leydig cells. *Int. J. Biol. Macromol.* **2020**, *153*, 128–137. [CrossRef] [PubMed]
46. Purcar, V.; Rădiţoiu, V.; Nichita, C.; Bălan, A.; Rădiţoiu, A.; Căprărescu, S.; Raduly, F.M.; Manea, R.; Şomoghi, R.; Nicolae, C.-A. Preparation and Characterization of Silica Nanoparticles and of Silica-Gentamicin Nanostructured Solution Obtained by Microwave-Assisted Synthesis. *Materials* **2021**, *14*, 2086. [CrossRef] [PubMed]
47. de Moraes, M.A.; Silva, M.F.; Weska, R.F.; Beppu, M.M. Silk fibroin and sodium alginate blend: Miscibility and physical characteristics. *Mater. Sci. Eng. C* **2014**, *40*, 85–91. [CrossRef]
48. Jantrawut, P.; Bunrueangtha, J.; Suerthong, J.; Kantrong, N. Fabrication and characterization of low methoxyl pectin/gelatin/carboxymethyl cellulose absorbent hydrogel film for wound dressing applications. *Materials* **2019**, *12*, 1628. [CrossRef]
49. Sinha, P.; Udhumansha, U.; Rathnam, G.; Ganesh, M.; Jang, H.T. Capecitabine encapsulated chitosan succinate-sodium alginate macromolecular complex beads for colon cancer targeted delivery: In vitro evaluation. *Int. J. Biol. Macromol.* **2018**, *117*, 840–850. [CrossRef]
50. Eltabakh, M.; Kassab, H.; Badawy, W.; Abdin, M.; Abdelhady, S. Active Bio-composite Sodium Alginate/Maltodextrin Packaging Films for Food Containing Azolla pinnata Leaves Extract as Natural Antioxidant. *J. Polym. Environ.* **2021**, 1–11. [CrossRef]
51. Gemeinder, J.L.P.; Barros, N.R.d.; Pegorin, G.S.A.; Singulani, J.d.L.; Borges, F.A.; Arco, M.C.G.D.; Giannini, M.J.S.M.; Almeida, A.M.F.; Salvador, S.L.d.S.; Herculano, R.D. Gentamicin encapsulated within a biopolymer for the treatment of *Staphylococcus aureus* and *Escherichia coli* infected skin ulcers. *J. Biomater. Sci. Polym. Ed.* **2021**, *32*, 93–111. [CrossRef]
52. Setapa, A.; Ahmad, N.; Mohd Mahali, S.; Mohd Amin, M.C.I. Mathematical Model for Estimating Parameters of Swelling Drug Delivery Devices in a Two-Phase Release. *Polymers* **2020**, *12*, 2921. [CrossRef] [PubMed]
53. Korsmeyer, R.W.; Gurny, R.; Doelker, E.; Buri, P.; Peppas, N.A. Mechanisms of solute release from porous hydrophilic polymers. *Int. J. Pharm.* **1983**, *15*, 25–35. [CrossRef]
54. Pires, A.; Westin, C.; Hernandez-Montelongo, J.; Sousa, I.; Foglio, M.; Moraes, A. Flexible, dense and porous chitosan and alginate membranes containing the standardized extract of *Arrabidaea chica* Verlot for the treatment of skin lesions. *Mater. Sci. Eng. C* **2020**, *112*, 110869. [CrossRef] [PubMed]

International Journal of
Molecular Sciences

MDPI

Article

Polypropylene Graft Poly(methyl methacrylate) Graft Poly(*N*-vinylimidazole) as a Smart Material for pH-Controlled Drug Delivery

Felipe López-Saucedo *, Jesús Eduardo López-Barriguete, Guadalupe Gabriel Flores-Rojas, Sharemy Gómez-Dorantes and Emilio Bucio *

Departamento de Química de Radiaciones y Radioquímica, Instituto de Ciencias Nucleares, Universidad Nacional Autónoma de México, Circuito Exterior, Ciudad Universitaria, Mexico City 04510, Mexico; jelbarrig@gmail.com (J.E.L.-B.); ggabofo@hotmail.com (G.G.F.-R.); shar.hj@gmail.com (S.G.-D.)
* Correspondence: felipelopezsaucedo@gmail.com (F.L.-S.); ebucio@nucleares.unam.mx (E.B.)

Abstract: Surface modification of polypropylene (PP) films was achieved using gamma-irradiation-induced grafting to provide an adequate surface capable of carrying glycopeptide antibiotics. The copolymer was obtained following a versatile two-step route; pristine PP was exposed to gamma rays and grafted with methyl methacrylate (MMA), and afterward, the film was grafted with *N*-vinylimidazole (NVI) by simultaneous irradiation. Characterization included Fourier transform infrared spectroscopy (FTIR), scanning electron microscope (SEM), thermogravimetric analysis (TGA), X-ray photoelectron spectroscopy (XPS), and physicochemical analysis of swelling and contact angle. The new material (PP-*g*-MMA)-*g*-NVI was loaded with vancomycin to quantify the release by UV-vis spectrophotometry at different pH. The surface of (PP-*g*-MMA)-*g*-NVI exhibited pH-responsiveness and moderate hydrophilicity, which are suitable properties for controlled drug release.

Keywords: grafting; polypropylene; gamma rays; methyl methacrylate; *N*-vinylimidazole; pH-responsiveness; vancomycin; release

Citation: López-Saucedo, F.; López-Barriguete, J.E.; Flores-Rojas, G.G.; Gómez-Dorantes, S.; Bucio, E. Polypropylene Graft Poly(methyl methacrylate) Graft Poly(*N*-vinylimidazole) as a Smart Material for pH-Controlled Drug Delivery. *Int. J. Mol. Sci.* **2022**, *23*, 304. https://doi.org/10.3390/ijms23010304

Academic Editor: Yury A. Skorik

Received: 26 November 2021
Accepted: 27 December 2021
Published: 28 December 2021

1. Introduction

The surface functionalization of polypropylene using conventional chemical methods presents well-known difficulties [1]. Such difficulties are attributed to the thermal stability and lack of reactivity from alkyl chains in the polymer. Nonetheless, the modification of polymer materials is often a prolific rewarding task because the addition of functional groups, whether in bulk or only in a specific part [2], is a standardized strategy to provide new physical and chemical properties to materials to boost their functionality [3].

Grafting polymerization has been demonstrated as an excellent option to modify PP [4], where different methods [5] have been tried to graft vinyl monomers, such as *N*-vinylimidazole (NVI) [6], methyl methacrylate (MMA) [7], *N*-vinylcaprolactam (NVCL) [8], or glycidyl methacrylate (GMA) [9], among others. However, there are issues related to the low reactivity of PP, which produces non-uniform grafting, low yields, waste, and residues [10].

Currently, alternative energy sources are becoming more relevant to carry out grafting polymerization, such as gamma rays [11], plasma [12], UV-light, and electron-beam [13]. Among these energy sources, high energy gamma rays of 1.17 and 1.33 MeV from ^{60}Co [14] are suitable to promote the homolytic rupture of stable C-H and C-C bonds from PP polymeric chains. Exposing PP to gamma rays produces free radicals (unstable), which are stabilized as peroxides and hydroperoxides under an oxidizing atmosphere [15]. The said process is named the "pre-irradiation oxidative method" and is used to induce grafting polymerization onto PP surfaces with vinyl monomers [2]. The grafting degree of vinyl monomers depends on different factors, such as the solvent, time, reaction temperature, and absorbed dose. In general terms, a good understanding of reactants ensures obtaining materials with the desired properties.

Therefore, gamma-ray-induced graft polymerization can provide new materials that meet multifunctional or multipurpose needs [3]. MMA is a methacrylic monomer used for multiple purposes with an aliphatic part (non-polar) and a carbonyl group (polar) [16]. In addition, NVI is a vinyl molecule with an N heteroatom ring used for diverse objectives [17,18]. Both carbonyl and imidazole groups can be employed in drug delivery systems, thanks to electrostatic and hydrophobic interactions drug-polymer. Hence, these polymers, PP, PMMA, and PNVI, may be implemented to design new materials with tailored properties for biomedical devices considering their biocompatibility [12,19,20].

In summary, this work presents grafting polymerization of MMA and NVI onto PP with the subsequent loading and release of vancomycin [21], where the pH-responsiveness of NVI chains [22] was studied as well as other physicochemical properties, such as swelling and contact angle.

2. Results

The materials were modified successfully with MMA by grafting polymerization using the pre-irradiation oxidative method. The grafting degree showed a dependence on the absorbed dose, temperature, monomer concentration, and reaction time, offering an excellent control on the yield and leading to the possibility of obtaining tailored grafted PP films. In the case of the grafting degree of NVI, this was carried out by the direct method and did not show a considerable grafting yield if compared to MMA graft, which was more quantitative.

2.1. Grafting

During the grafting process on PP films, it is possible to observe certain tendencies regarding the grafted acrylate. Grafting of MMA exhibited a linear slope in the absorbed dose experiment (5 to 25 kGy) and in the time reaction experiment (5 to 26 h) reaching a maximum grafting of 49.5% (25 kGy and 16 h) and 31% (5 kGy and 26 h), respectively (Figure 1a,b). However, the absorbed dose of 25 kGy could cause a detriment or deterioration on the PP matrix caused by polymer chain rupture, cross-linking, and increment of oxygenated groups [23]; for this reason, 5 kGy is the absorbed dose preferred.

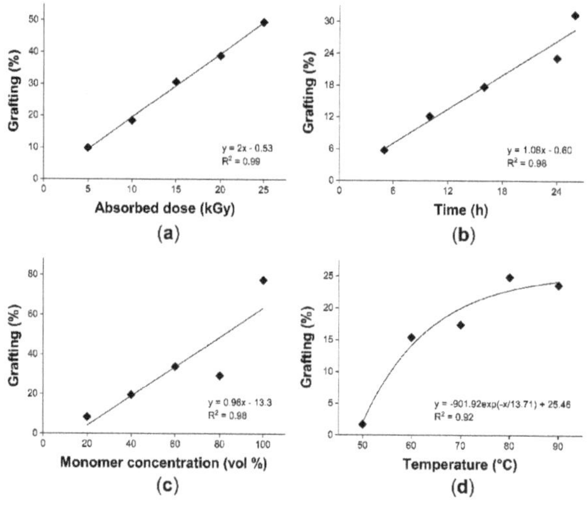

Figure 1. Grafting of MMA onto pristine PP: (**a**) Effect of absorbed dose (16 h, 65 °C, MMA 20 vol%); (**b**) reaction time (5 kGy, 65 °C, MMA 30 vol%); (**c**) monomer concentration (15 kGy, 16 h, 70 °C), and (**d**) reaction temperature (5 kGy, 16 h, MMA 30 vol%).

Graft by varying temperature and monomer concentration completed the grafting study. Regarding the effect of monomer concentration, it had a maximum grafting of 77.5% with a linear tendency (Figure 1c), but at the lower monomer concentrations (20%), the graft was adequate to incorporate a superficial modification. Finally, the results indicated that the minimum activation temperature for this system is about 50 °C, which is congruent with the temperature to activate peroxides. The grafting degree increased progressively up to 80 °C, but at 90 °C the graft slightly decreased, indicating that when the reaction took place at 90 °C, the homopolymerization was benefited, so the graft was affected (Figure 1d). Therefore, the results suggest a possible control on the grafting rate either by reaction time or reaction temperature, offering a reasonable percentage of functionalization using a low absorbed dose of 5 kGy and low monomer concentration, thus, ensuring lower damage in the properties of the matrix.

The grafting of NVI was carried out on PP-*g*-MMA with different grafting degrees from 8.5 to 77.5%. The NVI grafting degree was lower compared to the results obtained in the MMA grafting, which indicates a lower monomer reactivity. Since the MMA grafting degree of PP-*g*-MMA was higher, the grafting yield of NVI did not increase proportionally, obtaining yields ranging from 4 to 6.5%. Hence, it is understood that a slight modification with MMA is enough to promote the graft of NVI in a second step (Figure 2). In the following lines, the notation (PP-*g*-MMA)-*g*-NVI (x/y%) represents the binary grafted weight percent of "x" PMMA and "y" PNVI, respectively.

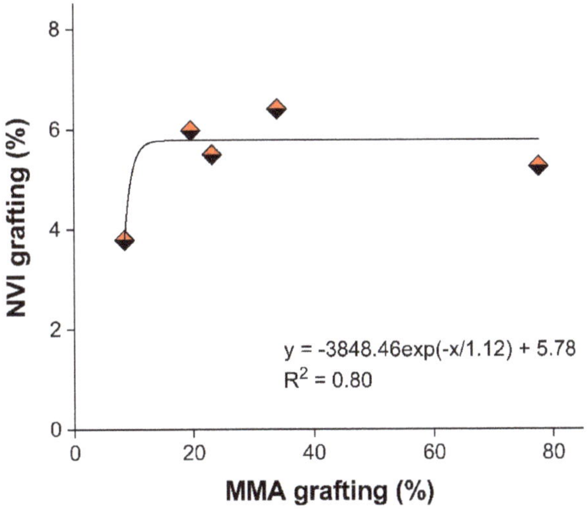

Figure 2. Results of NVI grafted on PP-*g*-MMA (50%); reaction conditions, absorbed dose 15 kGy, and room temperature (around 25 °C).

SEM microscopy was performed to analyze the surface morphology of PP-*g*-MMA (17%) and (PP-*g*-MMA)-*g*-NVI (19.5/6%) (Figure 3). Morphological changes due to the grafting polymerizations were observed, clearly indicating a surface copolymerization with an amorphous appearance, which is suitable for the adsorption of solids, as was found in this case.

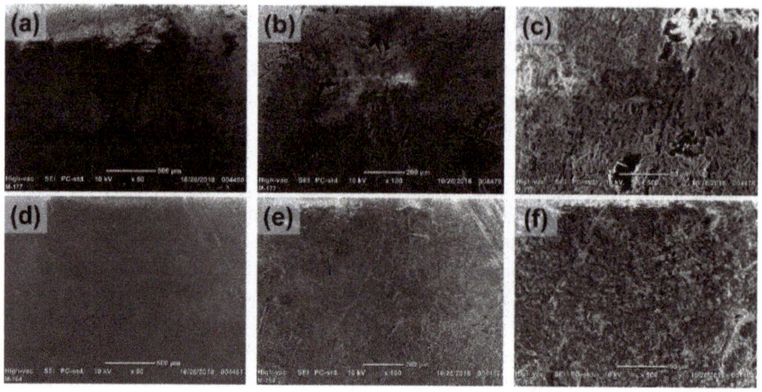

Figure 3. SEM images of (**a–c**) PP-*g*-MMA (17%) and (**d–f**) (PP-*g*-MMA)-*g*-NVI (19.5/6%), augmented from left to right ×50, ×100, and ×500.

2.2. Infrared Spectroscopy

For pristine PP, as a linear polymer constituted just of propylene units, strong bands of infrared corresponding to different modes of C-H vibration were displayed, which stretch in the region of 2949–2838 cm^{-1} and bend methyl (-CH_3) and methylene (-CH_2) groups at 1456 and 1375 cm^{-1}, respectively (Figure 4). Once the first copolymer was achieved, the spectrum of PP-*g*-MMA, besides the aliphatic bands, showed the characteristic carbonyl band around 1724 cm^{-1}, which appeared as a strong signal accompanied by the C-O stretching at 1145 and 1063 cm^{-1} [24]. After the second graft with NVI [25], in addition to the mentioned bands, there was an aromatic C-H stretching band at 3112 cm^{-1} and the characteristic bands of aromatic compounds in the fingerprint region between 900 and 650 cm^{-1} [26,27].

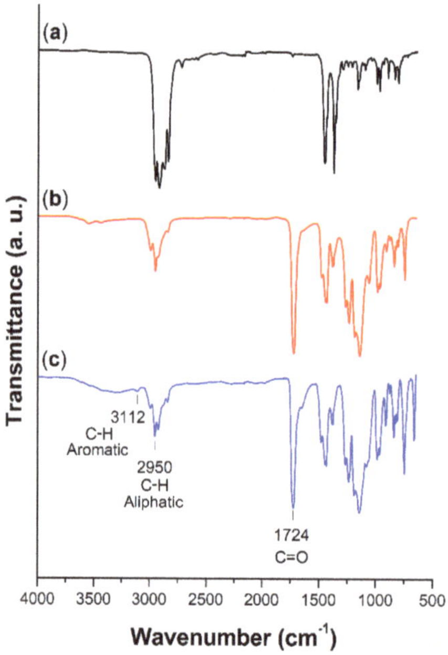

Figure 4. FTIR spectra: (**a**) non-irradiated PP, (**b**) PP-*g*-MMA (10%), and (**c**) (PP-*g*-MMA)-*g*-NVI (77.5/5%).

2.3. XPS Spectroscopy

XPS study determined the surface atomic compositions of pristine PP [28] and grafted films, confirming the existence of grafted PMMA [29] and PNVI [27] on the surface, as shown in Figure 5. The characteristic peaks of carbon (C 1 s at 285.0 eV), oxygen (O 1 s at 531.0 eV), and nitrogen (N 1 s at 399.4 eV) were detected in the scanning, and the atomic level relationship was obtained from the core level peak areas of C 1 s, O 1 s, and N 1 s, and multiplied by the corresponding sensitivity factors giving the results in Table 1.

Table 1. XPS results of pristine PP, PP-*g*-MMA (17%), and (PP-*g*-MMA)-*g*-NVI (77.5/5%): elemental composition used atomic sensitivity factor of C 1 s: 0.314, O 1 s: 0.733 and N 1 s: 0.499.

Film	Atomic (%)		
	C	O	N
PP	100	-	-
PP-*g*-MMA (17%)	75.28	24.51	-
(PP-*g*-MMA)-*g*-NVI (77.5/5%)	72.59	22.90	3.76

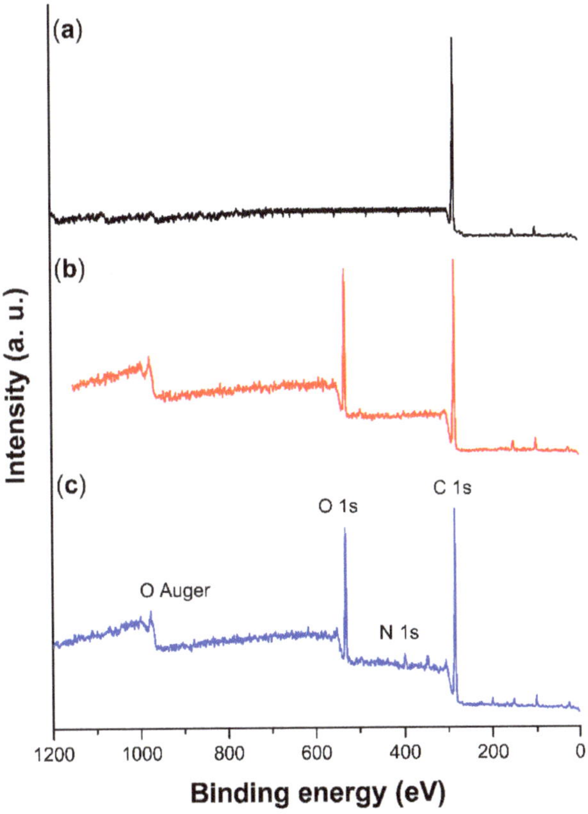

Figure 5. XPS scan of: (**a**) non-irradiated PP, (**b**) PP-*g*-MMA (17%), and (**c**) (PP-*g*-MMA)-*g*-NVI (77.5/5%).

2.4. Thermal Gravimetry Analysis

Thermograms of grafted films displayed a faster weight loss compared to the TGA curve observed in the pristine non-irradiated PP film, as the 10% weight loss indicates

(Figure 6). Decomposition temperature (Td) of pristine PP was higher than in the PP-*g*-MMA (25%) and (PP-*g*-MMA)-*g*-NVI (19.5/6%). The grafted films exhibited a multi-step decomposition, as is shown in the thermogram of PP-*g*-MMA (25%), where there were two decomposition stages, while in the (PP-*g*-MMA)-*g*-NVI (19.5/6%), there were three decomposition stages (Table 2). In conclusion, the study showed that pristine PP had better thermal stability in comparison to grafted films, but this difference is merely informative because grafted films worked well at load and release temperatures.

Table 2. Results of TGA (weight loss, decomposition temperature, residue) analyses of pristine non-irradiated PP film and films after single and binary grafting.

Film	10 wt% Loss (°C)	Td (°C)	Residue 800 °C (%)
PP	411.7	458.5	8.30
PP-*g*-MMA (25%)	358.5	370.3, 453.6	4.1
(PP-*g*-MMA)-*g*-NVI (19.5/6%)	303.4	291.32, 406.9, 455.9	2.3

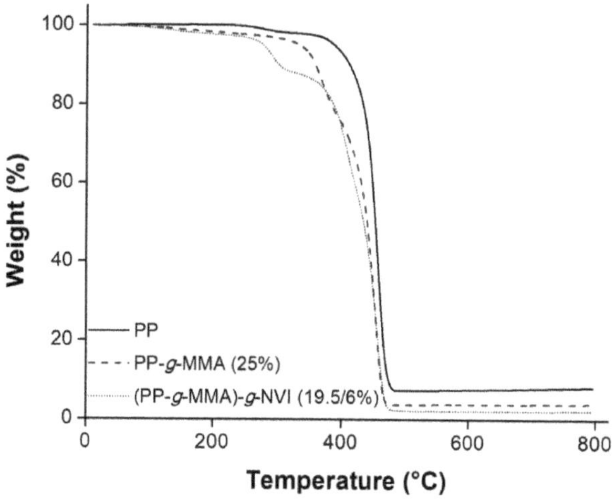

Figure 6. Thermogram runs under nitrogen atmosphere at 800 °C and heating rate 10 °C min^{-1}.

Multiple decomposition stages in the grafted films suggest a localized polymer composition forming a multilayer material. These zones are core, internal layer, and surface; nonetheless, the PP zone is in the nucleus and preserves its inherent thermal properties. This characteristic is found in a surface-grafting polymer [30].

2.5. Swelling and Critical pH

The unmodified and grafted PP films were put to swelling tests by immersion in different solvents for 24 h to determine their behavior in liquid mediums. Solvents were chosen according to their dielectric constant (ε), in order of polarity: water (78.5), dimethyl formaldehyde (DMF) (38.25), methanol (32.6), n-propanol (20.1), toluene (2.38), and n-hexane (1.89). The non-polar solvents n-hexane and toluene swelled the films more than the other solvents. One parameter for choosing a suitable solvent to graft NVI is its capability of swelling the film PP-*g*-MMA and as was expected, the highest swellings were achieved in toluene because both PP-*g*-MMA and (PP-*g*-MMA)-*g*-NVI have non-polar groups in their chains. These preliminary tests also helped to determine the viability of water for the load/release assays (Figure 7), although the water had the lowest swelling percentage followed by DMF, methanol, and n-propanol.

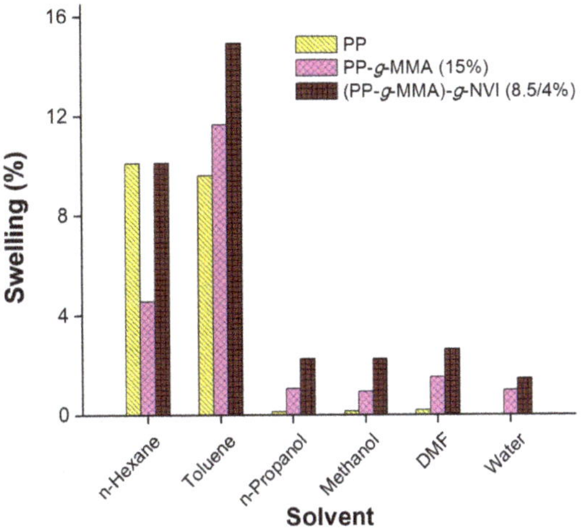

Figure 7. Swelling of the films in different solvents.

One of the characteristics of NVI-containing polymers is their pH response. This property is conferred by grafted PNVI, an electron donor polyelectrolyte (i.e., a Lewis base) that shrinks or expands by varying pH, as was verified with the study of swelling in a pH range of 3 to 11. In this case, two films of (PP-*g*-MMA)-*g*-NVI with different compositions were analyzed, as shown in Figure 8. The behavior was similar in both films since they were hydrophilic at acid pH and hydrophobic at alkaline pH, with inflection points at pH 6.9 for (PP-*g*-MMA)-*g*-NVI (20/6%) and pH 7.3 for (PP-*g*-MMA)-*g*-NVI (34/6.5%), respectively. The most significant difference was the higher swelling in the film with more PMMA grafted, so it is inferred that this polyacrylate conferred a more hydrophilic behavior to the film. Although swelling between 1 and 4% would seem low, the thickness of the films was 18 mm, and samples were above 250 mg, so even small weight changes, such as 0.1 mg, were detected. Furthermore, the swelling percentage was enough to load and release the vancomycin quantitatively (see Section 2.7), because in this case, the swelling occurred exclusively on the surface. Thus, it is concluded that the surface of the grafted films is moderately hydrophilic and pH-responsive can uptake water and molecules between their chains.

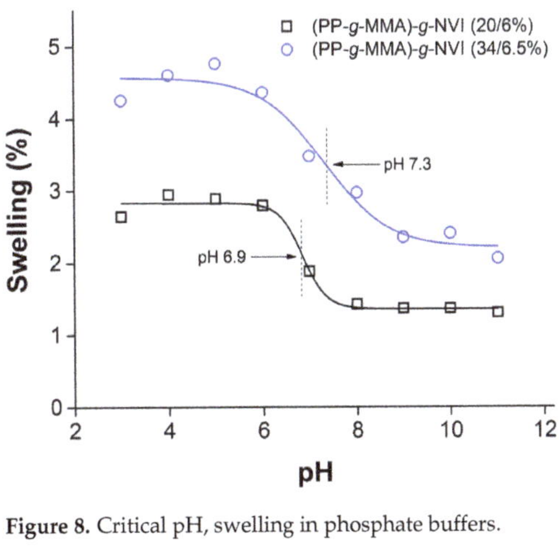

Figure 8. Critical pH, swelling in phosphate buffers.

2.6. Contact Angle

Once MMA is grafted onto the surface, the wettability was an important parameter because the acrylic chains are moderately hydrophilic, as swelling experiments showed (see Section 2.5). Thereby, the contact angle before and after graft can be interpreted in terms of surface energy, increasing in this case, which means that grafted films were more hydrophilic than the pristine PP films, even when the grafted chains had a hydrophobic moiety, the amorphous acrylic chains increased the wettability (Figure 9). The contact angles determined by the drop of water on the pristine PP film were 87.5 ± 0.9° at 1 min and 84.4 ± 0.9° at 5 min, and the highest compared with the grafted films, in which case the contact angle decreased as the MMA graft percentage increased. The change from hydrophobic to hydrophilic occurred since the first sample with the lowest graft that was PP-*g*-MMA (10%), with angles of 76 ± 1.7° at 1 min and 71.2 ± 2.3° at 5 min and decreased consecutively up to the last sample, PP-*g*-MMA (50%), exhibiting a contact angle of 64.5 ± 4.4° at 1 min and 59.4 ± 4.8° at 5 min, being the lowest angle and therefore the most hydrophilic.

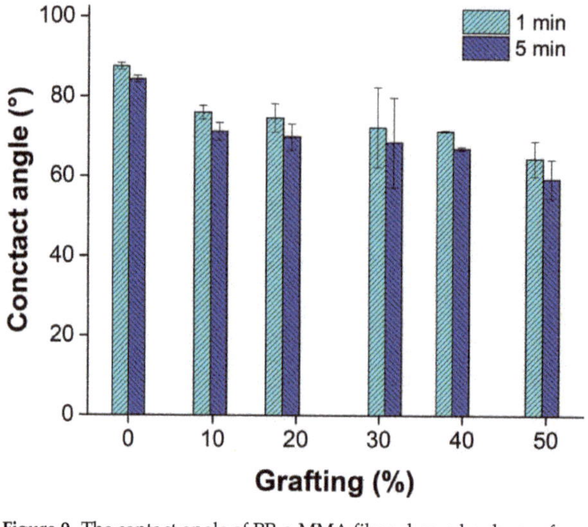

Figure 9. The contact angle of PP-*g*-MMA films showed a change from hydrophobic to hydrophilic.

2.7. Load and Release of Vancomycin

Drug loading was performed using the sample (PP-*g*-MMA)-*g*-NVI (23/5.5%) in an aqueous solution of vancomycin hydrochloride [2 mg mL^{-1}]. The main reasons for choosing water as solvent were its capability to dissolve the drug and its innocuousness. These properties are more relevant than the limited capability to swell the film [31].

Once the vancomycin was loaded, the release was in a controlled pH [32], in both acid (pH 4–6) and alkaline (pH 8–9) buffer medium to determine the release rate in simulated physiological conditions, considering the pH responsiveness of PNVI grafted [33], particularly at a pH close to that of the skin [34,35].

At neutral pH, the release rate was the highest, and the maximum reached within the first 2 h, which was 109.5 ± 4.3 µg cm^{-2} (Figure 10a), which was around 83% of the total loaded vancomycin. While in alkaline and acid pH, the release rate and the amount of vancomycin released decreased considerably (Figure 10b). It was found that even after 48 h, the release value at pH 8 was only 47.1 ± 2.5 µg cm^{-2}. In all buffers, unlike neutral pH, release rates were slower, and the maximum concentration was not reached at 48 h. When the kinetics were checked, it was found that they followed a different path than at neutral pH, given that under neutral conditions the release rate was around three times and in a fraction of the time (2 h). Therefore, there is an expedited diffusion at pH 7 and a prolonged release at pH 4, 5, 8, and 9.

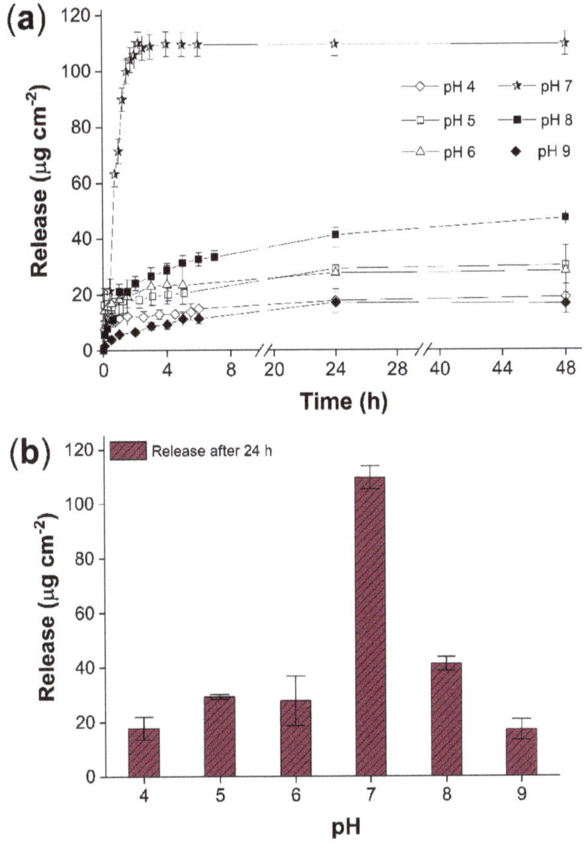

Figure 10. (PP-*g*-MMA)-*g*-NVI (23/5.5%): (**a**) Release of vancomycin in phosphate buffers 0.1 M at different pH and (**b**) release reached at 24 h.

3. Discussion

The difficulty of grafting NVI directly onto pristine PP was overcome by the easiness of grafting MMA. The difference among the reactivity of MMA and NVI is attributed to electronic effects, solvent, stability of intermediates (during chain reaction), and homopolymerization rate (reaction in competition during the copolymerization). The surface of PP grafted films exhibited changes in their hydrophilic/hydrophobic behavior and became able to load and release vancomycin in different pH conditions. These grafted materials are not limited to the vancomycin since the drug loading by swelling is a general method for delivery with many bioactive principles [36].

SEM analysis suggested that the graft of both copolymers took place on the surface of PP-*g*-MMA and (PP-*g*-MMA)-*g*-NVI films, which was supported by observing amorphous layers and by the fact that there was no significant change in the film's size. This information was consistent with infrared, where it was possible to observe the corresponding bands of the acrylic chains on the surface. While in XPS, elemental analysis detected the presence of O and N from the graft. Finally, TGA confirmed that grafted PP was thermally stable [37], even when the amorphous grafted binary and single copolymers decomposed earlier than pristine PP.

Regarding the swelling properties, grafted films showed a slight swelling on the polar solvents, which is a significant difference with pristine PP because the polar solvent absorbs between the alkyl backbones. The water drop contact angle showed that the surface of the PP-*g*-MMA was wet consistently as the contact angle decreased while the grafting degree increased. Overall, the swelling and contact angle results suggest an increase in the surface energy of the grafted films and are suitable to use in drug release systems [38].

The release rate of vancomycin on (PP-*g*-MMA)-*g*-NVI (23/5.5%) was studied, finding that this molecule, with several amines and one carboxylic group, at an acid or alkaline pH formed strong H interactions with the grafted chains, prolonging the release and decreasing the release rate [39]. At these conditions, the release could be conducted by a simultaneous equilibrium, with an interchange of ions from the system and medium [40]. However, the critical pH was reached at neutral conditions, where there was no excess of H^+ or OH^- ions, which eased the release due to the strong interaction among medium and grafted chains [41,42], yielding 83% of the total drug released in the first hours.

4. Materials and Methods

4.1. Materials

High-density polypropylene films (0.18 cm thickness) were from Goodfellow (Huntingdon, Cambridgeshire, UK). Vancomycin hydrochloride, MMA (99%), and NVI (99%) were purchased from Aldrich Chemical Co. (Saint. Louis, Missouri, USA), and monomers were purified by vacuum distillation. Boric acid, citric acid, trisodium orthophosphate, and solvents (including double distilled water) were acquired from Baker (Mexico City, Mexico).

4.2. Grafting Method

MMA was grafted by oxidative pre-irradiation method, and NVI was grafted by direct method [43], in both cases using a gamma-rays source of ^{60}Co Gammabeam 651-PT (UNAM, Mexico City, Mexico) at a dose rate of 8.4 kGy h^{-1}, the methods are described in detail in the next Sections 4.2.1 and 4.2.2.

4.2.1. Grafting Polymerization of MMA Using the Oxidative Pre-Irradiation Method

PP films of 3cm × 2cm × 0.18 cm (width, length, and thickness) were weighed (around 250 mg) and placed into open glass ampoules and exposed to gamma irradiation in the presence of air. Afterward, the solutions of MMA were prepared in methanol as the solvent at different concentrations (Table 3) and then added (5 mL) into the glass ampoules with the pre-irradiated PP film. The ampoules were degassed by freezing and thawing cycles with liquid nitrogen followed by a purged in the vacuum line; subsequently, the ampoules were sealed at vacuum. Then, the polymerization was initiated by heating in a water bath at different times and temperatures. Once completed the reaction time, the ampoules were open, and the films were rinsed in a water/ethanol mixture 50/50 vol% under constant stirring for 24 h. Finally, the samples were dried in a vacuum oven at 60 °C for 24 h. The grafting percentage was calculated according to Equation (1), using the weight of pristine (W_0) and grafted (W_g) film.

$$\text{Grafting (\%)} = 100[(W_g - W_0)/W_0] \quad (1)$$

Table 3. Reaction conditions of grafting polymerization of MMA by pre-irradiation oxidative and grafting degree.

Experiment	Dose (kGy)	Time (h)	Temperature (°C)	Concentration (vol%)	Grafting (%)
Dose	5	16	65	20	10
Dose	10	16	65	20	18
Dose	15	16	65	20	31
Dose	20	16	65	20	39

Table 3. *Cont.*

Experiment	Dose (kGy)	Time (h)	Temperature (°C)	Concentration (vol%)	Grafting (%)
Dose	25	16	65	20	49.5
Time	5	5	65	30	6
Time	5	10	65	30	12
Time	5	16	65	30	18
Time	5	24	65	30	23
Time	5	26	65	30	31
Concentration	15	16	70	20	8.5
Concentration	15	16	70	40	19.5
Concentration	15	16	70	60	34
Concentration	15	16	70	80	29.5
Concentration	15	16	70	100	77.5
Temperature	5	16	60	30	15.5
Temperature	5	16	70	30	17.5
Temperature	5	16	80	30	25
Temperature	5	16	90	30	23.5

4.2.2. Grafting of NVI on PP-*g*-MMA by Direct Method

PP-*g*-MMA films with different MMA grafting degree (Table 4) were weighed and placed into glass ampoules containing 6 mL of NVI in toluene (50 vol%). Then, oxygen was removed from the ampoules with freezing and thawing cycles (see Section 4.2.1), then the ampoules were sealed with a blowtorch and irradiated with an absorbed dose of 15 kGy at room temperature. Finally, the ampoules were open, and the grafted films were rinsed with methanol and dried into a vacuum oven at 60 °C for 24 h. The weight of films was recorded to calculate the grafting degree according to Equation (1).

Table 4. (PP-*g*-MMA)-*g*-NVI, reaction conditions of NVI grafting by the direct method.

MMA Grafting (%)	Dose (kGy)	Concentration (vol%)	Total Grafting MMA/NVI (%)
8.5	15	50	8.5/4
19.5	15	50	19.5/6
23	15	50	23/5.5
34	15	50	34/6.5
77.5	15	50	77.5/5

4.3. *Swelling Experiments*

The samples were placed in different solvents until they reached the limit swelling (maximum 24 h) at room temperature (around 25 °C). Excess solvent was removed with an absorbent paper. The solvents used for swelling were water, methanol, n-propanol, DMF, and n-hexane. The swelling percentage (%) was calculated according to Equation (2):

$$\text{Swelling (\%)} = 100[(W_s - W_d)/W_d] \quad (2)$$

where W_s and W_d are the weights of swollen and dried films, respectively.

The (PP-*g*-MMA)-*g*-NVI (23/5.5%) film was employed to determine the critical pH. The sample was put inside different phosphate buffers (pH 4–9) for 24 h to record the weight. After each measurement, the sample was rinsed with double distilled water and submerged in the next buffer. Equation (2) was also applied to calculate the swelling at different pH.

4.4. *Load and Release of Vancomycin*

Vancomycin was loaded to the (PP-*g*-MMA)-*g*-NVI (23/5.5%) film. A fresh dissolution of vancomycin hydrochloride [2 mg mL^{-1}] was prepared with double distilled water and poured into a vial containing the grafted film. The vial was stored in refrigeration at 4 °C for 48 h; then, the film was taken out, dried, and stored at room temperature (around 25 °C). The amount of vancomycin loaded was calculated by measuring the vancomycin released under sonication and replacing the solvent (double distilled water) until reaching absorbances close to 0, these absorbances represent the total vancomycin concentration, resulting 132.2 ± 0.8 μg cm^{-2}.

Release experiments were performed using the same film which was (PP-*g*-MMA)-*g*-NVI (23/5.5%) in 4 mL of sodium phosphate buffer (pH 4–9), 0.1 M, and at 37 °C. The releasing was monitored at different times, recording absorbances by spectrophotometry at 280 nm [44].

4.5. Instrumental

Infrared spectroscopy attenuated total reflection (FTIR-ATR) spectra of dry pristine and modified films were analyzed using a Perkin–Elmer Spectrum 100 spectrometer (Norwalk, CT, USA) with 16 scans.

X-ray photoelectron spectroscopy was performed in an ultra-high vacuum (UHV) system Scanning XPS microprobe PHI 5000 Versa Probe II (Chanhassen, MN, USA), with an excitation source of Al Kα monochromatic, energy 1486.6 eV, 100 μm beam diameter, and with a Multi-Channel Detector (MCD). The XPS spectra were obtained at 45° to the normal surface in pass energy mode (CAE) E0 = 117.40 and 11.75 eV. Peak positions were calibrated to Ag 3d5/2 photopeak at 368.20 eV, having a full width at half maximum of 0.56 eV, and the energy scale corrected using the C 1s peak brought to 285.0 eV.

A Kruss DSA 100 drop shape analyzer (Matthews, NC, USA) was employed to measure water droplet contact angle at 1 and 5 min in triplicates.

Scanning electron microscope (SEM) images were acquired by the Zeiss Evo LS15 instrument (Jena, Germany). Small pieces (1 cm length) of grafted samples were cut and directly analyzed under a high vacuum without using any coating.

Thermogravimetric analysis (TGA) data of weight loss and decomposition of pristine and modified films (around 10 mg) were analyzed under a heating rate of 10 °C min^{-1} and run from 20 to 800 °C in a TGA instrument Q50 TA Instruments (New Castle, DE, USA).

Ultraviolet-visible (UV-vis) spectrophotometer model Agilent 8453 (Waldbronn, Germany) was utilized to analyze the release of vancomycin at 280 nm, using quartz cuvettes (1 cm length).

5. Conclusions

Radiation-grafting was a convenient method to modify the surface of PP films. In the first step, the PP-*g*-MMA was obtained by the pre-irradiation oxidative method, and in the second step, the final material (PP-*g*-MMA)-*g*-NVI was achieved by simultaneous irradiation. The grafting of NVI endowed the surface with pH responsiveness and the chains were able to load vancomycin hydrochloride in aqueous dissolution (2 mg mL^{-1}). The release of vancomycin was pH dependent with a higher rate at pH 7 and more controlled release at non-neutral pH; the maximum amount of drug released at buffer pH 7 was 109.5 ± 4.3 μg cm^{-2} after 48 h. These findings suggest an active interaction in the equilibrium of the NVI chains-vancomycin-release medium. This type of superficial modification onto a non-reactive thermoplastic, such as the PP, provides a route to get more sophisticated materials and devices.

Author Contributions: Conceptualization, F.L.-S. and E.B.; methodology, F.L.-S.; software, F.L.-S. and J.E.L.-B.; validation, J.E.L.-B.; formal analysis, G.G.F.-R. and S.G.-D.; investigation, F.L.-S.; resources, E.B.; data curation, S.G.-D.; writing—original draft preparation, F.L.-S.; writing—review and editing, J.E.L.-B., G.G.F.-R., and S.G.-D.; visualization, G.G.F.-R. and S.G.-D.; supervision, E.B.; project administration, E.B.; funding acquisition, E.B. All authors have read and agreed to the published version of the manuscript.

Funding: This research was funded by DGAPA-UNAM, grant number IN202320 (Mexico).

Institutional Review Board Statement: Not applicable.

Informed Consent Statement: Not applicable.

Data Availability Statement: Not applicable.

Acknowledgments: FLS CVU 409872 thanks CONACyT. The authors would like to thank B. Leal from ICN-UNAM for his technical assistance in the irradiation experiments, to C. Flores from IIM-UNAM for his technical assistance in SEM, and to L. Huerta-Arcos from IIM-UNAM for his technical assistance in XPS.

Conflicts of Interest: The authors declare no conflict of interest.

References

1. Maddah, H.A. Polypropylene as a promising plastic: A review. *Am. J. Polym. Sci.* **2016**, *6*, 1–11. [CrossRef]
2. Minko, S. Grafting on solid surfaces: "grafting to" and "grafting from" methods. In *Polymer Surfaces and Interfaces: Characterization, Modification and Applications*; Stamm, M., Ed.; Springer: Berlin/Heidelberg, Germany, 2008; pp. 215–234. ISBN 978-3-540-73865-7.
3. Zhang, S.; Yu, P.; Zhang, Y.; Ma, Z.; Teng, K.; Hu, X.; Lu, L.; Zhang, Y.; Zhao, Y.; An, Q. Remarkably boosted molecular delivery triggered by combined thermal and flexoelectrical field dual stimuli. *ChemistrySelect* **2020**, *5*, 6715–6722. [CrossRef]
4. Contreras-García, A.; Bucio, E.; Concheiro, A.; Alvarez-Lorenzo, C. Polypropylene grafted with NIPAAm and APMA for creating hemocompatible surfaces that load/elute nalidixic acid. *React. Funct. Polym.* **2010**, *70*, 836–842. [CrossRef]
5. Flores-Rojas, G.G.; López-Saucedo, F.; Bucio, E.; Isoshima, T. Covalent immobilization of lysozyme in silicone rubber modified by easy chemical grafting. *MRS Commun.* **2017**, *7*, 904–912. [CrossRef]
6. Naguib, H.F.; Aly, R.O.; Sabaa, M.W.; Mokhtar, S.M. Gamma radiation induced graft copolymerization of vinylimidazole-acrylic acid onto polypropylene films. *Polym. Test.* **2003**, *22*, 825–830. [CrossRef]
7. Chung, T.C.; Rhubright, D.; Jiang, G.J. Synthesis of polypropylene-graft-poly(methyl methacrylate) copolymers by the borane approach. *Macromolecules* **1993**, *26*, 3467–3471. [CrossRef]
8. Kudryavtsev, V.N.; Kabanov, V.Y.; Yanul, N.A.; Kedik, S.A. Polypropylene modification by the radiation graft polymerization of N-vinylcaprolactam. *High Energy Chem.* **2003**, *37*, 382–388. [CrossRef]
9. Xu, X.; Zhang, L.; Zhou, J.; Wang, J.; Yin, J.; Qiao, J. Thermal behavior of polypropylene-g-glycidyl methacrylate prepared by melt grafting. *J. Macromol. Sci. Part B* **2015**, *54*, 32–44. [CrossRef]
10. Anastas, P.T.; Kirchhoff, M.M. Origins, current status, and future challenges of green chemistry. *Acc. Chem. Res.* **2002**, *35*, 686–694. [CrossRef]
11. Bucio, E.; Burillo, G. Radiation-induced grafting of sensitive polymers. *J. Radioanal. Nucl. Chem.* **2009**, *280*, 239–243. [CrossRef]
12. Saxena, S.; Ray, A.R.; Kapil, A.; Pavon-Djavid, G.; Letourneur, D.; Gupta, B.; Meddahi-Pellé, A. Development of a new polypropylene-based suture: Plasma grafting, surface treatment, characterization, and biocompatibility studies. *Macromol. Biosci.* **2011**, *11*, 373–382. [CrossRef] [PubMed]
13. Kumari, M.; Gupta, B.; Ikram, S. Characterization of N-isopropyl acrylamide/acrylic acid grafted polypropylene nonwoven fabric developed by radiation-induced graft polymerization. *Radiat. Phys. Chem.* **2012**, *81*, 1729–1735. [CrossRef]
14. Strohmaier, S.; Zwierzchowski, G. Comparison of (60)Co and (192)Ir sources in HDR brachytherapy. *J. Contemp. Brachytherapy* **2011**, *3*, 199–208. [CrossRef] [PubMed]
15. Clark, D.E. Peroxides and peroxide-forming compounds. *Chem. Health Saf.* **2001**, *8*, 12–22. [CrossRef]
16. Wypych, G. PMMA polymethylmethacrylate. In *Handbook of Polymers*, 2nd ed.; Wypych, G.B.T.-H., Ed.; ChemTec Publishing: Toronto, ON, Canada, 2016; pp. 467–471. ISBN 978-1-895198-92-8.
17. Fink, J.K. Poly(vinylimidazole). In *Handbook of Engineering and Specialty Thermoplastics*; Fink, J.K., Ed.; John Wiley & Sons: Hoboken, NJ, USA, 2011; pp. 251–291.
18. Camacho-Cruz, L.A.; Velazco-Medel, M.A.; Parra-Delgado, H.; Bucio, E. Functionalization of cotton gauzes with poly(N-vinylimidazole) and quaternized poly(N-vinylimidazole) with gamma radiation to produce medical devices with pH-buffering and antimicrobial properties. *Cellulose* **2021**, *28*, 3279–3294. [CrossRef]
19. de Magalhães Pereira, M.; Lambert Oréfice, R.; Sander Mansur, H.; Paz Lopes, M.T.; de Marco Turchetti-Maia, R.M.; Vasconcelos, A.C. Preparation and biocompatibility of poly (methyl methacrylate) reinforced with bioactive particles. *Mater. Res.* **2003**, *6*, 311–315. [CrossRef]
20. Wang, B.; Liu, H.-J.; Chen, Y. A biocompatible poly(N-vinylimidazole)-dot with both strong luminescence and good catalytic activity. *RSC Adv.* **2016**, *6*, 2141–2148. [CrossRef]
21. Szász, M.; Hajdú, M.; Pesti, N.; Domahidy, M.; Kristóf, K.; Zahár, Á.; Nagy, K.; Szabó, D. In vitro efficiency of vancomycin containing experimental drug delivery systems. *Acta Microbiol. Immunol. Hung.* **2013**, *60*, 461–468. [CrossRef] [PubMed]
22. Zavala-Lagunes, E.; Ruiz, J.C.; Varca, G.H.C.; Bucio, E. Synthesis and characterization of stimuli-responsive polypropylene containing N-vinylcaprolactam and N-vinylimidazole obtained by ionizing radiation. *Mater. Sci. Eng. C* **2016**, *67*, 353–361. [CrossRef] [PubMed]
23. Kawamura, Y. Effects of gamma irradiation on polyethylene, polypropylene, and polystyrene. In *Irradiation of Food and Packaging*; Komolprasert, V., Morehouse, K.M., Eds.; ACS Symposium Series; American Chemical Society: Washington, DC, USA, 2004; Volume 875, pp. 262–276, ISBN 0-8412-3869-3.

24. Duan, G.; Zhang, C.; Li, A.; Yang, X.; Lu, L.; Wang, X. Preparation and characterization of mesoporous zirconia made by using a poly (methyl methacrylate) template. *Nanoscale Res. Lett.* **2008**, *3*, 118–122. [CrossRef] [PubMed]
25. López-Saucedo, F.; Flores-Rojas, G.G.; Bucio, E.; Alvarez-Lorenzo, C.; Concheiro, A.; González-Antonio, O. Achieving antimicrobial activity through poly(N-methylvinylimidazolium) iodide brushes on binary-grafted polypropylene suture threads. *MRS Commun.* **2017**, *7*, 938–946. [CrossRef]
26. Ramasamy, R. Vibrational spectroscopic studies of imidazole. *Armen. J. Phys.* **2015**, *8*, 51–55.
27. Kuba, A.G.; Smolin, Y.Y.; Soroush, M.; Lau, K.K.S. Synthesis and integration of poly(1-vinylimidazole) polymer electrolyte in dye sensitized solar cells by initiated chemical vapor deposition. *Chem. Eng. Sci.* **2016**, *154*, 136–142. [CrossRef]
28. Lannon, J.M.; Meng, Q. Analysis of a poly(propylene)(PP) Homopolymer by XPS. *Surf. Sci. Spectra* **1999**, *6*, 79–82. [CrossRef]
29. Pignataro, B.; Fragalà, M.E.; Puglisi, O. AFM and XPS study of ion bombarded poly(methyl methacrylate). *Nucl. Instrum. Methods Phys. Res. Sect. B* **1997**, *131*, 141–148. [CrossRef]
30. Abudonia, K.S.; Saad, G.R.; Naguib, H.F.; Eweis, M.; Zahran, D.; Elsabee, M.Z. Surface modification of polypropylene film by grafting with vinyl monomers for the attachment of chitosan. *J. Polym. Res.* **2018**, *25*, 125. [CrossRef]
31. Alvarez-Lorenzo, C.; Bucio, E.; Burillo, G.; Concheiro, A. Medical devices modified at the surface by γ-ray grafting for drug loading and delivery. *Expert Opin. Drug Deliv.* **2010**, *7*, 173–185. [CrossRef]
32. Johnson, J.L.H.; Yalkowsky, S.H. Reformulation of a new vancomycin analog: An example of the importance of buffer species and strength. *AAPS PharmSciTech* **2006**, *7*, E33–E37. [CrossRef] [PubMed]
33. Obando-Mora, Á.; Acevedo-Gutiérrez, C.; Pérez-Cinencio, J.; Sánchez-Garzón, F.; Bucio, E. Synthesis of a pH- and thermo-responsive binary copolymer poly(N-vinylimidazole-co-N-vinylcaprolactam) grafted onto silicone films. *Coatings* **2015**, *5*, 758–770. [CrossRef]
34. Wagner, H. pH profiles in human skin: Influence of two in vitro test systems for drug delivery testing. *Eur. J. Pharm. Biopharm.* **2003**, *55*, 57–65. [CrossRef]
35. Hendi, A.; Umair Hassan, M.; Elsherif, M.; Alqattan, B.; Park, S.; Yetisen, A.K.; Butt, H. Healthcare applications of pH-sensitive hydrogel-based devices: A review. *Int. J. Nanomed.* **2020**, *15*, 3887–3901. [CrossRef]
36. Siepmann, J.; Siepmann, F. Swelling controlled drug delivery systems. In *Fundamentals and Applications of Controlled Release Drug Delivery*; Siepmann, J., Siegel, R., Rathbone, M., Eds.; Springer: Boston, MA, USA, 2012; pp. 153–170.
37. Quinelato, R.R.; Albitres, G.A.; Mariano, D.M.; Freitas, D.F.; Mendes, L.C.; Rodrigues, D.C.; Filho, M.F. Influence of polycaprolactone and titanium phosphate in the composites based upon recycled polypropylene. *J. Thermoplast. Compos. Mater.* **2020**, 1–21. [CrossRef]
38. Nasef, M.M.; Gupta, B.; Shameli, K.; Verma, C.; Ali, R.R.; Ting, T.M. Engineered bioactive polymeric surfaces by radiation induced graft copolymerization: Strategies and applications. *Polymers* **2021**, *13*, 3102. [CrossRef]
39. Ahmad Nor, Y.; Zhang, H.; Purwajanti, S.; Song, H.; Meka, A.K.; Wang, Y.; Mitter, N.; Mahony, D.; Yu, C. Hollow mesoporous carbon nanocarriers for vancomycin delivery: Understanding the structure–release relationship for prolonged antibacterial performance. *J. Mater. Chem. B* **2016**, *4*, 7014–7021. [CrossRef] [PubMed]
40. Zhuo, S.; Zhang, F.; Yu, J.; Zhang, X.; Yang, G.; Liu, X. pH-Sensitive biomaterials for drug delivery. *Molecules* **2020**, *25*, 5649. [CrossRef]
41. Shan, S.; Herschlag, D. The change in hydrogen bond strength accompanying charge rearrangement: Implications for enzymatic catalysis. *Proc. Natl. Acad. Sci. USA* **1996**, *93*, 14474–14479. [CrossRef] [PubMed]
42. Wood, J.L. pH-controlled hydrogen-bonding. *Biochem. J.* **1974**, *143*, 775–777. [CrossRef] [PubMed]
43. López-Saucedo, F.; Alvarez-Lorenzo, C.; Concheiro, A.; Bucio, E. Radiation-grafting of vinyl monomers separately onto polypropylene monofilament sutures. *Radiat. Phys. Chem.* **2017**, *132*, 1–7. [CrossRef]
44. Li, B.; Brown, K.V.; Wenke, J.C.; Guelcher, S.A. Sustained release of vancomycin from polyurethane scaffolds inhibits infection of bone wounds in a rat femoral segmental defect model. *J. Control Release* **2010**, *145*, 221–230. [CrossRef]

International Journal of *Molecular Sciences*

MDPI

Article

Heparin Enriched-WPI Coating on Ti6Al4V Increases Hydrophilicity and Improves Proliferation and Differentiation of Human Bone Marrow Stromal Cells

Davide Facchetti [1,2,*], Ute Hempel [3], Laurine Martocq [1], Alan M. Smith [4], Andrey Koptyug [5], Roman A. Surmenev [6,7], Maria A. Surmeneva [6,7] and Timothy E. L. Douglas [1,8,*]

1 Engineering Department, Lancaster University, Lancaster LA1 4YW, UK; l.martocq@lancaster.ac.uk
2 Department of Chemistry, Molecular Sciences Research Hub, Imperial College London, London W12 0BZ, UK
3 Institute of Physiological Chemistry, Technische Universität Dresden, 01307 Dresden, Germany; hempel-u@msx.tu-dresden.de
4 Department of Pharmacy, School of Applied Sciences, University of Huddersfield, Huddersfield HD1 3DR, UK; a.m.smith@hud.ac.uk
5 Department of Quality Technology, Mechanical Engineering & Mathematics, Mid Sweden University, 831 25 Ostersund, Sweden; andrey.koptyug@miun.se
6 Physical Materials Science and Composite Materials Centre, Research School of Chemistry & Applied Biomedical Sciences, National Research Tomsk Polytechnic University, 634050 Tomsk, Russia; rsurmenev@mail.ru (R.A.S.); surmenevamaria@mail.ru (M.A.S.)
7 Piezo- and Magnetoelectric Materials Research & Development Centre, Research School of Chemistry & Applied Biomedical Sciences, National Research Tomsk Polytechnic University, 634050 Tomsk, Russia
8 Materials Science Institute (MSI), Lancaster University, Lancaster, UK
* Correspondence: d.facchetti21@imperial.ac.uk (D.F.); t.douglas@lancaster.ac.uk (T.E.L.D.)

Citation: Facchetti, D.; Hempel, U.; Martocq, L.; Smith, A.M.; Koptyug, A.; Surmenev, R.A.; Surmeneva, M.A.; Douglas, T.E.L. Heparin Enriched-WPI Coating on Ti6Al4V Increases Hydrophilicity and Improves Proliferation and Differentiation of Human Bone Marrow Stromal Cells. *Int. J. Mol. Sci.* **2022**, *23*, 139. https://doi.org/10.3390/ijms23010139

Academic Editor: Yury A. Skorik

Received: 8 November 2021
Accepted: 19 December 2021
Published: 23 December 2021

Abstract: Titanium alloy (Ti6Al4V) is one of the most prominent biomaterials for bone contact because of its ability to bear mechanical loading and resist corrosion. The success of Ti6Al4V implants depends on bone formation on the implant surface. Hence, implant coatings which promote adhesion, proliferation and differentiation of bone-forming cells are desirable. One coating strategy is by adsorption of biomacromolecules. In this study, Ti6Al4V substrates produced by additive manufacturing (AM) were coated with whey protein isolate (WPI) fibrils, obtained at pH 2, and heparin or tinzaparin (a low molecular weight heparin LMWH) in order to improve the proliferation and differentiation of bone-forming cells. WPI fibrils proved to be an excellent support for the growth of human bone marrow stromal cells (hBMSC). Indeed, WPI fibrils were resistant to sterilization and were stable during storage. This WPI-heparin-enriched coating, especially the LMWH, enhanced the differentiation of hBMSC by increasing tissue non-specific alkaline phosphatase (TNAP) activity. Finally, the coating increased the hydrophilicity of the material. The results confirmed that WPI fibrils are an excellent biomaterial which can be used for biomedical coatings, as they are easily modifiable and resistant to heat treatments. Indeed, the already known positive effect on osteogenic integration of WPI-only coated substrates has been further enhanced by a simple adsorption procedure.

Keywords: WPI fibrils; Ti6Al4V; additive manufacturing; osseointegration; heparin; tinzaparin; osteoblast differentiation; coating; enriched

1. Introduction

Ti6Al4V is a well-known biomaterial for orthopaedic implants widely used due to its high specific strength and high corrosion resistance. One of the advantages of this material is that it can be produced using AM techniques, allowing the generation of implants with complicated shapes while keeping production costs reasonably low [1]. The stability and long-term success of an implant depends on its ability to osseointegrate after its stable fixation to the surrounding bone. One of the known methods of facilitating this process is by modifying the surface of the implant with a coating. Collagen is the most studied protein for this type of application, especially in its fibrillar form [2–4]. The advantages of using protein fibrils lie in their high surface/volume ratio, which facilitates their adsorption onto the substrate, and the possibility of incorporating other molecules on their surface such as phenols [4,5] and marine polysaccharides [5]. Heparin is a highly sulfated glycosaminoglycan (GAGs) widely used as a coating for implants due to its ability to accumulate and release growth factors (in the form of a crosslinked hydrogel), improve blood compatibility by reducing the inflammatory and coagulative response, and facilitate bone cells' adhesion, growth and osteogenic differentiation [6].

Previous studies demonstrated that WPI increases cell proliferation and osteogenic differentiation in soluble form both in cell culture medium [7] and as a hydrogel [8]. Additionally, this compound shows antibacterial effects [9]. WPI consists of more than 75% β-lactoglobulin (β-lg) and under conditions of acid hydrolysis (pH < 3, T > 80 °C) it forms uniform fibrils that are several micrometers long and a few nanometers thick. The fibrillation process occurs thanks to peptides that self-assemble into amyloid-like structures [10]. In the fibrillar form, they have supported the attachment, spreading and differentiating of human bone marrow stromal cells (hBMSC), which are cells with great clinical relevance for bone regeneration. WPI fibrils also proved to be resistant to autoclaving, making them an excellent option for biomedical coatings which need to be sterile [11].

In this study, we demonstrated the possibility of coating Ti6Al4V with heparin-enriched WPI fibrils in order to combine the positive effects of all the components, resulting in a system that is easily sterilizable and can lead to the production, via AM, of complex-shaped prostheses that are highly biocompatible. In addition, we evaluated the influence of the molecular weight of heparin on its ability to modulate the growth of hBMSC by comparing the effects of sodium heparin (MW: ~20,000 Da) and tinzaparin (LMWH; MW: ~8000 Da). For this purpose, we studied cell proliferation via metabolic activity by a metabolic assay and osteogenic differentiation via the activity of the tissue non-specific alkaline phosphatase enzyme (TNAP) at days, in culture after two and 11 days.

2. Results and Discussion

2.1. Fibrils Characterization

The fluorescence emission of Thioflavin T (ThT) is an indication of the presence of fibrils since the molecule binds to β-sheets that are present in the protein [12]. The internal structure of the fibrils consists largely of crossed β-sheets arranged perpendicularly to the main axis of the fibrils. Figure 1 shows the results of the ThT assay made to assess if the fibrillation process has been successful. As clearly shown by the higher emission of the acid-hydrolysis treated sample (WPI fibrils) compared to the untreated sample (WPI), the reaction led to an increase of the fluorescence. A signal was also present in the untreated sample. This is certainly due to the presence of a small number of β-sheets prior to heat treatment, in accordance with Akkermans et al. [10]. Thereby, given the high increase in β-sheet numbers after heat treatment, the fibrillation reaction was successful.

The effective presence of the fibrils has been confirmed by the scanning electron microscopy (SEM) images shown in Figure 2. As indicated by the black arrows, on the coated not sterile (NS) (Figure 2a,b) and sterilized (S) (Figure 2c,d) samples there were long and thin fibrils well dispersed onto the analyzed area. The morphology appeared similar to the structure expected after fibrillation at pH 2 [13]. Moreover, the images further confirmed the ability of these type of fibrils to withstand washing and sterilization, as shown by previous work, in which glass substrates were used [14]. However, the presence of tinzaparin and heparin could not be detected by SEM. Therefore, the images are not shown in the article but can be found in the Supplementary Materials (S9–S10).

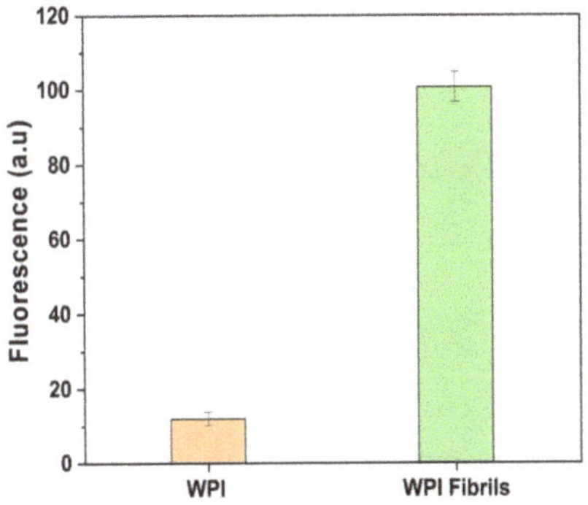

Figure 1. ThT test results of: 2.5% WPI at pH 2 (WPI); 2.5% WPI at pH 2 after 5 h incubation at 90 °C under stirring at 350 rpm (WPI Fibrils). λ_{ex} = 440 nm; λ_{em} = 486 nm. Error bars represent the standard deviation (SD).

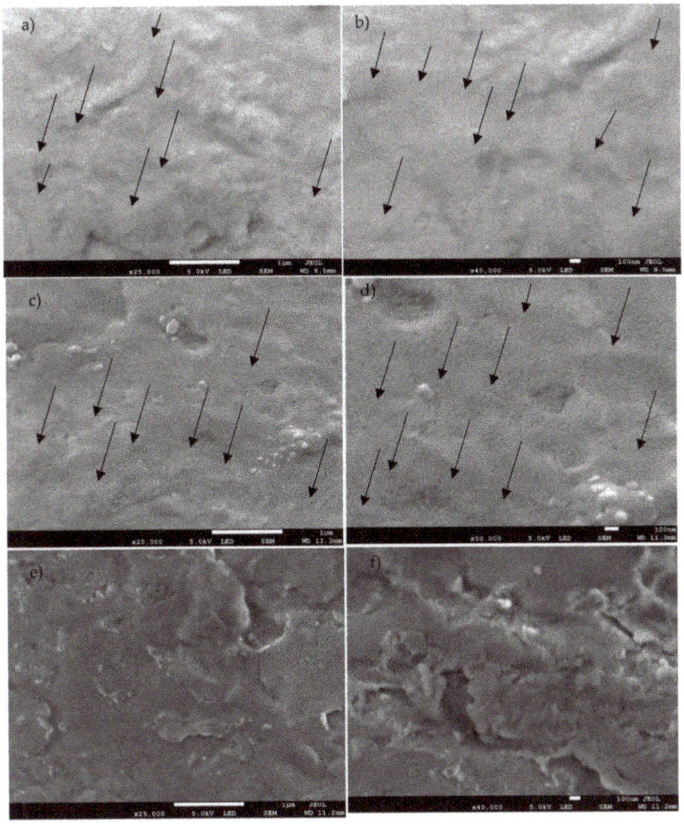

Figure 2. SEM images of Ti6Al4V WPI coated NS with 25,000× magnification (**a**) and 40,000× magnification (**b**); WPI coated S at 25,000× (**c**) and 50,000× (**d**); Ti6Al4V uncoated at 25,000× (**e**) and 40,000× (**f**). Black arrows indicate WPI fibrils. Scale bar: 1 μm for the 25,000× images and 0.1 μm for the more magnified.

2.2. Coating Characterization

Table 1 shows the name of the samples object of this study and the corresponding treatment.

Table 1. Samples name and treatment.

Sample Name	Treatment
Uncoated	Bare Ti6Al4V
WPI Coated NS (Not Sterile)	Ti6Al4V coated with WPI fibrils
WPI Coated S (Sterile)	Ti6Al4V coated with WPI fibrils and autoclaved
WPI Coated S + H (Sterile + Heparin)	Ti6Al4V coated with WPI fibrils, autoclaved and subsequently coated with Heparin
WPI Coated S + T (Sterile + Tinzaparin)	Ti6Al4V coated with WPI fibrils, autoclaved and subsequently coated with Tinzaparin (LMWH)

2.2.1. Contact Angle

Figure 3 shows the contact angle of a 5 μL-MilliQ water drop on the surface of Ti6Al4V discs coated with WPI fibrils at pH 2 treated in different ways. Each sample was coated following the protocol described in Section 3.2. The uncoated sample (bare Ti6Al4V) showed a value of 114.73 ± 3.61°. On the other hand, the coated material showed a lower angle of contact in each condition: the sample "WPI Coated NS" had an angle of contact of 53.49 ± 2.35° while the same sample but sterilized (WPI Coated S) showed a slight increase in the value with 80.40 ± 2.31°. Finally, the samples with heparin and tinzaparin showed a value in between those for the "WPI Coated S" and "WPI Coated NS" of 75.11 ± 4.85° and 59.33 ± 2.45° respectively. The large reduction in the contact angle between the uncoated and coated pre-sterilization sample demonstrated a decrease in the hydrophobicity of the Ti6Al4V following coating with WPI fibrils.

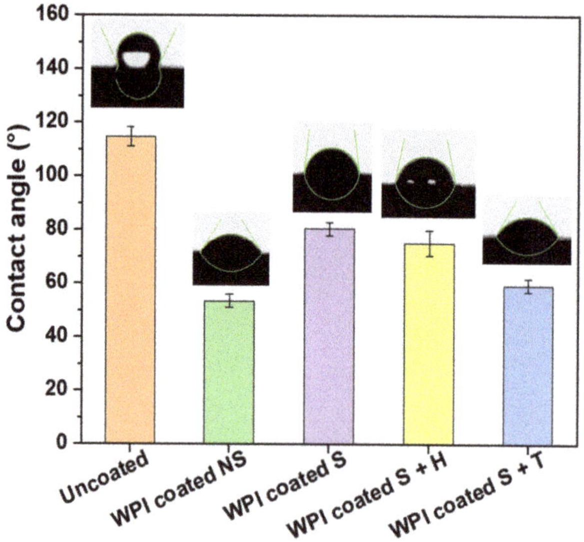

Figure 3. Contact angle measurements of Ti6Al4V discs: uncoated, coated with WPI fibrils at pH 2 not sterile (NS); coated with WPI fibrils at pH 2 and sterilized (S); coated and sterilized sample + heparin (S + H); coated and sterilized sample + tinzaparin (S + T). Error bars represent the standard deviation.

According to statistical analysis (Table 2), each coated sample differed significantly from uncoated samples. However, among the coated samples, unsterilized and tinzaparin-enriched samples did not differ significantly in hydrophilicity. The same behavior can be observed between sterilized and heparin-enriched samples, which do not differ significantly from each other, but do differ from all other treatments. These results are in agreement with previous studies with titanium alloy [11] and WPI fibrils [14]. In addition, the surface became more hydrophobic following sterilization. One explanation could be that a small percentage of the coating was lost because of the heat treatment; however, further analyses are required. On the other hand, the addition of heparin made the sterilized sample more hydrophilic (as expected) from the polar nature of the molecule and the results of other studies on heparin-enriched coatings [15,16]. It is also important to note that the addition of tinzaparin increased hydrophilicity more than heparin, giving a first proof of the presence of the LMWH molecule. The increase of hydrophilicity leads to considerable advantages because a low hydrophobicity is required in orthopedic prostheses, especially if in contact with blood, as it reduces implant friction allowing better control in surgeries, limiting thrombosis and increasing general biocompatibility. The result is a reduction in the cost and time of the procedures [15].

Table 2. Water-Contact angle test results on Ti6Al4V uncoated and coated in different conditions. Results are expressed as mean ± standard deviation (SD). Means without a common superscript differ ($p < 0.05$).

Sample	Mean ± SD (°)
Uncoated	114.73 ± 3.61^{a}
WPI Coated NS	53.49 ± 2.35^{be}
WPI Coated S	80.40 ± 2.31^{cd}
WPI Coated S + H	75.11 ± 4.85^{cd}
WPI Coated S + T	59.33 ± 2.45^{be}

2.2.2. X-ray Photoelectron Spectroscopy

X-ray photoelectron spectroscopy (XPS) results (Figure 4) showed that the superficial atomic composition of Ti6Al4V was: C (51%), O (34.8%), Ti (8.96%), Al (3.16%), N (2%). The WPI coated NS sample showed an increase in the amount of carbon and nitrogen, which are the main components of the coating. Consequently, the relative contents of oxygen and titanium decreased (22.6% and 3.3% respectively) accordingly. Weak signals of sodium (1.16%) and chlorine (0.71%) were recorded following sterilization (WPI coated S), which could be due to autoclave steam contamination since the instrument was loaded with tap water. Finally, it was possible to record the Sulphur signal (4.16%) in the sample further coated with heparin, indicating the actual presence of the molecule (survey and high-resolution spectra reported in Supplementary Figures S1–S8). However, with this technique, it was not possible to record the signal for the sample with tinzaparin. We assumed the presence of the LMWH molecule since in vitro investigations with hBMSC, e.g., the TNAP activity test on tinzaparin-coated samples, gave a higher result compared to the control and heparin-coated samples (see Section 2.4) and the CA measurement differed significantly among the heparin coated samples, as explained in the previous section. However, in future works it would be useful to develop a test to assess the presence of the molecule directly. In our preliminary work with WPI fibrils coatings, similar results were obtained with the same adsorption time on glass substrates [14], indicating that our protocol allows a good coating of the substrate with a thickness of at least 10 nm, which is considered to be the limit of detection of this technique.

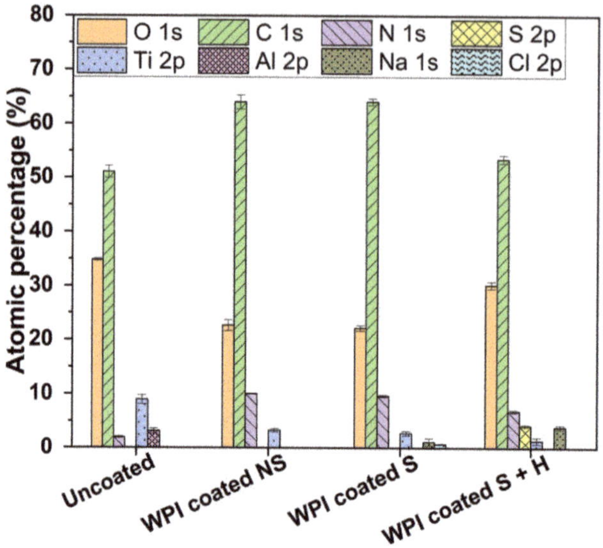

Figure 4. Atomic composition of uncoated and coated Ti6Al4V samples measured by XPS. Error bars represent the standard deviation.

2.3. *Influence of Different Ti6Al4V Coating on Metabolic Activity of hBMSC*

The MTS assay measures metabolic activity of cells as an indication for cell number and cell viability [17]. After two days in culture, the hBMSC number on coated Ti6Al4V seemed to be slightly lower than that on uncoated Ti6Al4V. In particular, the WPI-Tinzaparin coating tended to show the highest reduction of metabolic activity. However, the difference was not statistically significant (Figure 5). Hence, the coating had no impact on the number of cells that proliferate on the surface. In our previous study, similar results were found on glass-WPI coated samples showing that, even if the number of cells does not change after the coating, a better organization of the cytoskeleton occurs on the coated samples [14].

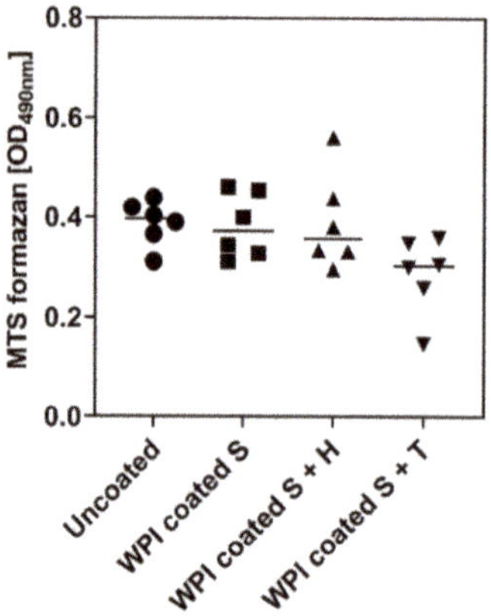

Figure 5. Metabolic activity of hBMSC on bare and coated Ti6Al4V. 6500 hBMSC/sample plated onto bare Ti6Al4V (uncoated) and Ti6Al4V coated with WPI + heparin, and WPI + tinzaparin and analysed after 48 h for metabolic activity. The results are shown as mean (bar) and individual values; $n = 3$. Statistically significant differences were not noted.

2.4. Influence of Different Ti6Al4V Coatings on TNAP Activity of hBMSC

For evaluation of the osteoinductive potential of different Ti6Al4V coatings, the activity of the TNAP enzyme in hBMSC cultured for 11 days was analyzed. Besides the deposition of calcium phosphate as a final differentiation marker for osteoblasts, TNAP activity is frequently used as an early osteogenic differentiation marker that indicates the potential of hBMSC to form hydroxyapatite [18,19]. Thus, TNAP activity is a prerequisite for bone mineralization since the ectoenzyme releases phosphate ions and therefore increases their concentration locally so that, finally, hydroxyapatite can be formed. The higher the activity of alkaline phosphatase (ALP), the more bone mineral is deposited and the better an implant can be integrated [20,21]. As highlighted in Figure 6, the coating of Ti6Al4V with WPI slightly increased TNAP activity. The addition of heparin and tinzaparin to the WPI significantly increased the activity of TNAP at day 11. The higher effect of tinzaparin could be related to its short chain length compared to heparin. These findings were consistent with our previous work which showed the significant effect of WPI coating on the activity of TNAP [14]. However, here we proved that this behavior can be further enhanced by adding LMWH to the system. Simann et al. [22] and Mathews et al. [23] both showed a higher activity of ALP in hBMSC cultured in the presence of heparin, supporting our results. Additionally, in our previous study, the positive effect of heparin on the early-stage osteoblast-differentiation occurred only when the molecule was bound to a protein substrate [24]. The results showed in this work gave a clear indication of a pro-osteogenic effect of LMWH-modified WPI implant coatings. These results are a promising starting point for future investigations involving more osteoblast features and osteogenic parameters as mineral deposition, formation of osteocalcin, osteoprotegerin, osteopontin, bone sialoprotein, etc. in order to elucidate the molecular mechanisms of such coatings and to characterize the effects of heparin and tinzaparin as coating components in more detail.

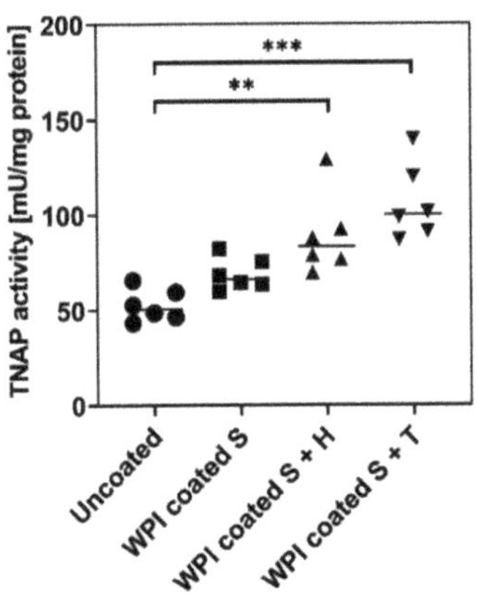

(a)

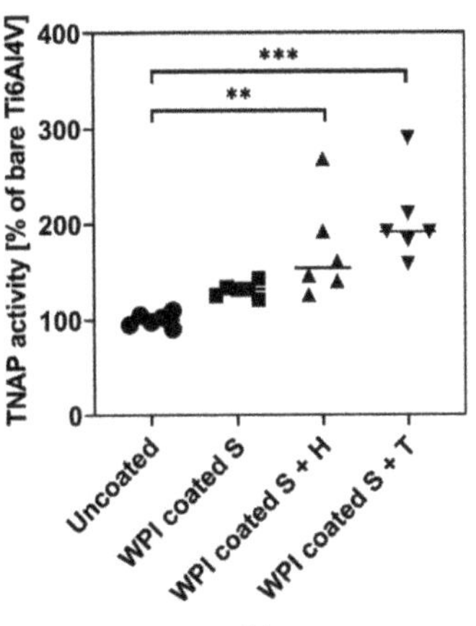

(b)

Figure 6. TNAP activity of hBMSC on bare and coated Ti6Al4V. 6500 hBMSC/sample plated in onto bare Ti6Al4V and Ti6Al4V coated WPI, WPI + heparin, and WPI + tinzaparin. From day four after plating, the cells were cultured with osteogenic supplements and analysed at day 11 for TNAP enzyme activity. The results are shown as mean (bar) and individual values; $n = 3$. Statistical significant differences versus bare Ti6Al4V are indicated with ** ($p < 0.01$) and *** ($p < 0.001$). (**a**) mU/mg protein; (**b**) % of bare Ti6Al4V.

3. Materials and Methods

3.1. Fibrils Production and Characterization

WPI fibrils were prepared according to the protocol described by Keppler et al. [25]): WPI (BiPro, Davisco Foods International Inc., Eden Prairie, MN, USA) was dissolved in Milli-Q water to a final concentration of 2.5 wt% and the pH was set to 2.0 with 2M HCl. 40 mL of protein solution were heated at 90 °C for five hours under stirring at 350 rpm to allow the fibrillation reaction to take place. At the end of the specified time, the solution was immediately cooled on ice to stop the reaction. The fibrils were stored in a refrigerator and proved to be stable for approximately four months.

Qualitative determination of the fibrils was carried out using the ThT colorimetric assay described by Loveday et al. [26]: 12 µL of the protein solution was added to 1 mL of ThT (Sigma-Aldrich, Schnelldorf, Germany) solution and fluorescence was measured with a plate reader (Infinite M200 PRO; Tecan, Reading, UK) after 1 min of incubation directly in a 96-well plate. Excitation and emission wavelengths were 420 nm and 486 nm, respectively. The analysis has been conducted against a blank sample with water and ThT.

3.2. Coating Protocol and Characterization

The Ti6Al4V discs (2 cm of diameter and 1 mm thick) additively manufactured in an A2 ARCAM EBM machine (ARCAM EBM, Mölnlycke, Sweden) were coated with fibrils by adsorption from the suspension. The substrates were left in contact with 1 mL of the fibril solution (2.5 wt%) for one hour, then rinsed three times with Milli-Q water to remove excess coating and left to air dry. For samples with heparin (or tinzaparin), the protocol was repeated on the fibril-coated samples using a 10 wt% heparin (or tinzaparin) solution.

The sterilization (121 °C, 15 min, 1 atm) was performed after the coating procedure with a Bench top Autoclave. Addition of heparin or tinzaparin by adsorption from solutions was performed in sterile conditions under a laminar flow hood using syringe (NORM-JECT 12 mL, Henke-Sass Wolf GmbH, Tuttlingen, Germany) and filters (28 mm Diameter Syringe Filters, 0.2 µm Pore SFCA-PF Membrane, Corning International, New York, NY, USA).

X-ray photoelectron spectroscopy (XPS) was performed to analyze the surface chemical composition on the samples using an Axis Supra spectrometer (Kratos Analytical Ltd., Manchester, UK) with a monochromatic Al Kα source (1.487 keV). Samples were mounted using carbon tape on a sample holder. An internal flood gun was used for neutralizing charging effects. Wide scans were recorded at a pass energy of 160 eV and a step size of 1 eV. Samples were measured at an emission angle of 0° (relative to the surface normal), power of 225 W (15 kV × 15 mA) and an analysis area of 700 × 300 µm. Three different locations on each coating's type were analyzed. Spectra were analyzed with CasaXPS software (version 2.3.22, Casa Software Ltd., Devon, UK). All binding energies were referenced to the C-C component of the C1s spectrum at 284.8 eV to compensate for the surface charging effects. The curve fitting procedure of the components was performed using Gaussian-Lorentzian function and a linear background.

Contact angle measurements were achieved with a home-built system consisting of a light, a stand and a camera connected to a computer. The images were analysed with ImageJ software with the drop analysis plugin [27].

The SEM characterization was performed as follows: samples were mounted on standard aluminum pin stubs using double sided conductive carbon adhesive dots. They were subsequently sputter coated with approximately 5 nm of gold (at 20 mA for 60 s, 1x 10-2 mBar, under argon) using a system by Quorum Technologies Ltd., Lewes, UK, Q150RES. Finally, samples were imaged using a Jeol JSM-7800F Field Emission Scanning Electron Microscope (FEG-SEM) using the lower secondary electron detector.

3.3. Isolation and Cultivation of Human Bone Marrow Stromal Cells (hBMSC)

hBMSC were isolated from bone marrow aspirates obtained from donors at the Bone Marrow Transplantation Center of the University Hospital Dresden. The cells were characterized as described in the work by [28]. The donors (males, average age 25 ± 3 years.)

were duly informed about the procedures and gave their full consent. The study was approved by the local ethics commission (ethic vote No. EK466112016).

For the in vitro experiments, 5000 hBMSC were seeded in an 80 µL-droplet of cell culture medium (Dulbecco's minimal essential medium (DMEM; Merck-Millipore, Darmstadt, Germany), supplemented with 10% heat-inactivated fetal calf serum and antibiotics (Sigma-Aldrich,) onto the surface of each sample (Ø 10 mm, 1 mm height, 0.78 cm^2). Two hours after plating, DMEM was added to cover the samples with medium. At day four after plating, cell culture medium was replaced by osteogenic differentiation medium (DMEM with 10% heat-inactivated fetal calf serum and antibiotics supplemented with 10 mM β-glycerol phosphate, 300 µM ascorbate, and 10 nM dexamethasone (Sigma-Aldrich) as described previously [29]. The medium was changed twice per week.

3.4. Determination of Metabolic Activity

The metabolic activity of hBMSC cells was determined by the MTS assay (Cell Titer96 AQ_{ueous} One Solution Proliferation Assay; Promega, Germany) at day two after plating. The cell culture medium was replaced by fresh medium containing 10% of MTS dye solution. After 2 h of incubation at 37 °C in a humidified CO_2 incubator, 80 µL of medium was transferred into a 96-well plate and the absorbance of the formed formazan dye was measured photometrically at 490 nm.

3.5. Determination of Tissue Non-Specific Alkaline Phosphatase (TNAP) Enzyme Activity

At day 11 after seeding, hBMSC were analysed for TNAP enzyme activity. TNAP enzyme activity was determined from cell lysates (TNAP lysis buffer: 1.5 M Tris-HCl, pH 10 containing 1 mM $ZnCl_2$, 1 mM $MgCl_2$ and 1% Triton X-100; Sigma-Aldrich, Germany) with p-nitrophenylphosphate (Sigma-Aldrich, Germany) as a substrate, as previously described [30]. TNAP activity was calculated from a linear calibration curve ($r > 0.99$) prepared with p-nitrophenolate. Protein concentration of the lysate was determined with RotiQuant protein assay (Roth GmbH, Karlsruhe, Germany) and was calculated from a linear calibration curve ($r > 0.99$) obtained with bovine serum albumin (Serva, Heidelberg, Germany). Specific TNAP activity is given in mU/mg protein.

3.6. Statistical Analysis

Cell experiments were performed with cells from three different donors ($n = 3$) each in duplicate. The results were presented as mean ± standard error of the mean (SEOM). Statistical significance was analyzed with GraphPad Prism 8.4 software (Statcon, Witzenhausen, Germany) by ANOVA analysis with Bonferroni's post-test. The contact angle was measured on four water drops for each sample. One-way ANOVA with Tukey's post-test was carried out with the software IBM SPSS Statistics Version 27.

4. Conclusions

Coating Ti6Al4V with WPI fibrils obtained at pH 2 enriched with heparin and tinzaparin proved to be a successful strategy to create a viable substrate for hBMSC. This substrate for the cells promotes osteogenic differentiation by improving the quality of the differentiated cells, as evidenced by the increase in the TNAP activity. In particular, this work showed that enriching the coating with heparin and tinzaparin improves the aforementioned effect considerably. Specifically, it seems that tinzaparin has the highest impact on the TNAP activity. In any case, further investigations of the molecular mechanism are needed to further elucidate this behavior. As far as the coating protocol is concerned, one hour of adsorption time was sufficient to successfully coat Ti6Al4V substrates. However, the presence of some uncoated areas was detected. Additionally, XPS can be used as a method to evaluate the presence of heparin and proteins on the surface of the material. Nonetheless, it seems that this is not a suitable method for the detection of tinzaparin, and it advisable to use a different approach in future work. The coating increased the hydrophilicity of the material with a higher extent when enriched with tinzaparin compared to heparin.

This behavior, together with the statistically significant difference seen between the heparin and tinzaparin coated samples in the TNAP activity test, provides further evidence of the presence of the LMWH molecule.

Supplementary Materials: The following are available online at https://www.mdpi.com/article/10.3390/ijms23010139/s1.

Author Contributions: Conceptualization, T.E.L.D.; methodology, T.E.L.D., U.H., L.M. and A.K.; formal analysis, D.F., U.H. and L.M.; investigation, D.F., U.H., L.M., A.M.S. and A.K.; resources, T.E.L.D., U.H., A.M.S., M.A.S. and R.A.S.; writing—original draft preparation, D.F.; writing—review and editing, D.F., U.H., L.M., A.M.S., A.K., M.A.S., R.A.S. and T.E.L.D.; supervision, T.E.L.D., R.A.S. and M.A.S.; project administration, T.E.L.D.; funding acquisition, T.E.L.D., D.F., L.M., A.M.S., R.A.S. and M.A.S. All authors have read and agreed to the published version of the manuscript.

Funding: This research was funded by EPSRC "A novel coating technology based upon polyatomic ions from plasma" grant number EP/S004505/1 (L.M.)". The University of Milan, Italy, is thanked for financial support for a research stay (D.F.).

Institutional Review Board Statement: The study was conducted in accordance with the Declaration of Helsinki, and approved by the Institutional Ethics Committee of TU Dresden (protocol code EK466112016 and approval date 3 November 2016).

Informed Consent Statement: Not applicable.

Data Availability Statement: Not applicable.

Acknowledgments: RAS and MAS acknowledge the support from Ministry of Science and Higher Education of the Russian Federation (grant agreement #075-15-2021-588 from 1 June 2021).

Conflicts of Interest: The authors declare no conflict of interest. The funders had no role in the design of the study; in the collection, analyses, or interpretation of data; in the writing of the manuscript, or in the decision to publish the results.

Abbreviations

AM	Additive Manufacturing
ALP	Alkaline phosphatase
CA	Contact angle
DMEM	Dulbecco's minimal essential medium
FEG-SEM	Field Emission Scanning Electron Microscope
GAGs	Glycosaminoglycans
hBMSC	human bone marrow stromal cells
LMWH	Low molecular weight heparin
NS	Not sterile
S	Sterile
SD	Standard deviation
SEOM	Standard error of the mean
SEM	Scanning Electron Microscopy
ThT	Thioflavin T
TNAP	Tissue non-specific alkaline phosphatase
WPI	Whey protein isolate
β-lg	β-lactoglobulin

References

1. Sidambe, A.T. Biocompatibility of Advanced Manufactured Titanium Implants—A Review. *Materials* **2014**, *7*, 8168–8188. [CrossRef] [PubMed]
2. Douglas, T.; Heinemann, S.; Mietrach, C.; Hempel, U.; Bierbaum, S.; Scharnweber, D.; Worch, H. Interactions of collagen types I and II with chondroitin sulfates A-C and their effect on osteoblast adhesion. *Biomacromolecules* **2007**, *8*, 1085–1092. [CrossRef] [PubMed]

3. Douglas, T.; Hempel, U.; Mietrach, C.; Viola, M.; Vigetti, D.; Heinemann, S.; Bierbaum, S.; Scharnweber, D.; Worch, H. Influence of collagen-fibril-based coatings containing decorin and biglycan on osteoblast behavior. *J. Biomed. Mater. Res. A* **2008**, *84*, 805–816. [CrossRef] [PubMed]
4. Mieszkowska, A.; Beaumont, H.; Martocq, L.; Koptyug, A.; Surmeneva, M.A.; Surmenev, R.A.; Naderi, J.; Douglas, T.E.L.; Gurzawska-Comis, K.A. Phenolic-Enriched Collagen Fibrillar Coatings on Titanium Alloy to Promote Osteogenic Differentiation and Reduce Inflammation. *Int. J. Mol. Sci.* **2020**, *21*, 6406. [CrossRef]
5. Norris, K.; Mishukova, O.I.; Zykwinska, A.; Colliec-Jouault, S.; Sinquin, C.; Koptioug, A.; Cuenot, S.; Kerns, J.G.; Surmeneva, M.A.; Surmenev, R.A.; et al. Marine Polysaccharide-Collagen Coatings on Ti6Al4V Alloy Formed by Self-Assembly. *Micromachines* **2019**, *10*, 68. [CrossRef]
6. Ferreira, A.M.; Gentile, P.; Toumpaniari, S.; Ciardelli, G.; Birch, M.A. Impact of Collagen/Heparin Multilayers for Regulating Bone Cellular Functions. *ACS Appl. Mater. Interfaces* **2016**, *8*, 29923–29932. [CrossRef]
7. Douglas, T.E.L.; Vandrovcová, M.; Kročilová, N.; Keppler, J.K.; Zárubová, J.; Skirtach, A.G.; Bačáková, L. Application of whey protein isolate in bone regeneration: Effects on growth and osteogenic differentiation of bone-forming cells. *J. Dairy Sci.* **2018**, *101*, 28–36. [CrossRef]
8. Gupta, D.; Kocot, M.; Tryba, A.M.; Serafim, A.; Stancu, I.C.; Jaegermann, Z.; Pamuła, E.; Reilly, G.C.; Douglas, T.E. Novel naturally derived whey protein isolate and aragonite biocomposite hydrogels have potential for bone regeneration. *Mater. Des.* **2020**, *188*, 108408. [CrossRef]
9. Keppler, J.K.; Martin, D.; Garamus, V.M.; Berton-Carabin, C.; Nipoti, E.; Coenye, T.; Schwarz, K. Functionality of whey proteins covalently modified by allyl isothiocyanate. Part 1 physicochemical and antibacterial properties of native and modified whey proteins at pH 2 to 7. *Food Hydrocoll.* **2017**, *65*, 130–143. [CrossRef]
10. Akkermans, C.; Venema, P.; van der Goot, A.J.; Gruppen, H.; Bakx, E.J.; Boom, R.M.; van der Linden, E. Peptides are building blocks of heat-induced fibrillar protein aggregates of beta-lactoglobulin formed at pH 2. *Biomacromolecules* **2008**, *9*, 1474–1479. [CrossRef]
11. Douglas, T.E.; Hempel, U.; Żydek, J.; Vladescu, A.; Pietryga, K.; Kaeswurm, J.A.; Buchweitz, M.; Surmenev, R.A.; Surmeneva, M.A.; Cotrut, C.M.; et al. Pectin coatings on titanium alloy scaffolds produced by additive manufacturing: Promotion of human bone marrow stromal cell proliferation. *Mater. Lett.* **2018**, *227*, 225–228. [CrossRef]
12. Krebs, M.R.H.; Bromley, E.H.C.; Donald, A.M. The binding of thioflavin-T to amyloid fibrils: Localisation and implications. *J. Struct. Biol.* **2005**, *149*, 30–37. [CrossRef]
13. Heyn, T.R.; Garamus, V.M.; Neumann, H.R.; Uttinger, M.J.; Guckeisen, T.; Heuer, M.; Selhuber-Unkel, C.; Peukert, W.; Keppler, J.K. Influence of the polydispersity of pH 2 and pH 3.5 beta-lactoglobulin amyloid fibril solutions on analytical methods. *Eur. Polym. J.* **2019**, *120*, 109211. [CrossRef]
14. Rabe, R.; Hempel, U.; Martocq, L.; Keppler, J.K.; Aveyard, J.; Douglas, T.E.L. Dairy-Inspired Coatings for Bone Implants from Whey Protein Isolate-Derived Self-Assembled Fibrils. *Int. J. Mol. Sci.* **2020**, *21*, 5544. [CrossRef]
15. Wang, X.H.; Li, D.P.; Wang, W.J.; Feng, Q.L.; Cui, F.Z.; Xu, Y.X.; Song, X.H. Covalent immobilization of chitosan and heparin on PLGA surface. *Int. J. Biol. Macromol.* **2003**, *33*, 95–100. [CrossRef]
16. Chen, J.L.; Li, Q.L.; Chen, J.Y.; Chen, C.; Huang, N. Improving blood-compatibility of titanium by coating collagen–heparin multilayers. *Appl. Surf. Sci.* **2009**, *255*, 6894–6900. [CrossRef]
17. Hempel, U.; Hefti, T.; Dieter, P.; Schlottig, F. Response of human bone marrow stromal cells, MG-63, and SaOS-2 to titanium-based dental implant surfaces with different topography and surface energy. *Clin. Oral Implants Res.* **2013**, *24*, 174–182. [CrossRef]
18. Pittenger, M.F. Mesenchymal stem cells from adult bone marrow. *Methods Mol. Biol.* **2008**, *449*, 27–44. [CrossRef]
19. Pittenger, M.F.; Mackay, A.M.; Beck, S.C.; Jaiswal, R.K.; Douglas, R.; Mosca, J.D.; Moorman, M.A.; Simonetti, D.W.; Craig, S.; Marshak, D.R. Multilineage potential of adult human mesenchymal stem cells. *Science* **1999**, *284*, 143–147. [CrossRef]
20. Hempel, U.; Matthäus, C.; Preissler, C.; Möller, S.; Hintze, V.; Dieter, P. Artificial matrices with high-sulfated glycosaminoglycans and collagen are anti-inflammatory and pro-osteogenic for human mesenchymal stromal cells. *J. Cell. Biochem.* **2014**, *115*, 1561–1571. [CrossRef]
21. Jafary, F.; Hanachi, P.; Gorjipour, K. Osteoblast Differentiation on Collagen Scaffold with Immobilized Alkaline Phosphatase. *Int. J. Organ Transplant. Med.* **2017**, *8*, 195–202.
22. Simann, M.; Schneider, V.; Le Blanc, S.; Dotterweich, J.; Zehe, V.; Krug, M.; Jakob, F.; Schilling, T.; Schütze, N. Heparin affects human bone marrow stromal cell fate: Promoting osteogenic and reducing adipogenic differentiation and conversion. *Bone* **2015**, *78*, 102–113. [CrossRef]
23. Mathews, S.; Mathew, S.A.; Gupta, P.K.; Bhonde, R.; Totey, S. Glycosaminoglycans enhance osteoblast differentiation of bone marrow derived human mesenchymal stem cells. *J. Tissue Eng. Regen. Med.* **2014**, *8*, 143–152. [CrossRef]
24. Vogel, S.; Arnoldini, S.; Möller, S.; Schnabelrauch, M.; Hempel, U. Sulfated hyaluronan alters fibronectin matrix assembly and promotes osteogenic differentiation of human bone marrow stromal cells. *Sci. Rep.* **2016**, *6*, 36418. [CrossRef]
25. Keppler, J.K.; Heyn, T.R.; Meissner, P.M.; Schrader, K.; Schwarz, K. Protein oxidation during temperature-induced amyloid aggregation of beta-lactoglobulin. *Food Chem.* **2019**, *289*, 223–231. [CrossRef]
26. Loveday, S.M.; Wang, X.L.; Rao, M.A.; Anema, S.G.; Singh, H. β-Lactoglobulin nanofibrils: Effect of temperature on fibril formation kinetics, fibril morphology and the rheological properties of fibril dispersions. *Food Hydrocoll.* **2012**, *27*, 242–249. [CrossRef]

27. Rueden, C.T.; Schindelin, J.; Hiner, M.C.; DeZonia, B.E.; Walter, A.E.; Arena, E.T.; Eliceiri, K.W. ImageJ2: ImageJ for the next generation of scientific image data. *BMC Bioinform.* **2017**, *18*, 529. [CrossRef]
28. Oswald, J.; Boxberger, S.; Jørgensen, B.; Feldmann, S.; Ehninger, G.; Bornhäuser, M.; Werner, C. Mesenchymal stem cells can be differentiated into endothelial cells in vitro. *Stem Cells* **2004**, *22*, 377–384. [CrossRef]
29. Hempel, U.; Müller, K.; Preissler, C.; Noack, C.; Boxberger, S.; Dieter, P.; Bornhäuser, M.; Wobus, M. Human Bone Marrow Stromal Cells: A Reliable, Challenging Tool for In Vitro Osteogenesis and Bone Tissue Engineering Approaches. *Stem Cells Int.* **2016**, *2016*, 7842191. [CrossRef]
30. Hempel, U.; Hintze, V.; Möller, S.; Schnabelrauch, M.; Scharnweber, D.; Dieter, P. Artificial extracellular matrices composed of collagen I and sulfated hyaluronan with adsorbed transforming growth factor β1 promote collagen synthesis of human mesenchymal stromal cells. *Acta Biomater.* **2012**, *8*, 659–666. [CrossRef]

International Journal of *Molecular Sciences*

MDPI

Article

Synthesis and Characterization of Novel Succinyl Chitosan-Dexamethasone Conjugates for Potential Intravitreal Dexamethasone Delivery

Natallia V. Dubashynskaya [1], Anton N. Bokatyi [1], Alexey S. Golovkin [2], Igor V. Kudryavtsev [3], Maria K. Serebryakova [3], Andrey S. Trulioff [3], Yaroslav A. Dubrovskii [2,4] and Yury A. Skorik [1,*]

1 Institute of Macromolecular Compounds of the Russian Academy of Sciences, Bolshoi VO 31, 199004 St. Petersburg, Russia; dubashinskaya@gmail.com (N.V.D.); qwezakura@yandex.ru (A.N.B.)
2 Almazov National Medical Research Centre, Akkuratova 2, 197341 St. Petersburg, Russia; golovkin_a@mail.ru (A.S.G.); dubrovskiy.ya@gmail.com (Y.A.D.)
3 Institute of Experimental Medicine, Akademika Pavlova 12, 197376 St. Petersburg, Russia; igorek1981@yandex.ru (I.V.K.); m-serebryakova@yandex.ru (M.K.S.); trulioff@gmail.com (A.S.T.)
4 Research and Training Center of Molecular and Cellular Technologies, St. Petersburg State Chemical Pharmaceutical University, Prof. Popova 14, 197376 St. Petersburg, Russia
* Correspondence: yury_skorik@mail.ru

Abstract: The development of intravitreal glucocorticoid delivery systems is a current global challenge for the treatment of inflammatory diseases of the posterior segment of the eye. The main advantages of these systems are that they can overcome anatomical and physiological ophthalmic barriers and increase local bioavailability while prolonging and controlling drug release over several months to improve the safety and effectiveness of glucocorticoid therapy. One approach to the development of optimal delivery systems for intravitreal injections is the conjugation of low-molecular-weight drugs with natural polymers to prevent their rapid elimination and provide targeted and controlled release. This study focuses on the development of a procedure for a two-step synthesis of dexamethasone (DEX) conjugates based on the natural polysaccharide chitosan (CS). We first used carbodiimide chemistry to conjugate DEX to CS via a succinyl linker, and we then modified the obtained systems with succinic anhydride to impart a negative ζ-potential to the polymer particle surface. The resulting polysaccharide carriers had a degree of substitution with DEX moieties of 2–4%, a DEX content of 50–85 μg/mg, and a degree of succinylation of 64–68%. The size of the obtained particles was 400–1100 nm, and the ζ-potential was −30 to −33 mV. In vitro release studies at pH 7.4 showed slow hydrolysis of the amide and ester bonds in the synthesized systems, with a total release of 8–10% for both DEX and succinyl dexamethasone (SucDEX) after 1 month. The developed conjugates showed a significant anti-inflammatory effect in TNFα-induced and LPS-induced inflammation models, suppressing CD54 expression in THP-1 cells by 2- and 4-fold, respectively. Thus, these novel succinyl chitosan-dexamethasone (SucCS-DEX) conjugates are promising ophthalmic carriers for intravitreal delivery.

Keywords: dexamethasone; succinyl chitosan; intravitreal delivery systems; anti-inflammatory activity

Citation: Dubashynskaya, N.V.; Bokatyi, A.N.; Golovkin, A.S.; Kudryavtsev, I.V.; Serebryakova, M.K.; Trulioff, A.S.; Dubrovskii, Y.A.; Skorik, Y.A. Synthesis and Characterization of Novel Succinyl Chitosan-Dexamethasone Conjugates for Potential Intravitreal Dexamethasone Delivery. *Int. J. Mol. Sci.* **2021**, 22, 10960. https://doi.org/10.3390/ijms222010960

Academic Editor: Antonella Piozzi

Received: 11 September 2021
Accepted: 7 October 2021
Published: 11 October 2021

1. Introduction

Inflammatory diseases of the posterior segment of the eye such as diabetic retinopathy, glaucoma, age-related macular degeneration, macular edema, and uveitis are serious medical challenges. These diseases affect the retina and choroid, resulting in visual impairment and blindness in millions of patients globally [1,2]. However, delivery of ocular drugs to the posterior segment of the eye is difficult due to the presence of anatomical and physiological ophthalmic barriers, including nasolacrimal drainage and the corneal barrier, and because of the non-target absorption of drugs by the conjunctiva [1,3–6]. Therefore, traditional dosage forms for both topical (eye drops) and systemic administration (enteral and parenteral drugs) are ineffective for the treatment of retinal diseases [7,8].

Intravitreal injections enable the delivery of anti-inflammatory agents specifically to the target sites of the posterior segment of the eye, but this medical procedure is invasive, and frequent injections can have severe side effects, including infections and retinal detachment [9,10]. For this reason, an intravitreal dosage form must have a prolonged release profile of active pharmaceutical substances over several months while still maintaining the drug concentration at an adequate therapeutic level [8,11,12]. Current intravitreal drugs include implants (e.g., FDA approved Ozurdex [13]) and various types of polymeric nanoparticles, but many of these implants and nanoparticles consist of non-degradable polymers, and this limits their wide clinical application for controlled drug release [2,3,14–17]. The use of various nanoparticles for retinal delivery does not ensure the desired months-long release profile and can also lead to increased intraocular pressure and visual impairment [11]. In addition, the physical size of both implants and particles can preclude targeted delivery specifically to the retinal cells [7].

One approach to the development of optimal intravitreal delivery systems is the conjugation of low-molecular-weight active pharmaceutical substances with natural polymers [1]. The diffusion rate of a drug in the vitreous humor and its elimination from the eye depend on the size of the molecule [18]. Thus, the pharmacological effects of drugs injected into the vitreous humor can be prolonged by increasing their molecular size [19], as this prevents the rapid elimination seen with low-molecular-weight components while also ensuring a targeted and controlled drug release. In addition to size, the particle surface charge is a crucial factor for successful intravitreal delivery [20–22].

The vitreous humor consists of a negatively charged three-dimensional matrix based on collagen and hyaluronic acid [14,23]. This structure allows the free movement of negatively charged and neutral particles in the vitreous humor, while the mobility of positively charged particles is strongly limited by their interactions with the anionic components of the vitreous gel [24–26]. Covalent conjugation with macromolecular compounds (especially non-toxic, biocompatible, and biodegradable natural polymers) can increase the size of a drug molecule, thereby increasing its residence time in the vitreous humor [27,28]. Altiok et al. [29,30] conjugated an anti-vascular endothelial growth factor (anti-VEGF) drug (sFlt-1) with polyanionic hyaluronic acid to decrease sFlt-1 clearance and increase drug retention time in the vitreous. This resulted in a tenfold increase in the drug half-life with no change in pharmacological activity. Famili et al. [31] developed hyaluronic acid-fragment antigen-binding (Fab) bioconjugates for anti-VEGF therapies. They found that conjugation of Fab with negatively charged hyaluronic acid significantly slowed in vivo clearance from rabbit vitreous humor after intravitreal injection. Compared with free Fab (the half-life in the vitreous humor was 2.8 days), Fab conjugated with hyaluronic acid with molecular weights of 40 kDa, 200 kDa, and 600 kDa cleared with half-lives of 7.6, 10.2 and 18.3 days, respectively.

Unfortunately, polysaccharide-based conjugates for intravitreal delivery have rarely been used [32]. The aim of the present study was therefore to synthesize conjugates of the anti-inflammatory drug dexamethasone (DEX) with the natural polysaccharide chitosan (CS) to explore its potential application as an intravitreal delivery system. CS was chosen as the natural biopolymer because it is easily modified by amino groups to form both positively and negatively charged water-soluble derivatives [33]. CS and its derivatives have demonstrated safety and satisfying biocompatibility following intravitreal injections on in vivo models [34–37]. DEX was chosen as the test drug because it is a high-efficacy synthetic glucocorticosteroid (7 times more potent than prednisolone) and one of the most frequently used anti-inflammatory drugs for the treatment of eye disease, including inflammatory diseases of both the anterior (e.g., keratitis, blepharitis, allergic conjunctivitis, and dry eye) and posterior (e.g., choroiditis, uveitis, age-related macular degeneration, diabetic macular edema, and diabetic retinopathy) segments [38–40].

2. Results and Discussion

2.1. Synthesis and Characterization of the Succinyl Chitosan-Dexamethasone Conjugates (SucCS-DEX)

Conjugation of drug molecules with different polymers through linkers with different stabilities (amide, hydrazone, or ester linkages) provides prolonged release by controlled hydrolysis of the formed chemical bonds [41]. We used a succinyl linker to introduce the carboxyl function to DEX for subsequent conjugation with CS amino groups [42]. The resulting succinyl DEX (SucDEX) was characterized by ^{1}H nuclear magnetic resonance (^{1}H NMR) spectroscopy (Figure S1).

SucCS-DEX was prepared by a two-step synthesis (Figure 1). SucDEX was first conjugated to CS by carbodiimide chemistry. The resulting intermediate product (CS-DEX) was isolated and characterized by Fourier transform infrared (FTIR) spectroscopy (Figure S2), ^{1}H NMR spectroscopy (Figure 2), and elemental analysis (Table 1). The CS-DEX-20 sample was poorly redispersed in water, so we did not use it in further tests. The degrees of substitution with DEX moieties (DS_{DEX}) of CS-DEX ranged from 2 to 4% (Table 1).

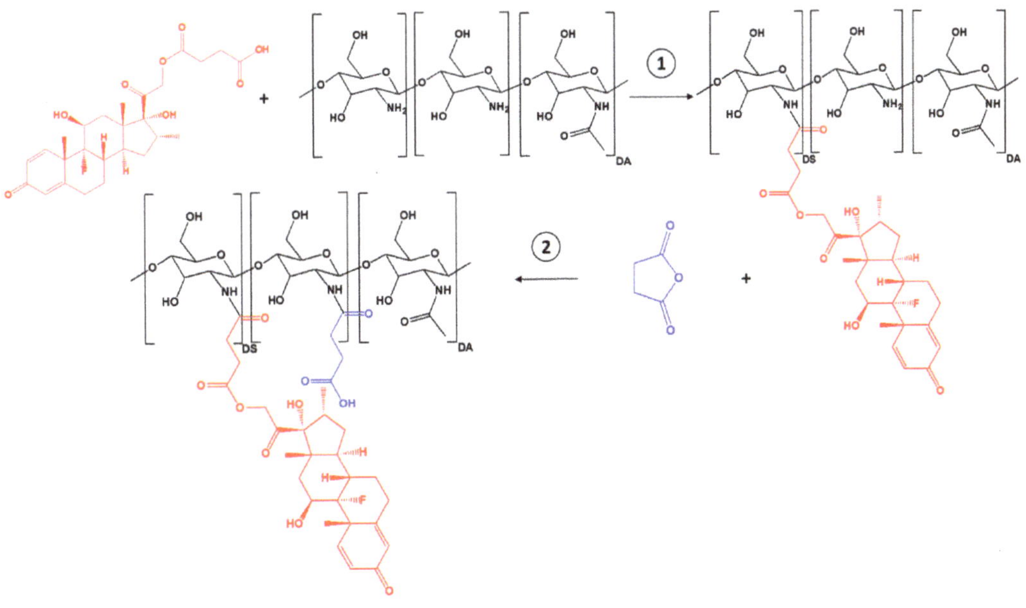

Figure 1. Synthesis scheme for SucCS-DEX.

Table 1. Synthesis conditions and characterization of CS-DEX conjugates.

Sample	Molar Ratio of Reagents				ω (%)		DS_{DEX} ** (%) by EA	DS_{DEX} (%) by NMR
	CS	EDC *	NHS *	SucDEX	C	N		
CS-DEX-5	1	0.05	0.05	0.05	40.07	6.749	1.7	1.8
CS-DEX-10	1	0.1	0.1	0.1	40.09	6.424	3.1	2.9
CS-DEX-20	1	0.2	0.2	0.2	39.27	6.145	3.8	3.9
CS	-	-	-	-	41.55	7.449	-	-

* EDC—1-ethyl-3-(3-dimethylaminopropyl) carbodiimide hydrochloride, NHS—N-hydroxysuccinimide. ** SD_{DEX} was calculated from elemental analysis data using the following formula: $\left(\frac{\omega(C)CS-DEX}{\omega(N)CS-DEX} - \frac{\omega(C)CS}{\omega(N)CS}\right)\frac{M(N)}{M(C)\ n} \times 100\%$, where ω is the mass fraction of the element in the sample, M is the molar mass of the element, n = 26 (the number of C atoms in SucDEX).

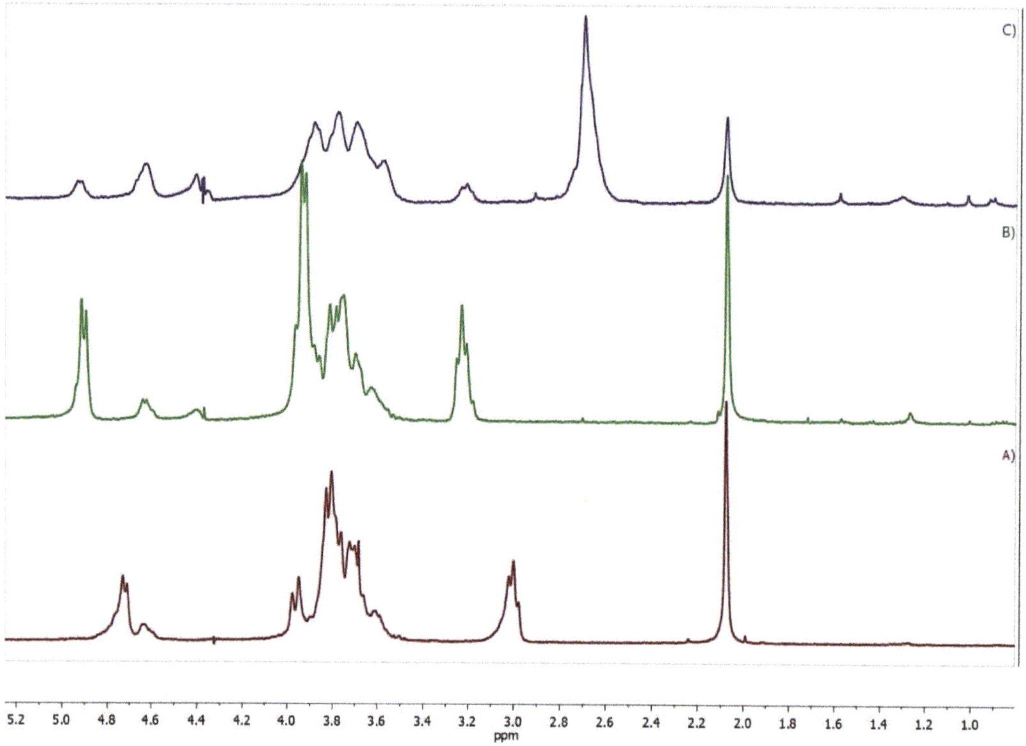

Figure 2. ^{1}H NMR spectra (400 MHz, D_2O/trifluoroacetic acid) of (**A**) CS, (**B**) CS-DEX-10, and (**C**) SucCS-DEX-10.

The ^{1}H NMR spectrum of CS-DEX (Figure 2B) revealed all the signals of the initial CS, including the signal of the acetamide protons (2.08 ppm), H-2 of the glucosamine unit (3.23 ppm), a multiplet at 3.5–4.1 ppm (H-3–H-6 and H-2 of the N-acetylglucosamine unit), and the signals of the H-1 anomeric protons at 4.64 and 4.92 ppm. The spectrum also contained signals of the DEX protons at 0.75–3.0 ppm. The signals H-2–H-6 of the glucosamine ring were chosen as the reference signals, since no DEX protons are present in this region. The DS was calculated from the DEX proton signals at 0.85–1.6 ppm ($I(DEX)_{0.85-1.6}$), which corresponds to 14 DEX protons using the following equation $DS_{DEX} = \frac{6\, I(DEX)_{0.85-1.6}}{14\, I(H-2-H-6)}$. The DS determined by the NMR method was in agreement with the elemental analysis data (Table 1).

The synthesized conjugates self-assembled in aqueous media into submicron particles with a positive ζ-potential due to the presence of the protonated CS amino groups on the surface. Therefore, the second step of conjugate synthesis was the succinylation of the positively charged particles at the amino group (Figure 1, step 2); this resulted in negatively charged final particles suitable for intravitreal administration. The negative ζ-potential prevents the particles from undergoing polyelectrolyte interactions with oppositely charged ions and thus provides the stability and mobility of the conjugates in the vitreous humor environment that consists of polyanionic hyaluronic acid [43]. Succinic anhydride (SA) was chosen as the surface modifier for CS since it is biotransformed in the body into succinic acid, a natural endogenous metabolite of the Krebs cycle, and is therefore non-toxic [44]. Succinylated CS, in addition to its high safety profile and low toxicity, is also less biodegradable than CS, so it is an excellent polymer platform for prolonged glucocorticoid delivery systems and has a release profile for the active pharmaceutical substance of several months [33,45,46]. The resulting compound was characterized by elemental analysis (Table 2) and ^{1}H NMR spectroscopy (Figure 2).

Table 2. Characterization of SucCS-DEX.

Sample	ω (%)		DS_{Suc} * (%) by EA	DS_{Suc} (%) by NMR	DS_{DEX} (%) by NMR	DEX Content (μg/mg)
	C	N				
SucCS-DEX-5	31.58	3.868	65	64	1.8	50
SucCS-DEX-10	30.40	3.575	66	68	3.0	85

* DS_{Suc} was calculated from elemental analysis data using the following formula: $DS_{Suc} = \left(\frac{\omega(C)SucCS-DEX}{\omega(N)SucCS-DEX} - \frac{\omega(C)CS-DEX}{\omega(N)CS-DEX}\right)\frac{M(N)}{M(C)\,n} \times 100\%$, where ω is the mass fraction of the element in the sample, M is the molar mass of the element, and n = 4 (the number of C atoms in SA).

The ^{1}H NMR spectrum of SucCS-DEX (Figure 2C) as compared to the CS-DEX spectrum (Figure 2B) shows the appearance of a signal of methylene protons of the succinyl substituent at 2.7 ppm. The degree of succinylation (DS_{Suc}) was calculated from the integral intensity of this signal (the number of protons is 4) using the following equation: $DS_{Suc} = \frac{6\,I(CH_2)}{4\,I(H-2-H-6)} \times 100\%$. DS_{Suc}, determined by the NMR method, which agreed with the elemental analysis data. The DS_{DEX} of SucCS-DEX determined by the NMR method (Table 2) was in agreement with that of CS-DEX (Table 1), which indicates the absence of hydrolysis of the SucDEX substituent in the CS-DEX succinylation process. The DS_{Suc} in the SucCS-DEX was about 64–68%. The DEX content in SucCS-DEX samples was determined by UV spectroscopy (Figure S3) at 242 nm (Table 2). The spectrum of SucCS was also recorded to confirm that the polymer itself had no absorption at 242 nm.

The physicochemical characteristics (the hydrodynamic size and the ζ-potential) of the conjugates are presented in Table 3. The synthesized conjugates were capable of self-assembly in aqueous media and formed submicron-sized particles (400–1100 nm). The ζ-potential was 14–23 mV for the non-succinylated samples and −30 to −33 mV for the succinylated particles. The spherical shape of the particles was confirmed by scanning electron microscopy (SEM) (Figure 3). The average particle size on SEM images was 200–600 nm, which did not conflict with the dynamic light scattering data.

Table 3. Physicochemical characteristics of CS-DEX and SucCS-DEX (mean ± standard deviation, n = 3).

Sample	2Rh (nm)	ζ-Potential (mV)
CS-DEX-5	816 ± 268	22.5 ± 0.5
SucCS-DEX-5	916 ± 326	−32.1 ± 0.5
CS-DEX-10	700 ± 252	14.9 ± 0.8
SucCS-DEX-10	950 ± 330	−30.9 ± 0.7

2.2. *In Vitro DEX Release from the SucCS-DEX Conjugates*

Anionic conjugates for intravitreal delivery were studied to determine the DEX release pathway (i.e., whether by amide or ester bond hydrolysis). The pharmacological activity depends on the chemical structure of the substance and the presence of substituents; therefore, we needed to determine which form of DEX (native DEX or SucDEX) is released from these polymer carriers. The mass spectrometry data showed that the hydrolysis of SucCS-DEX results in the formation of both SucDEX and DEX (the extracted ion chromatograms of the DEX forms released from SucCS-DEX-10 for 30 days are shown in Figure S4).

Both the DEX and SucDEX release kinetics for the SucCS-DEX-5 and SucCS-DEX-10 samples in phosphate buffered saline (PBS) at 37 °C are shown in Figures 4 and 5, respectively. Under the conditions studied, the hydrolysis rate of the succinyl linker at the amide bond was higher than at the ester bond, which resulted in the favorable formation of SucDEX over DEX.

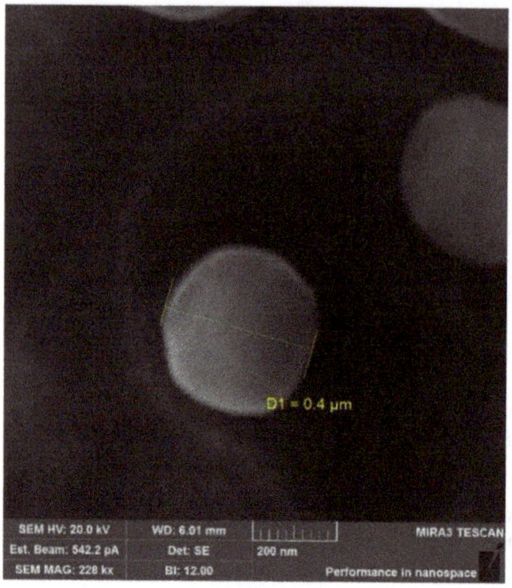

Figure 3. SEM images of the SucCS-DEX-10 particles.

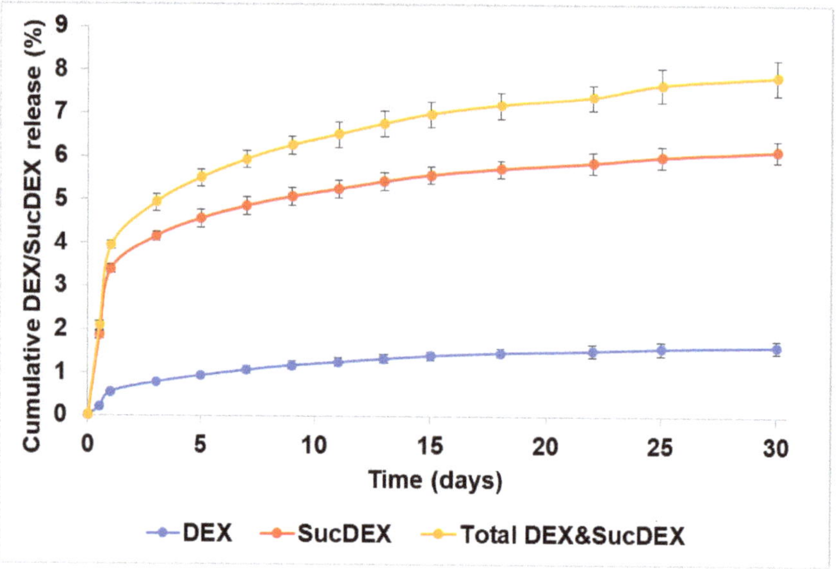

Figure 4. Release of DEX and SucDEX from the SucCS-DEX-5 particles; $n = 3$, error bars represent one standard deviation.

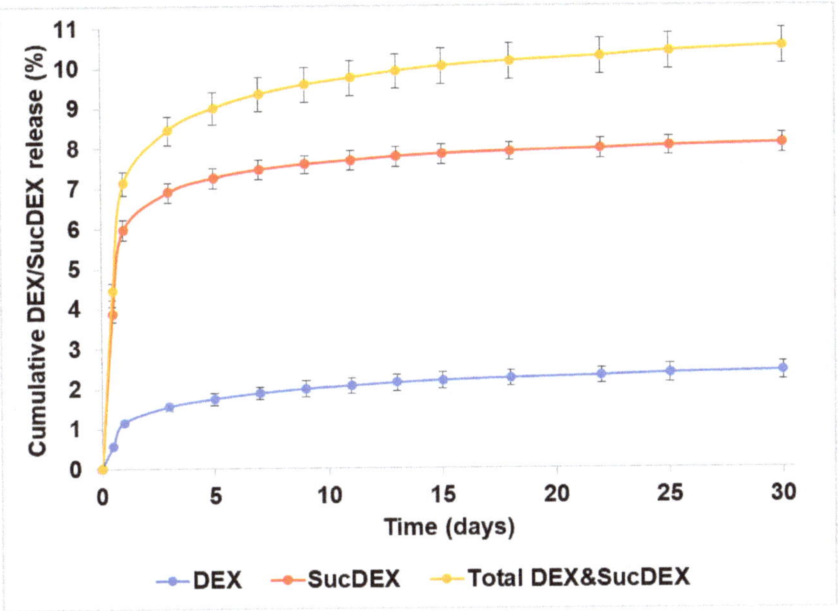

Figure 5. Release of DEX and SucDEX from the SucCS-DEX-10 particles; $n = 3$, error bars represent one standard deviation.

Thus, the obtained conjugates prolonged the release of both DEX and SucDEX by a total of 8–10% for at least a month at physiological pH. Polymeric conjugates of DEX with this release profile are attractive candidates for use as intravitreal delivery systems.

2.3. *Anti-Inflammatory Activity of the SucCS-DEX Conjugates*

We needed to confirm that the two DEX forms (DEX and SucDEX) released from the SucCS-DEX conjugate retained comparative anti-inflammatory activities to that of native DEX. SucCS with DS of 65% [33] was used as a control. We first tested the influence of SucCS-DEX on in vitro THP-1 cell viability. In the absence of LPS or TNFα stimulation, all tested compounds at concentrations of 1 and 10 μg/mL significantly reduced the relative number of viable THP-1 cells (Table 4); however, no significant differences were detected between all samples under the inflammatory conditions.

Table 4. The influence of DEX, SucDEX, SucCS, and SucCS-DEX-10 solutions on THP-1 cell viability under inflammatory conditions (mean ± SE).

Samples	Concentration (μg/mL)	w/o LPS or TNF	LPS (1 μg/mL)	TNFα (2 ng/mL)
Negative control ($n = 11$)	-	95.8 ± 0.3	92.40 ± 0.7	87.8 ± 1.6
DEX ($n = 6$)	1	94.0 ± 0.8 *	91.49 ± 1.2	87.1 ± 0.9
	10	92.6 ± 0.8 *	89.54 ± 1.5	85.9 ± 1.4
SucDEX ($n = 6$)	1	94.7 ± 0.5	91.94 ± 1.1	88.6 ± 0.8
	10	94.0 ± 0.6 *	88.8 ± 1.8	84.6 ± 1.1
SucCS ($n = 6$)	1	94.2 ± 0.5 *	92.9 ± 0.7	88.8 ± 1.1
	10	94.4 ± 0.6 *	92.1 ± 0.9	89.5 ± 0.9
SucCS-DEX-10 ($n = 6$)	1	94.4 ± 0.5 *	91.2 ± 0.8	87.33 ± 1.3
	10	91.8 ± 0.9 *	91.4 ± 0.7	85.23 ± 1.2

*—the differences versus the negative control samples are significant, according to the Mann–Whitney U-test with $p < 0.05$.

We then tested SucCS-DEX-10 for its anti-inflammatory activity. We used the two standard in vitro models of THP-1 cell activation, using TNFα or LPS to induce CD54 expression. CD54 (another name is intercellular adhesion molecule-1 or ICAM-1) is a cell-surface adhesion glycoprotein that is expressed on the surface of endothelial and immune system cells, including monocytes. Proinflammatory cytokines stimulation led to augmentation of CD54 expression by human alveolar macrophages [47] and peripheral blood neutrophil [48]. THP-1 cell activation results in increased expression of CD54 on its surface.

Recombinant TNFα was added at a final concentration of 2 ng/mL and incubated for 24 h. The effect on CD54 expression was studied by flow cytometry (Table 5). We detected a reduction in the TNFα-induced expression of CD54 in THP-1 cells in the presence of SucCS-DEX-10, but expression was elevated compared to samples not induced with TNFα.

Table 5. The influence of DEX, SucDEX, SucCS, and SucCS-DEX-10 solutions on CD54 expression in THP-1 cells under TNFα-stimulated inflammatory conditions, mean ± SE.

Sample	Concentration (μg/mL)	w/o TNF	TNFα (2 ng/mL)	TNF vs. w/o TNF, *p*
Negative control (*n* = 11)	-	0.51 ± 0.02	3.3 ± 0.5	<0.001
DEX (*n* = 6)	1	0.47 ± 0.03	1.9 ± 0.2	<0.001
	10	0.55 ± 0.05	2.01 ± 0.10	<0.001
SucDEX (*n* = 6)	1	0.46 ± 0.01	2.0 ± 0.2	<0.001
	10	0.46 ± 0.02	2.1 ± 0.2	<0.001
SucCS (*n* = 6)	1	0.43 ± 0.06	2.2 ± 0.4	<0.001
	10	0.43 ± 0.06	2.4 ± 0.4	<0.001
SucCS-DEX-10 (*n* = 6)	1	0.48 ± 0.02	1.2 ± 0.3 *	0.030
	10	0.49 ± 0.02	1.04 ± 0.19 **	0.018

* and **—differences with 2 ng/mL TNF-treated control sample are significant according to the Mann–Whitney U-test with $p = 0.004$ and $p = 0.008$, respectively.

We also tested the anti-inflammatory activity of SucCS-DEX-10 in LPS-treated THP-1 cells (Table 6). We observed a significant decrease in LPS-induced CD54 expression in THP-1 cells in the presence of SucCS-DEX-10, but no differences between LPS-treated and untreated samples.

Table 6. The influence of DEX, SucDEX, SucCS, and SucCS-DEX-10 solutions on CD54 expression on THP-1 cells under the LPS-stimulated inflammatory conditions, mean ± SE.

Sample	Concentration (μg/mL)	w/o LPS	LPS (1 μg/mL)	LPS vs. w/o LPS, *p*
Negative control (*n* = 11)	-	0.51 ± 0.02	2.6 ± 0.9	<0.001
DEX (*n* = 6)	1	0.47 ± 0.03	2.1 ± 0.9	0.012
	10	0.55 ± 0.05	3.0 ± 1.2	<0.001
SucDEX (*n* = 6)	1	0.46 ± 0.01	1.4 ± 0.5	<0.001
	10	0.46 ± 0.02	1.8 ± 0.8	<0.001
SucCS (*n* = 6)	1	0.43 ± 0.06	1.1 ± 0.2	0.016
	10	0.43 ± 0.06	1.0 ± 0.3 *	0.039
SucCS-DEX-10, (*n* = 6)	1	0.48 ± 0.02	0.69 ± 0.16 **	0.231
	10	0.49 ± 0.02	0.54 ± 0.12 **	0.693

* and **—differences with the 1 μg/mL LPS-treated control sample are significant according to the Mann-Whitney U-test, with $p = 0.047$ and $p = 0.016$, respectively.

Thus, despite the minor cytotoxic effect observed for all the tested samples in THP-1 cells cultured under standard conditions, we observed a significant anti-inflammatory effect of SucCS-DEX-10. This effect was demonstrated in both infectious (LPS-induced) and sterile (TNF-induced) models of inflammation. An interesting feature was that the anti-inflammatory effect was dose-independent in both models and was the same and significant at concentrations of 1 and 10 μg/mL. SucCS-DEX-10 significantly decreased the expression of CD54 in the THP-1 cells, and this can be interpreted as an inhibition of cell activation under inflammatory conditions. This effect was detected only for SucCS-DEX-10.

3. Materials and Methods

3.1. Materials and Reagents

In this work, we used CS from crab shells (Bioprogress, Russia) with a viscosity average molecular weight (Mη) of 37,000 and a degree of acetylation (DA) of 26% [49]. The intrinsic viscosity of CS was determined by viscometry using an Ubbelohde capillary viscometer (Design Bureau Pushchino, Russia) at 20 °C with 0.33 M acetic acid/0.3 M NaCl as the solvent. The Mη of CS was calculated using the Mark–Houwink equation: $[\eta] = 3.41 \times 10^{-3} \times M\eta^{1.02}$ [50]; $[\eta] = 1.56$ dL/g.

DEX, PBS, SA, tetrahydrofuran, 1-ethyl-3-(3-dimethylaminopropyl) carbodiimide hydrochloride (EDC), N-hydroxysuccinimide (NHS), sodium hydrogen carbonate, hydrochloric acid, trifluoroacetic acid, and deuterium oxide (D_2O) (99.9 atom %D) were obtained from Sigma-Aldrich Co. (St. Louis, MO, USA). Pyridine was dried over sodium hydroxide and distilled before use. Other reagents and solvents were obtained from commercial sources and were used without further purification.

3.2. Synthesis of SucDEX

SucDEX was prepared according to the following procedure, with some modifications [42]: DEX 0.2 g (0.51 mmol) and SA 0.255 g (2.55 mmol) were dissolved in 2.2 mL pyridine. The reaction mixture was left for 7 days at room temperature, and then the product was precipitated with 30 mL 1 M HCl for purification from pyridine. The resulting white precipitate was separated by centrifugation, washed 2 times with deionized water, and dried at 60 °C for 3 h.

3.3. Synthesis of the SucCS-DEX Conjugates

The CS (0.1 g, 0.53 mmol of N) was first dissolved in 5 mL 0.1 M HCl, followed by the addition of a certain amount of SucDEX dissolved in 0.5 mL tetrahydrofuran. EDC and NHS were then added to activate the carboxyl group of SucDEX, and the reaction mixture was stirred for 72 h at 20 °C (the molar ratio of the SucDEX, EDC, and NHS is presented in Table 1). The resulting product was precipitated with acetone, washed twice with acetone, filtered, dissolved in distilled water, dialyzed against distilled water for 3 days, and lyophilized. The synthesized CS-DEX (0.1 g, 0.45 mmol N) was then dispersed in 10 mL distilled water, SA was added at a 5-fold molar ratio (relative to the CS), and the reaction mixture was stirred at room temperature for approximately 7 h. Sodium hydrogen carbonate was added to the reaction mixture to bring it to pH 7–8. The resulting product was dialyzed for 3 days against distilled water and then freeze-dried.

3.4. Characterization of Conjugates

The ^1H NMR spectra were recorded using a Bruker Avance 400 instrument (Bruker, Billerica, MA, USA). Samples of SucDEX (10 mg) were dissolved in DMSO-d_6, and the spectrum was recorded at 20 °C. Samples of CS-DEX and SucCS-DEX were prepared by dissolving 5 mg of conjugates in D_2O. To protonate all amino groups, 5 μL trifluoroacetic acid was added to the solution. The spectra were recorded at 70 °C using a zgpr pulse sequence with suppression of residual H_2O.

The FTIR spectra were obtained in the attenuated total reflection mode using a Vertex-70 FTIR spectrometer (Bruker, Billerica, MA, USA) equipped with a ZnSe total reflection attachment (PIKE Technologies, Fitchburg, WI, USA). Elemental analysis (EA) was performed on a Vario EL CHN analyzer (Elementar, Langenselbold, Germany). The hydrodynamic diameter (2Rh) was measured by dynamic light scattering and the ζ-potential by electrophoretic light scattering on a Photocor Compact-Z device (Photocor, Moscow, Russia) equipped with a He-Ne laser (659.7 nm, 25 mV). The measurements were performed at a temperature of 20 °C with a scattering angle of 90°.

Particle morphology was studied by SEM using a Tescan Mira 3 instrument (Tescan, Brno, Czech Republic). Images were obtained in the secondary electron mode (SE) with an accelerating voltage of 20 kV; the distance between the sample and detector was 6 mm. The studied samples were placed on double-sided carbon tape and dried in a vacuum oven for 48 h prior to SEM observations.

3.5. Determination of DEX Content in the SucCS-DEX Conjugates

The DEX content in the SucCS-DEX was determined spectrophotometrically. To do this, the DEX concentration was determined in nanosuspensions (1 mg of SucCS-DEX per 20 mL double distilled water) at a wavelength of 242 nm using a calibration curve according to the equation y = 89.817x, R^2 = 0.9923, 10 mm quartz cuvette, UV-visible spectrophotometer UV-1700 Pharma Spec (Shimadzu, Kyoto, Japan). The DEX content was then calculated per 1 mg of the SucCS-DEX.

3.6. In Vitro DEX Release from the SucCS-DEX Conjugates

A 1 mg sample of SucCS-DEX was dispersed in PBS (4 mL, pH 7.4) and incubated at 37 °C for 30 days. At the selected time intervals, the nanosuspension was ultracentrifuged at 4500 rpm using a Vivaspin®Turbo4 5000 molecular weight cut-off centrifugal concentrator and a replenishing the volume of the dissolution medium with fresh buffer. The amount and chemical structure of the released DEX or SucDEX was determined by ultra-high performance liquid chromatography—mass spectrometry. Chromatographic separation was undertaken using an Elute UHPLC (Bruker Daltonics GmbH, Bremen, Germany), equipped with a Millipore Chromolith Performance/PR-18e, C18 analytical column (100 mm × 2 mm, Merck, Darmstadt, Germany) with a Chromolith® RP-18 endcapped 5-3 guard cartridges (Merck, Darmstadt, Germany), operated under a flow rate of 300 μL/min. Mobile phases were as follows: A = water: acetonitrile: formic acid—100:1:0.1 (*v*/*v*/*v*) and B = water: acetonitrile: formic acid—10:90:0.1 (*v*/*v*/*v*). Elution gradient was as follows: 0→1 min 40%B→50%B (linear gradient); 1→1.2 min 50%B→90%B (linear gradient); 1.2→2.2 min 90%B (isocratic); 2.2→2.4 min 90%B→40%B (linear gradient); 2.4→4 min 40%B (isocratic). The total run time of the method was 4 min; the injection volume was 5 μL. Mass spectra were obtained on a Maxis Impact Q-TOF mass spectrometer (Bruker Daltonics GmbH, Germany) equipped with an electrospray ionization (ESI) source (Bruker Daltonics GmbH, Germany). The instrument was operated in positive ionization mode with an electrospray voltage of 4.5 kV and a MS scanning range of 50–1000 m/z. The obtained mass spectra were analyzed using the DataAnalysis® and TASQ® software (Bruker Daltonics GmbH, Germany).

3.7. Anti-Inflammatory Activity of the SucCS-DEX Conjugates

THP-1 cells (human monocytic leukemia cells) were seeded in 50 mL plastic flasks (Sarstedt, Nümbrecht, Germany) at 37 °C in a humidified atmosphere containing 5% CO_2, and were maintained in RPMI-1640 culture medium (Biolot, St. Petersburg, Russia) supplemented with 10% fetal bovine serum (FBS; Gibco, Thermo Fisher Scientific Inc., Bartlesville, OK, USA), 50 μg/mL gentamicin (Biolot, Russia), and 2 mM L-glutamine (Biolot, Russia), at a cell density of 0.5–1 × 10^6 cells/mL with a medium change every 2–3 days.

For experiments, 200 μL of cell culture medium containing 1 × 10^5 cells in suspension were seeded into 96-well flat-bottom culture plates (Sarstedt, Germany). The THP-1 cells were activated in vitro by adding 2 ng/mL TNFα (BioLegend Inc., San Diego, CA, USA) or lipopolysaccharides (LPS) from *Escherichia coli* (Sigma-Aldrich, Merck KGaA, Darmstadt, Germany) to each well (unactivated cells served as a negative control). The test compounds (SucDEX, Suc-CS and Suc-DEX-CS-10) were added to the wells at final concentrations of 10 and 1 μg/mL (relative to DEX) and incubated with THP-1 cells for 24 h. The cells were then transferred to 75 × 12 mm flow cytometry tubes (Sarstedt, Germany) and washed with 4 mL sterile PBS (centrifugation at 300× *g* for 5 min). The washed cells were then resuspended in 50 μL fresh PBS, stained with mouse antibodies against human CD54

(Beckman Coulter Inc., Indianapolis, IN, USA) in the dark for 15 min, and washed again. Finally, the washed cells were resuspended in 100 μL fresh PBS and stained for 5 min with 250 nM PO-PRO-1 iodide (Invitrogen, Thermo Fisher Scientific, USA) and 3 μM DRAQ7 (Beckman Coulter Inc., USA), as described previously [17]. At least 10,000 single THP-1 cells per each sample were acquired. Flow cytometry data were obtained with a Navios™ flow cytometer (Beckman Coulter, USA) equipped with 405, 488, and 638 nm lasers and analyzed using Navios Software v.1.2 and Kaluza™ software v.2.0 (Beckman Coulter, USA). The data were presented as the percentage of viable cells per sample ± standard error (SE), and the intensity of CD54 expression was ultimately measured as mean fluorescence intensity (MFI) on the cell surface of viable THP-1 cells.

4. Conclusions

We designed a two-step synthesis of SucCS-DEX conjugates for potential application as a prolonged release intravitreal delivery system. We first used carbodiimide chemistry to conjugate CS to DEX via a succinyl linker to form a hydrolysable amide bond. The resulting conjugates had a DS_{DEX} of 2–4% (DEX content of 50–85 μg/mg), a size of 450–1000 nm, and a positive ζ-potential of 14–23 mV. Next, we modified the surface of the synthesized particles with SA to obtain a negatively charged system to level out the unwanted electrostatic interaction of CS with the vitreous contents following intravitreal injection.

As a result, we produced particles 400–1100 nm in size with a ζ-potential of −30 to −33 mV and a DS_{Suc} of 64–68%. The DEX content in the succinylated conjugates did not change compared to the initial conjugates, indicating an optimal selection of synthesis conditions and satisfactory reliability of the proposed procedure. In vitro tests showed that the developed conjugates sustained the release of the active pharmaceutical substance in the form of both DEX and SucDEX (8–10% in total for 1 month). Obviously, the vitreous environment has specific conditions (e.g., high viscosity and the presence of enzymes) that will affect the hydrolysis and release of the drug from this type of conjugate. Nevertheless, we expect that the achieved release profile will maintain therapeutic concentrations of DEX in the vitreous for several months due to the low biodegradation of succinylated CS and the sufficiently large particle size, thereby reducing side effects and the need for frequent injections. In addition, the developed conjugates demonstrated significant anti-inflammatory effects in both sterile (TNFα-induced) and infectious (LPS-induced) models of inflammation, as confirmed by 2- and 4-fold suppression, respectively, of CD54 expression in THP-1 cells. Based on the current results showing an improved release profile and the anti-inflammatory potential of the designed polymeric systems, we intend to expand this research to in vivo experiments aimed at creating an intravitreal DEX delivery system with improved pharmacological characteristics.

Supplementary Materials: The following are available online at https://www.mdpi.com/article/10.3390/ijms222010960/s1.

Author Contributions: Conceptualization, N.V.D. and Y.A.S.; methodology, N.V.D., A.S.G., Y.A.D. and I.V.K.; investigation, N.V.D., A.N.B., Y.A.D., I.V.K., M.K.S. and A.S.T.; writing—original draft preparation, N.V.D., A.N.B. and I.V.K.; writing—review and editing, Y.A.S.; supervision, Y.A.S.; project administration Y.A.S. All authors have read and agreed to the published version of the manuscript.

Funding: This research received no external funding.

Institutional Review Board Statement: Not applicable.

Informed Consent Statement: Not applicable.

Data Availability Statement: The data are contained within the article and Supplementary Materials.

Acknowledgments: The authors are grateful to A.V. Dobrodumov for conducting the NMR studies and to E.N. Vlasova for performing the FTIR spectroscopy analysis.

Conflicts of Interest: The authors declare no conflict of interest. The funders had no role in the design of the study; in the collection, analyses, or interpretation of data; in the writing of the manuscript, or in the decision to publish the results.

References

1. Bhattacharya, M.; Sadeghi, A.; Sarkhel, S.; Hagström, M.; Bahrpeyma, S.; Toropainen, E.; Auriola, S.; Urtti, A. Release of functional dexamethasone by intracellular enzymes: A modular peptide-based strategy for ocular drug delivery. *J. Control. Release* **2020**, *327*, 584–594. [CrossRef]
2. Liu, J.; Zhang, X.; Li, G.; Xu, F.; Li, S.; Teng, L.; Li, Y.; Sun, F. Anti-angiogenic activity of bevacizumab-bearing dexamethasone-loaded plga nanoparticles for potential intravitreal applications. *Int. J. Nanomed.* **2019**, *14*, 8819–8834. [CrossRef] [PubMed]
3. Yavuz, B.; Bozdağ Pehlivan, S.; Sümer Bolu, B.; Nomak Sanyal, R.; Vural, İ.; Ünlü, N. Dexamethasone–pamam dendrimer conjugates for retinal delivery: Preparation, characterization and in vivo evaluation. *J. Pharm. Pharmacol.* **2016**, *68*, 1010–1020. [CrossRef]
4. Raghava, S.; Hammond, M.; Kompella, U.B. Periocular routes for retinal drug delivery. *Expert Opin. Drug Deliv.* **2004**, *1*, 99–114. [CrossRef] [PubMed]
5. Urtti, A.; Salminen, L.; Miinalainen, O. Systemic absorption of ocular pilocarpine is modified by polymer matrices. *Int. J. Pharm.* **1985**, *23*, 147–161. [CrossRef]
6. Battaglia, L.; Gallarate, M.; Serpe, L.; Foglietta, F.; Muntoni, E.; del Pozo Rodriguez, A.; Aspiazu, M.A.S. Ocular delivery of solid lipid nanoparticles. In *Lipid Nanocarriers for Drug Targeting*; Elsevier: Amsterdam, The Netherlands, 2018; pp. 269–312.
7. Urtti, A.; Pipkin, J.D.; Rork, G.; Repta, A. Controlled drug delivery devices for experimental ocular studies with timolol 1. In vitro release studies. *Int. J. Pharm.* **1990**, *61*, 235–240. [CrossRef]
8. Dubashynskaya, N.V.; Poshina, D.N.; Raik, S.V.; Urtti, A.; Skorik, Y.A. Polysaccharides in ocular drug delivery. *Pharmaceutics* **2020**, *12*, 22. [CrossRef]
9. Cox, J.T.; Eliott, D.; Sobrin, L. Inflammatory complications of intravitreal anti-vegf injections. *J. Clin. Med.* **2021**, *10*, 981. [CrossRef]
10. Melo, G.B.; da Cruz, N.F.S.; Emerson, G.G.; Rezende, F.A.; Meyer, C.H.; Uchiyama, S.; Carpenter, J.; Shiroma, H.F.; Farah, M.E.; Maia, M.; et al. Critical analysis of techniques and materials used in devices, syringes, and needles used for intravitreal injections. *Prog. Retin. Eye Res.* **2021**, *80*, 100862. [CrossRef]
11. Li, Q.; Weng, J.; Wong, S.N.; Thomas Lee, W.Y.; Chow, S.F. Nanoparticulate drug delivery to the retina. *Mol. Pharm.* **2020**, *18*, 506–521. [CrossRef]
12. Kim, H.M.; Woo, S.J. Ocular drug delivery to the retina: Current innovations and future perspectives. *Pharmaceutics* **2021**, *13*, 108. [CrossRef]
13. Mehta, H.; Gillies, M.; Fraser-Bell, S. Perspective on the role of ozurdex (dexamethasone intravitreal implant) in the management of diabetic macular oedema. *Ther. Adv. Chronic Dis.* **2015**, *6*, 234–245. [CrossRef] [PubMed]
14. Nayak, K.; Misra, M. A review on recent drug delivery systems for posterior segment of eye. *Biomed. Pharmacother.* **2018**, *107*, 1564–1582. [CrossRef]
15. Ryu, M.; Nakazawa, T.; Akagi, T.; Tanaka, T.; Watanabe, R.; Yasuda, M.; Himori, N.; Maruyama, K.; Yamashita, T.; Abe, T.; et al. Suppression of phagocytic cells in retinal disorders using amphiphilic poly (γ-glutamic acid) nanoparticles containing dexamethasone. *J. Control. Release* **2011**, *151*, 65–73. [CrossRef] [PubMed]
16. Manickavasagam, D.; Oyewumi, M.O. Critical assessment of implantable drug delivery devices in glaucoma management. *J. Drug Deliv.* **2013**, *2013*, 895013. [CrossRef]
17. Dubashynskaya, N.V.; Golovkin, A.S.; Kudryavtsev, I.V.; Prikhodko, S.S.; Trulioff, A.S.; Bokatyi, A.N.; Poshina, D.N.; Raik, S.V.; Skorik, Y.A. Mucoadhesive cholesterol-chitosan self-assembled particles for topical ocular delivery of dexamethasone. *Int. J. Biol. Macromol.* **2020**, *158*, 811–818. [CrossRef] [PubMed]
18. Del Amo, E.M.; Vellonen, K.S.; Kidron, H.; Urtti, A. Intravitreal clearance and volume of distribution of compounds in rabbits: In silico prediction and pharmacokinetic simulations for drug development. *Eur. J. Pharm. Biopharm.* **2015**, *95*, 215–226. [CrossRef]
19. Shatz, W.; Hass, P.E.; Mathieu, M.; Kim, H.S.; Leach, K.; Zhou, M.; Crawford, Y.; Shen, A.; Wang, K.; Chang, D.P.; et al. Contribution of antibody hydrodynamic size to vitreal clearance revealed through rabbit studies using a species-matched fab. *Mol. Pharm.* **2016**, *13*, 2996–3003. [CrossRef]
20. Peynshaert, K.; Devoldere, J.; De Smedt, S.C.; Remaut, K. In vitro and ex vivo models to study drug delivery barriers in the posterior segment of the eye. *Adv. Drug Deliv. Rev.* **2018**, *126*, 44–57. [CrossRef]
21. Pitkanen, L.; Pelkonen, J.; Ruponen, M.; Ronkko, S.; Urtti, A. Neural retina limits the nonviral gene transfer to retinal pigment epithelium in an in vitro bovine eye model. *AAPS J.* **2004**, *6*, 72–80. [CrossRef]
22. Mains, J.; Wilson, C.G. The vitreous humor as a barrier to nanoparticle distribution. *J. Ocul. Pharmacol. Ther.* **2013**, *29*, 143–150. [CrossRef]
23. Bishop, P.N. Structural macromolecules and supramolecular organisation of the vitreous gel. *Prog. Retin. Eye Res.* **2000**, *19*, 323–344. [CrossRef]
24. Thakur, S.S.; Barnett, N.L.; Donaldson, M.J.; Parekh, H.S. Intravitreal drug delivery in retinal disease: Are we out of our depth? *Expert Opin. Drug Deliv.* **2014**, *11*, 1575–1590. [CrossRef]

25. Pitkanen, L.; Ruponen, M.; Nieminen, J.; Urtti, A. Vitreous is a barrier in nonviral gene transfer by cationic lipids and polymers. *Pharm. Res.* **2003**, *20*, 576–583. [CrossRef]
26. Martens, T.F.; Vercauteren, D.; Forier, K.; Deschout, H.; Remaut, K.; Paesen, R.; Ameloot, M.; Engbersen, J.F.; Demeester, J.; De Smedt, S.C.; et al. Measuring the intravitreal mobility of nanomedicines with single-particle tracking microscopy. *Nanomedicine* **2013**, *8*, 1955–1968. [CrossRef] [PubMed]
27. Shatz, W.; Aaronson, J.; Yohe, S.; Kelley, R.F.; Kalia, Y.N. Strategies for modifying drug residence time and ocular bioavailability to decrease treatment frequency for back of the eye diseases. *Expert Opin. Drug Deliv.* **2019**, *16*, 43–57. [CrossRef] [PubMed]
28. Jo, D.H.; Kim, J.H.; Lee, T.G.; Kim, J.H. Size, surface charge, and shape determine therapeutic effects of nanoparticles on brain and retinal diseases. *Nanomed. Nanotechnol. Biol. Med.* **2015**, *11*, 1603–1611. [CrossRef]
29. Altiok, E.I.; Santiago-Ortiz, J.L.; Svedlund, F.L.; Zbinden, A.; Jha, A.K.; Bhatnagar, D.; Loskill, P.; Jackson, W.M.; Schaffer, D.V.; Healy, K.E. Multivalent hyaluronic acid bioconjugates improve sflt-1 activity in vitro. *Biomaterials* **2016**, *93*, 95–105. [CrossRef]
30. Altiok, E.I.; Browne, S.; Khuc, E.; Moran, E.P.; Qiu, F.; Zhou, K.; Santiago-Ortiz, J.L.; Ma, J.-x.; Chan, M.F.; Healy, K.E. Sflt multivalent conjugates inhibit angiogenesis and improve half-life in vivo. *PLoS ONE* **2016**, *11*, e0155990. [CrossRef]
31. Famili, A.; Crowell, S.R.; Loyet, K.M.; Mandikian, D.; Boswell, C.A.; Cain, D.; Chan, J.; Comps-Agrar, L.; Kamath, A.; Rajagopal, K. Hyaluronic acid–antibody fragment bioconjugates for extended ocular pharmacokinetics. *Bioconjugate Chem.* **2019**, *30*, 2782–2789. [CrossRef]
32. Dubashynskaya, N.V.; Bokatyi, A.N.; Skorik, Y.A. Dexamethasone conjugates: Synthetic approaches and medical prospects. *Biomedicines* **2021**, *9*, 341. [CrossRef]
33. Skorik, Y.A.; Kritchenkov, A.S.; Moskalenko, Y.E.; Golyshev, A.A.; Raik, S.V.; Whaley, A.K.; Vasina, L.V.; Sonin, D.L. Synthesis of n-succinyl-and n-glutaryl-chitosan derivatives and their antioxidant, antiplatelet, and anticoagulant activity. *Carbohydr. Polym.* **2017**, *166*, 166–172. [CrossRef]
34. Pandit, J.; Sultana, Y.; Aqil, M. Chitosan coated nanoparticles for efficient delivery of bevacizumab in the posterior ocular tissues via subconjunctival administration. *Carbohydr. Polym.* **2021**, *267*, 118217. [CrossRef] [PubMed]
35. Elsaid, N.; Jackson, T.L.; Elsaid, Z.; Alqathania, A.; Somavarapu, S. Plga microparticles entrapping chitosan-based nanoparticles for the ocular delivery of ranibizumab. *Mol. Pharm.* **2016**, *13*, 2923–2940. [CrossRef] [PubMed]
36. Chaharband, F.; Daftarian, N.; Kanavi, M.R.; Varshochian, R.; Hajiramezanali, M.; Norouzi, P.; Arefian, E.; Atyabi, F.; Dinarvand, R. Trimethyl chitosan-hyaluronic acid nano-polyplexes for intravitreal vegfr-2 sirna delivery: Formulation and in vivo efficacy evaluation. *Nanomed. Nanotechnol. Biol. Med.* **2020**, *26*, 102181. [CrossRef]
37. Ugurlu, N.; Asik, M.D.; Cakmak, H.B.; Tuncer, S.; Turk, M.; Cag il, N.; Denkbas, E.B. Transscleral delivery of bevacizumab-loaded chitosan nanoparticles. *J. Biomed. Nanotechnol.* **2019**, *15*, 830–838. [CrossRef]
38. Villanueva, J.R.; Villanueva, L.R.; Navarro, M.G. Pharmaceutical technology can turn a traditional drug, dexamethasone into a first-line ocular medicine. A global perspective and future trends. *Int. J. Pharm.* **2017**, *516*, 342–351. [CrossRef]
39. Wong, C.W.; Wong, T.T. Posterior segment drug delivery for the treatment of exudative age-related macular degeneration and diabetic macular oedema. *Br. J. Ophthalmol.* **2019**, *103*, 1356–1360. [CrossRef]
40. Behar-Cohen, F. Recent advances in slow and sustained drug release for retina drug delivery. *Expert Opin. Drug Deliv.* **2019**, *16*, 679–686. [CrossRef] [PubMed]
41. Balasso, A.; Subrizi, A.; Salmaso, S.; Mastrotto, F.; Garofalo, M.; Tang, M.; Chen, M.; Xu, H.; Urtti, A.; Caliceti, P. Screening of chemical linkers for development of pullulan bioconjugates for intravitreal ocular applications. *Eur. J. Pharm. Sci.* **2021**, *161*, 105785. [CrossRef]
42. Suzuki, T.; Sato, E.; Tada, H.; TOJIMA, Y. Examination of local anti-inflammatory activities of new steroids, hemisuccinyl methyl glycolates. *Biol. Pharm. Bull.* **1999**, *22*, 816–821. [CrossRef]
43. Koo, H.; Moon, H.; Han, H.; Na, J.H.; Huh, M.S.; Park, J.H.; Woo, S.J.; Park, K.H.; Kwon, I.C.; Kim, K.; et al. The movement of self-assembled amphiphilic polymeric nanoparticles in the vitreous and retina after intravitreal injection. *Biomaterials* **2012**, *33*, 3485–3493. [CrossRef]
44. Chen, S.W.; Xin, Q.; Kong, W.X.; Min, L.; Li, J.F. Anxiolytic-like effect of succinic acid in mice. *Life Sci.* **2003**, *73*, 3257–3264. [CrossRef] [PubMed]
45. Kato, Y.; Onishi, H.; Machida, Y. Biological characteristics of lactosaminated n-succinyl-chitosan as a liver-specific drug carrier in mice. *J. Control. Release* **2001**, *70*, 295–307. [CrossRef]
46. Kato, Y.; Onishi, H.; Machida, Y. Evaluation of n-succinyl-chitosan as a systemic long-circulating polymer. *Biomaterials* **2000**, *21*, 1579–1585. [CrossRef]
47. Fattal-German, M.; Le Roy Ladurie, F.; Lecerf, F.; Berrih-Aknin, S. Expression of ICAM-1 and TNFα in human alveolar macrophages from lung-transplant recipients. *Ann. N. Y. Acad. Sci.* **1996**, *796*, 138–148. [CrossRef]
48. Yang, L.; Froio, R.M.; Sciuto, T.E.; Dvorak, A.M.; Alon, R.; Luscinskas, F.W. ICAM-1 regulates neutrophil adhesion and transcellular migration of TNF-α-activated vascular endothelium under flow. *Blood* **2005**, *106*, 584–592. [CrossRef] [PubMed]

49. Raik, S.V.; Poshina, D.N.; Lyalina, T.A.; Polyakov, D.S.; Vasilyev, V.B.; Kritchenkov, A.S.; Skorik, Y.A. N-[4-(n,n,n-trimethylammonium)benzyl]chitosan chloride: Synthesis, interaction with DNA and evaluation of transfection efficiency. *Carbohydr. Polym.* **2018**, *181*, 693–700. [CrossRef]
50. Pogodina, N.; Pavlov, G.; Bushin, S.; Mel'nikov, A.; Lysenko, Y.B.; Nud'ga, L.; Marsheva, V.; Marchenko, G.; Tsvetkov, V. Conformational characteristics of chitosan molecules as demonstrated by diffusion-sedimentation analysis and viscometry. *Polym. Sci. USSR* **1986**, *28*, 251–259. [CrossRef]

International Journal of
Molecular Sciences

MDPI

Article

Polycondensed Peptide Carriers Modified with Cyclic RGD Ligand for Targeted Suicide Gene Delivery to Uterine Fibroid Cells

Anna Egorova [1], Sofia Shtykalova [1], Marianna Maretina [1], Alexander Selutin [2], Natalia Shved [1], Dmitriy Deviatkin [1], Sergey Selkov [2], Vladislav Baranov [1] and Anton Kiselev [1,*]

1 Department of Genomic Medicine, D.O. Ott Research Institute of Obstetrics, Gynecology and Reproductology, Mendeleevskaya Line 3, 199034 Saint Petersburg, Russia; egorova_anna@yahoo.com (A.E.); sofia.shtykalova@gmail.com (S.S.); marianna0204@gmail.com (M.M.); natashved@mail.ru (N.S.); dimi02121@gmail.com (D.D.); baranov@vb2475.spb.edu (V.B.)

2 Department of Immunology and Intercellular Interactions, D.O. Ott Research Institute of Obstetrics, Gynecology and Reproductology, Mendeleevskaya Line 3, 199034 Saint Petersburg, Russia; a_selutin@yahoo.com (A.S.); selkovsa@mail.ru (S.S.)

* Correspondence: ankiselev@yahoo.co.uk

Abstract: Suicide gene therapy was suggested as a possible strategy for the treatment of uterine fibroids (UFs), which are the most common benign tumors inwomen of reproductive age. For successful suicide gene therapy, DNAtherapeutics should be specifically delivered to UF cells. Peptide carriers are promising non-viral gene delivery systems that can be easily modified with ligands and other biomolecules to overcome DNA transfer barriers. Here we designed polycondensed peptide carriers modified with a cyclic RGD moiety for targeted DNA delivery to UF cells. Molecular weights of the resultant polymers were determined, and inclusion of the ligand was confirmed by MALDI-TOF. The physicochemical properties of the polyplexes, as well as cellular DNA transport, toxicity, and transfection efficiency were studied, and the specificity of $\alpha v\beta 3$ integrin-expressing cell transfection was proved. The modification with the ligand resulted in a three-fold increase of transfection efficiency. Modeling of the suicide gene therapy by transferring the HSV-TK suicide gene to primary cells obtained from myomatous nodes of uterine leiomyoma patients was carried out. We observed up to a 2.3-fold decrease in proliferative activity after ganciclovir treatment of the transfected cells. Pro- and anti-apoptotic gene expression analysis confirmed our findings that the developed polyplexes stimulate UF cell death in a suicide-specific manner.

Keywords: DNA delivery; peptide-based carriers; gene therapy; thymidine kinase; uterine fibroids; integrins; polycondensation

Citation: Egorova, A.; Shtykalova, S.; Maretina, M.; Selutin, A.; Shved, N.; Deviatkin, D.; Selkov, S.; Baranov, V.; Kiselev, A. Polycondensed Peptide Carriers Modified with Cyclic RGD Ligand for Targeted Suicide Gene Delivery to Uterine Fibroid Cells. *Int. J. Mol. Sci.* **2022**, *23*, 1164. https://doi.org/10.3390/ijms23031164

Academic Editor: Yury A. Skorik

Received: 31 December 2021
Accepted: 18 January 2022
Published: 21 January 2022

1. Introduction

Gene therapy is a treatment by delivery of nucleic acid constructs into cells with the aim of repairing, adding, or removing a genetic sequence, or targeting genetic information processes [1]. Gene therapy can be used to treat a wide range of inherited and acquired diseases. According to the latest data from The Journal of Gene Medicine, most gene therapy clinical trials focus on the treatment of both malignant and benign tumors (http://www.abedia.com/wiley; accessed on 1 September 2021).

The most common benign tumors in women of reproductive age are uterine fibroids (UFs). UFs can cause profuse uterine bleeding, pelvic pain, complications inpregnancy, and infertility, and they arethe most common cause of myomectomy [2,3]. There are three major types of UF—submucosal, subserosal, and intramural fibroids [4]. The latter grow within the muscular uterine wall and imply certain difficulties and increased traumatism of the healthy myometrium during myomectomy, which makes this approach undesirable for treatment of intramural fibroids in comparison with hormonal therapy [5]. In fact, in the case of UFs with an intramural location in infertile women, the surgical treatment is controversial as there is no evidence that a myomectomy increases spontaneous or IVF-assisted fertility [4,6]. The fibroids are easily accessible by various endoscopic methods that make this disease a promising target for in situ gene therapy [7]. Delivery of apoptosis-inducing genes can be an effective approach to UF gene therapy. Most studies on suicide gene therapy use the delivery of the herpes simplex virus thymidine kinase (HSV-TK) gene followed by treatment with prodrugs, such as acyclovir or ganciclovir (GCV) [8]. A significant advantage of this approach is the so-called "bystander effect", the phenomenon based on the fact that phosphorylated GCV enters neighboring non-transfected cells through intercellular contacts widely spread in UF cells compared to normal myometrium; this in fact increases the efficiency of therapy without negative effects on healthy tissues [9,10]. Thus, the proposed gene therapy approach may be beneficial for UF treatment, especially for fibroids with an intramural location.

One of the main reasons for the limited widespread application of gene therapy is the lack of simultaneously effective, safe, and inexpensive methods for nucleic acid (NA) delivery. It is known that in most cases unprotected NAs are not able to overcome extracellular and intracellular barriers independently and easily degrade [11]. To date, the adenovirus (Ad) vectors are the most studied as delivery vehicles for UF gene therapy. The various modifications of Ad vectors are being developed to increase the efficiency and specificity of gene delivery into UF cells in vitro and in vivo [12–15]. However, the use of Ad vectors for systemic gene delivery is associated with the risk of the virus capsid proteins' interaction with blood components whichcan cause a systemic inflammatory response [16].

Non-viral vectors are being actively developed as an alternate delivery method in order to improve safety, reduce the difficulties in the production and storage of gene therapeutics, and also lower their cost. In terms of efficiency, they are still inferior to viral vectors, but these compounds are actively being developed in order to demonstrate their promise as successful gene delivery vectors for clinical use. The first application of the non-viral approach for HSV-TK gene delivery into UF cells was demonstrated earlier by Niu and colleagues using calcium phosphate-facilitated transfection. The authors emphasized the advantage of non-viral vector application associated with the lack of immune response to pDNA delivery and the susceptibility of UF cells to non-viral gene transfer [17].

Today, cationic lipids and polymers are the most promising non-viral gene delivery systems. Cationic lipids require extensive formulation work to optimize combinations and concentrations of liposomal components, whereas cationic polymers tend to complex with nucleic acids into smaller and more uniform nanoparticles. Cationic peptide carriers based on the cell-penetrating peptides (CPP) have additional advantages, such as biodegradability and better interaction with cellular membranes, enhancing transfection efficiency [18–21]. Arginine-rich CPPs form stable complexes with nucleic acids and promote their intracellular uptake due to hydrogen bond formation with various functional groups of cell surface glycosaminoglycans [22,23]. The structure of peptide-based carriers can be easily modified by the inclusion of different functional amino acids. For example, the inclusion of histidine provides the release of complexes into the cytoplasm after endosome destruction due to the "proton sponge" effect [24,25], whereas the inclusion of terminal cysteines enables the formation of higher molecular weight polypeptide-like polymers cross-linked by intermolecular disulfide bonds [26]. In the cytoplasm, the disulfide bonds are degraded by glutathione with the subsequent NA release, while the carrier toxicity is decreased to the level of low-molecular weight peptides [27]. There are two approaches to the formation of cross-linked polypeptide-like polymers: matrix polymerization, where interpeptide disulfide bonds are formed during NA complexation, and oxidative polycondensation, using dimethyl sulfoxide (DMSO) as a catalytic agent [28,29]. A number of studies have shown an advantage in transfection efficiency of the polycondensed peptide carriers compared to the matrix-polymerized carriers [26,29,30]. Recently, we observed a significant decrease in the survival of GCV-treated primary UF cells after delivery of the HSV-TK gene mediated by the polycondensed peptide carrier, polyR [31].

For successful gene therapy implementation, the NA complexes have to enter the cell specifically. In tumor gene therapy, the widespread approach is to modify carriers with the RGD motif (arginine–glycine–aspartic acid) for binding cell surface $\alpha v\beta 3$ integrins. This type of integrin is overexpressed on the surface of tumor cells compared to normal tissues [32]. Importantly, $\alpha v\beta 3$ integrins are widely expressed in UF cells but not in normal myometrium, which makes them promising for targeted UF gene therapy [33]. Previously, an increase in efficiency and specificity of UF cell transduction after Ad vector modification with the cyclic RGD (arginine–glycine–aspartic acid) ligand has been demonstrated [12–14]. In addition, we showed recently that combining the matrix-polymerized arginine-rich peptide carriers with the iRGD-modified peptides resulted in an increase in cell transfection efficiency and successful suicide gene therapy of UF primary cells [34].

Here, we present the development and detailed characterization of polycondensed peptide carriers modified with a cyclic RGD moiety during the polycondensation process. Molecular weights of the resultant polymers and inclusion of the ligand were determined by MALDI-TOF mass spectrometry. The physicochemical properties of the obtained peptide/DNA complexes (the size and zeta-potential of the nanoparticles, the efficiency of DNA complex formation, etc.), as well as the quantitative assessment of the complexes' transport into cells, toxicity, transfection efficiency, and specificity were studied in vitro. Modeling of the suicide gene therapy by the HSV-TK gene transfer into primary cells obtained from myomatous nodes of patients with UFs, followed by an assessment of the proliferative activity, apoptosis of the UF cells, along with apoptosis marker expression, was carried out.

2. Results and Discussion

2.1. Carrier Design and Molecular Weight Characterization

In this work, we obtained the R6p-cRGD carrier via oxidative polycondensation of arginine–histidine–cysteine-rich peptide R6 monomers with the inclusion of cyclic RGD-ligand in the reaction mixture (Table 1). The combination of arginine, histidine, and cysteine in the composition of the peptide-based gene delivery systems was proven to facilitate the effective entry of NA complexes into the cells and provide endosome disruption [35–37]. The cysteine thiol group in the composition of the cyclic RGDligand was protected by a 3-nitro-2-pyridinesulfenyl group (Npys), which acts as an activating group for disulfide bond formation, as it is displaced by the free thiol during oxidative polycondensation of R6 monomers [38]. Previously, we have shown that the peptide-based polyplexes require not less than 50% modification by the RGDligand to achievespecific gene delivery into $\alpha v\beta 3$ integrin-expressing cells [37]. Due to this fact, we included a cyclic RGD moiety in the reaction mixture at the rate of one ligand molecule per two R6 monomers.

Table 1. Design and composition of the synthesized peptides and peptide carriers.

	Name	Composition
Monomers	R6	CHRRRRRRHC
	cRGD	C(*Npys*)RGDy
Carriers	R6p	$(CHRRRRRRHC)_n$
	R6p-cRGD	cRGD-$(CHRRRRRRHC)_n$-cRGD

The molecular weight of the R6p and R6p-cRGD carriers formed by oxidative polycondensation was determined with MALDI-TOF mass spectrometry (Tables S1 and S2). It was found that the molecular weight of formed R6p polypeptides varied within 2.87–27.42 kDa, which corresponds to polymers with a degree polymerization of monomers from 2 to 19. MALDI-TOF analysis of R6p-cRGD also revealed a mixture of oligomers corresponding to 1–18 peptide monomers modified with the cRGD ligand. It should be noted that the absence or partial cRGD modification was also observed (Table S2). Nevertheless, it can be concluded that the addition of the cRGD ligand bearing a Npys-protected thiol group during polycondensation results in successful carrier modification. Moreover, despite the fact that cRGD is essentially a chain breaker during polycondensation, we did not observe a significant decrease in the degree of polymerization of the R6p-cRGD carrier.

2.2. DNA-Binding and DNA Protection Properties of the Carrier

The degree of DNA binding with the polycondensed carriers was assessed by ethidium bromide displacement assay (Figure 1). An increase of the carriers' concentration gradually resulted in an increase in the density of complexes. The EtBr fluorescence intensity in peptide/DNA complexes sharply decreased at anN/P ratio of three(up to 3–8%) compared to that of naked DNA (100%).

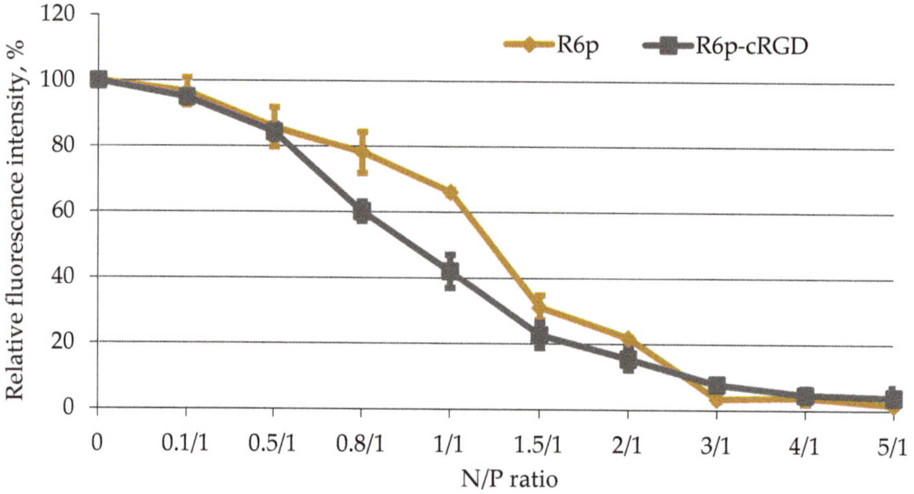

Figure 1. EtBr displacement assay of DNA complexes with R6p and R6p-cRGD carriers. Values are the mean ± SD of the mean of triplicates.

The DNA-binding capacity of carriers directly influences nuclease protection properties [39]. DNA integrity in the complexes was estimated using a DNase I protection assay. The studied polycondensed carriers were able to protect DNA from nuclease degradation at the same N/P ratio of three which corresponds to full DNA condensation (Figure 2).

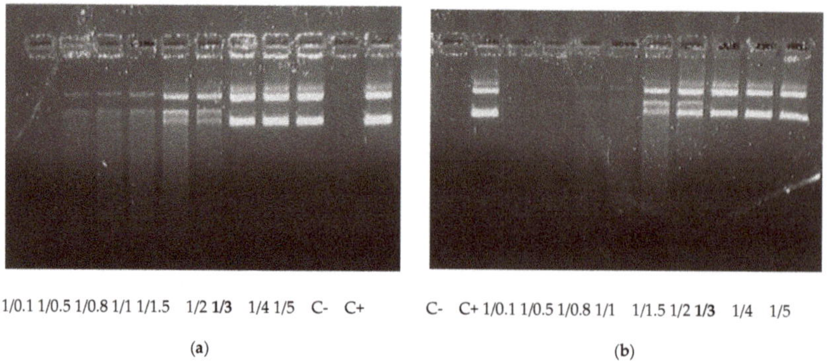

Figure 2. DNase I protection assay of DNAcomplexes formed with R6p (**a**) and R6p-cRGD (**b**) carriers. Charge ratio (CR) in **bold** indicates full DNA protection. C− = 'naked' plasmid DNA treated with DNase I; C+ = untreated plasmid DNA.

Thus, the obtained results showed no effect of the ligand inclusion inthe R6p-cRGD carrier composition on its DNA-binding and protection abilities. This result may have a positive effect on the specificity of DNA delivery because no interference between DNA-binding and receptor-binding properties could be expected.

2.3. Size and ζ-Potential of the Carrier/DNA Complexes

The size of peptide/DNA complexes is one of the key parameters affecting the mechanism of cell penetration [40]. For example, 100–200 nm-size complexes enter the cell via clathrin-mediated endocytosis, while larger ones penetrate into the cell by macropinocytosis [41]. Caveolin-mediated endocytosis, or potocytosis, may be used when the particle size does not exceed 100 nm [42]. Previous studies of non-viral carrier/DNA complexes revealed that particles of 70–200 nm have the longest duration of bloodstream circulation and the highest efficiency of tumor penetration [43]. We determined the size and ζ-potential of carrier/DNA complexes at N/P ratios 8/1 and 12/1, which provide full DNA binding/protection and should be colloidally stable. It was found that the size of the studied peptide/DNA complexes varied in the range of 100–200 nm (Table 2), and thus in the optimal range for cell membrane penetration via clathrin-mediated endocytosis. Importantly, cRGD ligand-modified nanoparticles were shown to internalize through the clathrin-mediated pathway [44].

Table 2. Size and ζ-potential of the carrier/DNA complexes.

Carrier	N/P Ratio	Size (nm) ± SD	ζ-Potential (mV) ± SD
R6p	8/1	120.9 ± 0.17	35.6 ± 0.21
R6p	12/1	110.3 ± 9.12	35.5 ± 0.81
R6p-cRGD	8/1	104.5 ± 0.15	32.0 ± 1.17
R6p-cRGD	12/1	178.4 ± 24.32	31.5 ± 0.61

The surface charge density of the carrier/DNA complex plays an important role in gene delivery efficiency [45]. To begin with, the positive charge of the carrier is substantial for DNA condensation, and the positive surface charge of the formed complex may increase cellular uptake due to electrostatic interaction with the negatively charged cell membrane [46]. The studied peptide/DNA polyplexes were found to be positively charged; their ζ-potential varied within a range from +30 to +36 mV (Table 2). This may contribute to increased cellular uptake and, as a result, to enhanced gene expression.

2.4. DTT and DS Treatment of the Complexes

DNA release is an essential step for the manifestation of its bioactivity [47]. R6p and R6p-cRGD carriers contain disulfide bonds designed to facilitate DNA release into the cytoplasm. In order to determine the rate of the DNA release after disulfide bond reduction, we used dithiothreitol (DTT) at a concentration of 200 mM to mimic a cytoplasmic glutathione interaction [48]. Before DTT treatment, we ensured complete DNA condensation by the tested carriers at a charge ratio of 8/1. DTT treatment for 1 h resulted in a 2–7-fold increase in fluorescence intensity, which corresponds to DNA release from the polyplexes (Figure 3). However, the relative level of fluorescence after incubation with DTT compared to free DNA (100%) was not very high, which indicates that electrostatic interaction is more important for the stability of complexes between DNA and arginine-rich carriers [34]. Additional data on DNA release were obtained after treatment of the complexes with negatively charged glycosaminoglycan in three-fold charge excess (Figure 4).

Glycosaminoglycans (GAGs), such as heparan sulfate (HS), chondroitin sulfate (CS), and dextran sulfate (DS), influence several aspects of peptide-mediated DNA transfer. It has been shown that GAGs may play a protective role against cytotoxicity associated with polycations [49]. On the other hand, they are capable of disrupting peptide/DNA complex integrity in the extracellular environment and significantly inhibit gene transfection [50,51]. After 24 h of DS treatment, an increase in the fluorescence intensity was detected, due to the relaxation of these complexes and a decrease in their density (Figure 4). For R6p and R6p-cRGD polyplexes, the fluorescence increased from 0.3% to 36.3% and from 2.3% to 63.3%, respectively. Notably, the cRGD-modified carrier released DNA more efficiently than the R6p carrier (Figure 4).

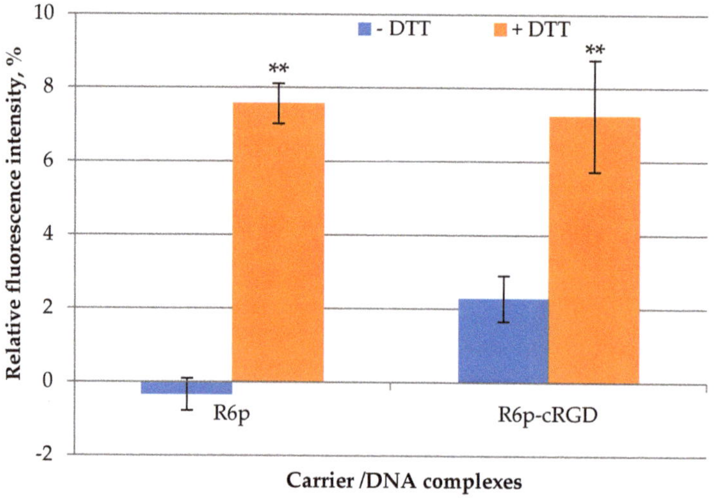

Figure 3. DNA release after DTT treatment of DNAcomplexes with R6p and R6p-cRGD carriers formed at anN/P ratio 8/1. Values are the mean ± SEM of the mean of triplicates. ** $p < 0.01$ compared to untreated complexes.

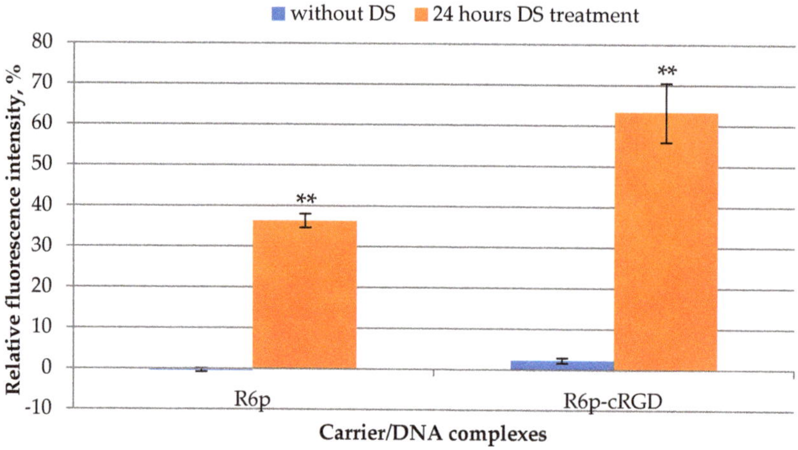

Figure 4. Relaxation of DNAcomplexes with R6p and R6p-cRGD carriers (N/P ratio 8/1) after 24 h of DS treatment in three-fold charge excess. Values are the mean ± SEM of the mean of triplicates. ** $p < 0.01$ compared to untreated complexes.

A delicate balance holds between DNA condensation and release. Too weak DNA binding might be insufficient to protect DNA against extra- and intracellular nuclease degradation, whereas too tight DNA binding might slow down or prevent DNA release from the carrier and hinder gene transcription [52]. A better understanding of nucleic acid binding mechanisms would be beneficial in designing more efficient gene delivery systems.

2.5. Cytotoxicity of Peptide/DNA Complexes

Gene transfection usually correlates with the perturbation of cellular membranes, which may elicit unwanted cytotoxicity [53,54]. Moreover, the relatively high net positive charge density of the carrier/DNA complex (>30 mV) also might induce cellular toxicity [55]. It is important to develop gene delivery carriers that have reduced toxicity without a decrease in transfection efficiency. One strategy is the incorporation of biodegradable linkers in the carrier composition [56,57]. The inclusion of the disulfide interpeptide bonds in R6p and R6p-cRGD polymers resulted in their cytoplasmic reduction, which was confirmed by DNA release after DTT treatment (Figure 4). A decrease in carrier molecular weight and, as a result, the reduction of positive charge density could decrease cytotoxicity. The cytotoxicity of R6p/DNA and R6p-cRGD/DNA complexes was determined in αvβ3-positive PANC-1 cells (Figure 5). Naked DNA was used as a negative control, and PEI/DNA complexes at an N/P ratio of 8/1 were used as a positive control. For the R6p/DNA complexes, cell viability exhibited a slightly decreasing trend with the increase of the N/P ratio and was significantly lower than that of intact cells and naked DNA. In the case of R6p/DNA complexes at anN/P of 12/1, the numberof living cells was reduced to 58% and was lower compared to PEI/DNA complexes. However, the R6p-cRGD/DNA complexes had relatively low cytotoxicity in PANC-1 cells at all tested N/P ratios; the cell viability after the transfection was comparable with that of naked DNA.

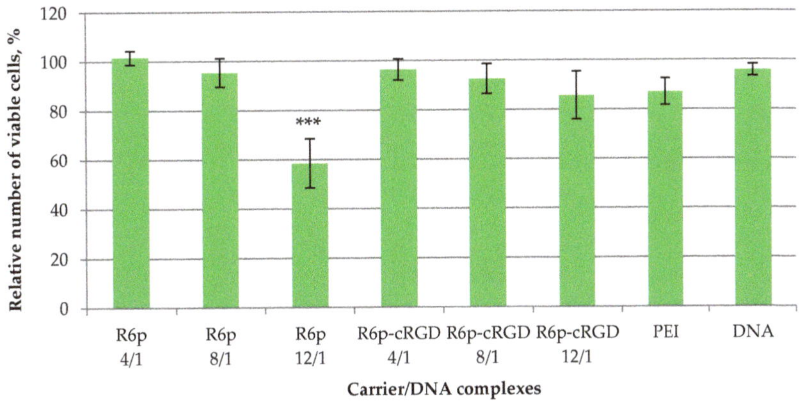

Figure 5. Cytotoxicity after transfection of DNA complexes formed with RGD1-R6, RGD0-R6, and R6 carriers at charge ratios 4/1, 8/1, and 12/1. Values are the mean ± SD of the mean of triplicates. *** $p < 0.001$ compared to intact cells.

2.6. Cellular Uptake

The intracellular uptake of DNA condensed by R6p or R6p-cRGD was determined by flow cytometry (Figure 6). To prove targeted DNA delivery with the R6p-cRGD carrier we used αvβ3-negative 293T cells and αvβ3-positive PANC-1 cells, which express a different level of αvβ3integrins on their surface [37]. The normalized fluorescence intensity of YOYO-1-labeled DNA taken up by cells 2 h after transfection is demonstrated in Figure 6. The complexes were studied at an 8/1 charge ratio because they do not exhibit toxicity at this N/P ratio, and no decrease in cellular uptake can be expected due to possible negative effects on the cell membrane (Figure 5). In PANC-1 cells, R6p-cRGD/DNA complexes showed higher intracellular uptake efficiency, 9.3times more efficient than R6p/DNA complexes. The data obtained are consistent with the other studies that proved an increase in the penetrating ability of the polyplexes modified with the αvβ3integrin ligand [58]. For 293T cells, differences in the cell penetration efficiency of ligand-modified and unmodified polyplexes were also found. However, R6p-cRGD/DNA polyplexes entered the cell only 3.7 times more efficiently than R6p/DNA complexes. Thus, we observed a direct correlationbetween the intracellular uptake efficiency of complexes and the presence of αvβ3 integrins on cell surfaces. It should be noted, that the cellular uptake efficiency depends not only on the presence of the ligand part in the carrier but also on polyplex size and other physicochemical properties [42]. Importantly, the size of R6p/DNA and R6p-cRGD/DNA complexes lies in the range of 100–200 nm (Table 2) and is optimal for clathrin-mediated endocytosis. Thus, it can be concluded that the polyplexes likely do not differ in terms of penetration, which further confirms the specificity of R6p-cRGD/DNA polyplexes uptake by receptor-mediated endocytosis via αvβ3 integrins.

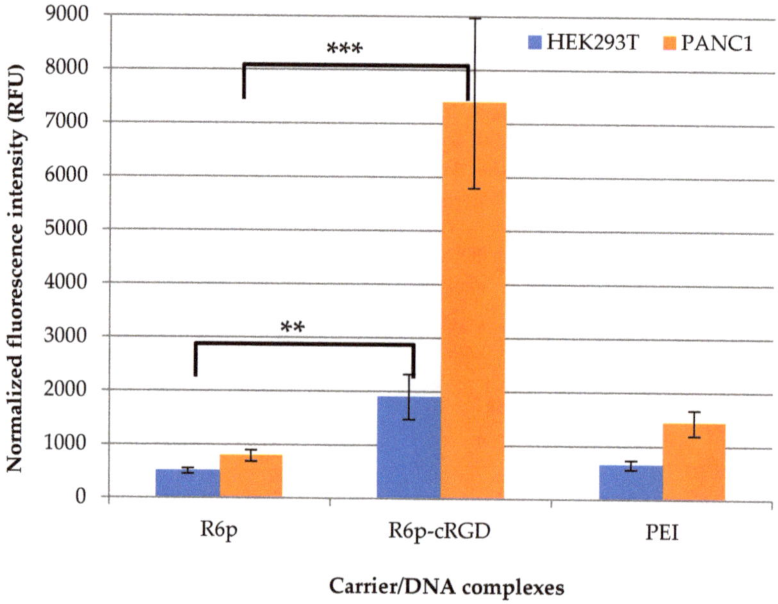

Figure 6. Normalized fluorescence intensity of PANC-1 and 293T cells after uptake of R6p/DNA and R6p-cRGD/DNA complexes at 8/1 charge ratios labeled with YOYO-1. ** $p < 0.01$, *** $p < 0.001$, when compared with R6p/DNA polyplexes.

2.7. Transfection Studies

In order to determine the transfection efficiency of R6p/DNA and R6p-cRGD/DNA complexes in αvβ3-positive PANC-1 cells, pCMV-lacZ was used as a reporter gene plasmid. PANC-1 cells were treated with the polyplexes at N/P ratios of 4/1, 8/1, and 12/1, with pCMV-lacZ only as a negative control and PEI/DNA complexes as a positive one (Figure 7a). At a 4/1 charge ratio, a significant difference in the transfection efficacy of ligand-modified and non-modified polyplexes was not found. As we showed previously at this charge ratio, iRGD-modified polyplexes could not provide ligand-mediated cellular penetration and subsequently could not result in enhanced transfection efficacy [37]. Beta-galactosidase activity after cell transfection with R6p-cRGD/DNA polyplexes at 8/1 and 12/1 N/P ratios was 2–2.5 times higher than that with R6p/DNA complexes; moreover, in most cases the transfection efficacy of ligand-modified polyplexes exceeded that of PEI/DNA complexes. These results demonstrate an increased transfection activity of R6p-cRGD/DNA polyplexes that was additionally proved using pEXPR-IBA5-eGFP plasmid transfer. In this instance, the transfection efficiency of PANC-1 cells mediated by the complexes was evaluated by flow cytometry analysis and estimated as a percentage (%) of transfected cells (Figure 7b). The R6p-cRGD/DNA transfection resulted in 43% of GFP-positive PANC-1 cells compared with 13.5% of GFP-positive cells after using DNA R6p/DNA complexes. Moreover, the efficacy of R6p-cRGD/DNA complexes was higher than that of PEI polyplexes (24%). The result demonstrated the high efficiency of R6p-cRGD polypeptides for DNA delivery in αvβ3-positive PANC-1 cells, and, thus, indirectly confirmed the specificity of gene transfer by the cRGD-modified carrier.

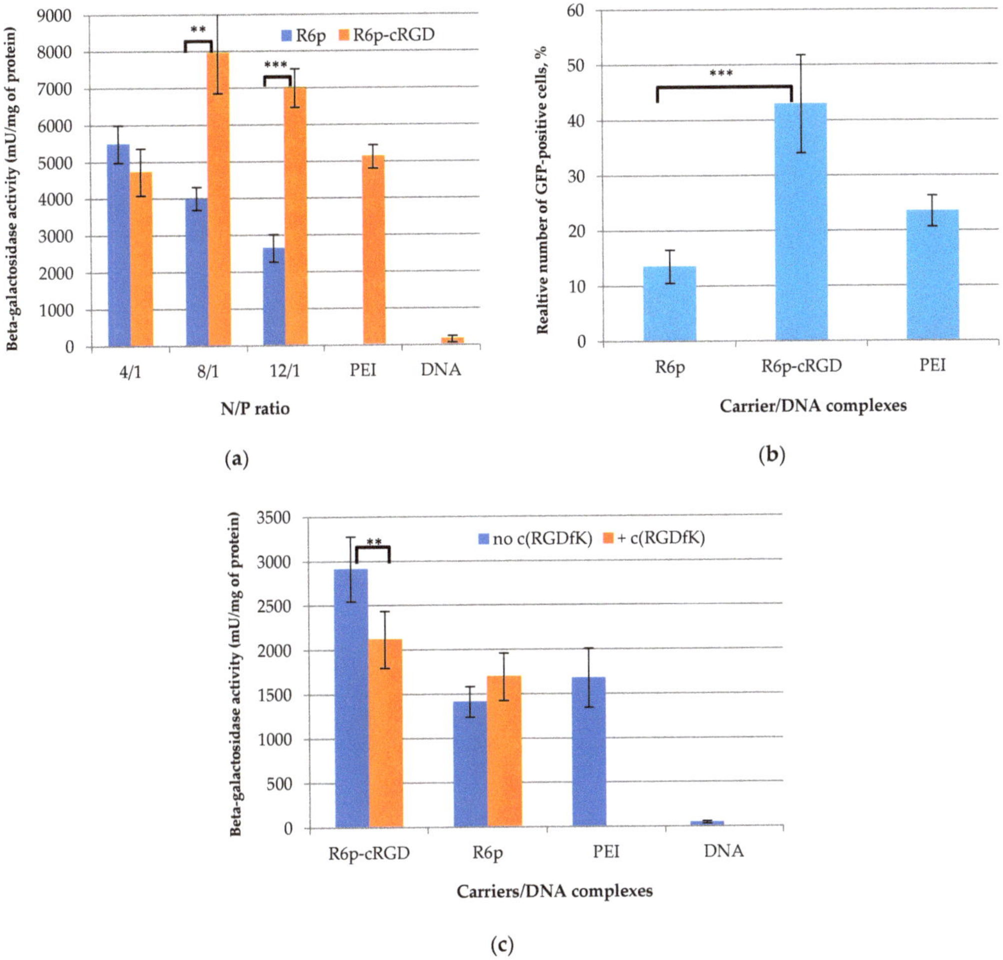

Figure 7. Transfection efficacy evaluation in the PANC-1 cells: (**a**) the cells were transfected with R6p/DNA and R6p-cRGD/DNA complexes at charge ratios of 4/1, 8/1, and 12/1, 'naked' plasmid pCMV-lacZ alone, and PEI/DNA complexes; (**b**) the cells were transfected with complexes of the pEXPR-IBA5-eGFP plasmid with R6p and R6p-cRGD carriers at a charge ratio of 8/1 withPEI/DNA polyplexes as a control; (**c**) the cells were transfected in the presence of the free cyclo(RGDfK) ligand with R6p/DNA and R6p-cRGD/DNA complexes at a charge ratio of 8/1, with thepCMV-lacZ plasmid alone and PEI/DNA polyplexes used as controls. *lacZ* gene expression was calculated as milliunits (mU) of beta-galactosidase per milligram of total cell extract protein. GFP gene expression was presented as a percentage of GFP-positive cells determined by flow cytometry assay. Values are the mean ± SD of the mean of triplicates. ** $p < 0.01$, *** $p < 0.001$, when compared with R6p/DNA polyplexes (**a**,**b**) and after free cyclo(RGDfK) ligand addition in transfection medium (**c**).

In order to further prove the selectivity of the R6p-cRGD-mediated transfection of αvβ3-positive cells, PANC-1 cells were pre-treated with free cyclo(RGDfK) peptide ligand. To perform competitive studies PANC-1 cells were transfected with R6p/DNA and R6p-cRGD/DNA complexes at an N/P ratio of 8/1 in the presence of a 10-fold excess of cyclo(RGDfK) (Figure 7c). The cell pre-treatment resulted in a 27% decrease in the efficacy of the R6p-cRGD/DNA polyplexes to the level of R6p/DNA complexes. However, the transfection efficiency after cell treatment with R6p/DNA polyplexes was not changed in the presence of the cyclo(RGDfK) ligand. Thus, we demonstrated the direct involvement of the cRGD ligand in the complexes' transfection via αvβ3 integrin binding.

To sum up, the obtained results proved that the cRGD-modified peptide carriers are favorable for transfection of αvβ3-positive cell lines, due to the specific recognition of the ligand-modified polyplexes by these cells that is beneficial for cellular uptake and the transfection process. So, the polycondensed cRGD-modified carrier can be applied as a targeted gene delivery vehicle for gene therapy purposes. Previously, it was shown that αvβ3 integrin targeting increases UF cells transfection efficacy compared to normal myometrial cells, indicating that an RGD moiety can selectively target uterine fibroids [13].

2.8. Assessment of Suicide Gene Therapy Effect in GSV-Treated Uterine Fibroid Cells after R6p-cRGD/pPTK1 Polyplex Transfection

There are no conservative and sufficiently safe approaches to UF therapy for women who want to keep their fertile potential [59]. The HSV-TK/GCV suicide gene therapy seems to be one of the most promising therapeutic strategies for UF treatment as a localized and organ-preserving method. The strategy has been actively developed for UF therapy and its successful application has been demonstrated in in vitro and in vivo studies [7]. The experimentally observed "bystander effect" arising from a large number of gap-junctions in uterine leiomyoma cells greatly enhanced the efficiency of this gene therapy approach [10,17]. It seems important that 48% of human UF cell death occurred when only 5% of the cells were transfected with the HSV-TK gene [10]. In addition, the reduced level of these junctions in normal myometrium compared to UFs will allow safety in suicide gene therapy for surrounding tissues. Thus, the studied approach could be a basis for future clinical trials of uterine leiomyoma.

UF cells overexpress αvβ3 integrins on their surface; our previous data demonstrated that up to 73% of UF cells can contain these receptors [34]. Importantly, when both viral and non-viral vehicles were modified with the RGD motif, a significant increase in the efficiency of suicide gene therapy was demonstrated [13–15,34].

Herein, the suicide effect of R6p-cRGD/pPTK1 polyplexes at N/P ratios of 8/1 and 12/1 with 0.7 μg and 0.35 μg of DNA per well was evaluated in primary UF cells, obtained from patients after myomectomy. The cells were treated with GCV at a concentration of 50 μg /mL. Assessment of the suicide gene therapy effect included determination of cell proliferation efficiency using the AlamarBlue assay (Figure 8a) and the amount of living cells by the trypan blue exclusion method (Figure 8b). For a visual demonstration of the obtained results, we present micrographs of UF cells (Figure 9). Further, the evaluation of relative apoptotic and necrotic cell number by ApoDETECT annexin V-FITC kit (Figure 10), and pro- and anti-apoptotic factor expression analysis were carried out (Figure 11). The RGD1-R6/pCMV-lacZ polyplexes were used to prove suicide gene therapy effects rather than those associated with polyplex cytotoxicity. Intact cells and naked DNAs treated with GCV also served as controls.

An AlamarBlue assay conducted 4 days after GCV treatment demonstrated the effect of suicide gene therapy for R6p-cRGD/pPTK complexes (Figure 8a). Cell transfection with naked DNA, both pPTKandpCMV-lacZ, showed no decrease in cell viability in the presence of GCV. The therapeutic effect was observed in UF cells treated with GCV after R6p-cRGD/pPTK transfection at 8/1 and 12/1 charge ratios; we registered a 1.7–2.3-fold decrease in the cells' proliferative activity compared to R6p-cRGD/pCMV-lacZ polyplexes. It should be noted that no suicide effect occurred when UF cells were transfected with PEI/pPTK1 complexes. The cytotoxicity caused by R6p-cRGD/pCMV-lacZ complexes with 0.35 μg of DNA did not exceed that of PEIpolyplexes. However, the DNA dose escalation resulted in an increase of polyplex cytotoxicity. Nevertheless, R6p-cRGD/pPTK polyplexes at an N/P ratio of 8/1 and a DNA dose of 0.35 μg were less toxic and showed high suicide efficiency at the level of 57% (Figure 8a).

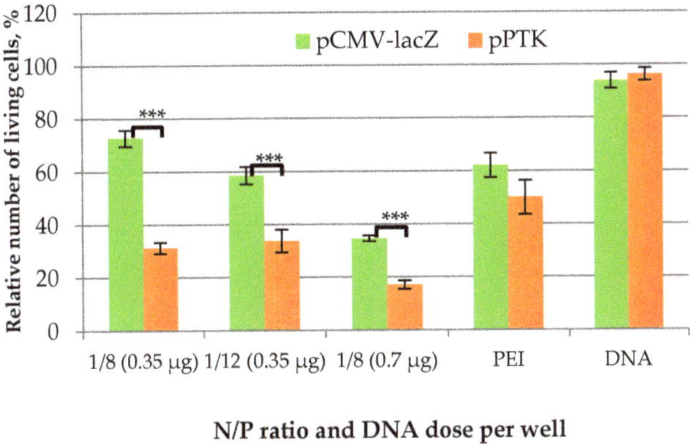

(a)

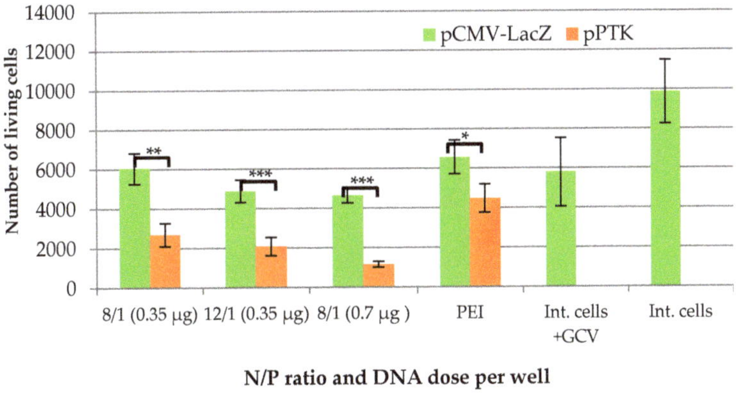

(b)

Figure 8. UF cell viability (**a**) and the number of living UF cells (**b**) after HSV thymidine kinase expression and GCV treatment as determined by Alamar Blue assay (**a**) and Trypan Blue exclusion assay (**b**). Values are the mean ± SEM of the mean of triplicates. * $p < 0.05$, ** $p < 0.01$, *** $p < 0.001$, compared to pCMV-LacZpolyplexes.

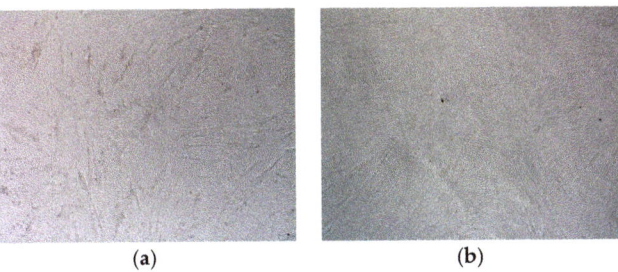

(**a**)

(**b**)

Figure 9. *Cont.*

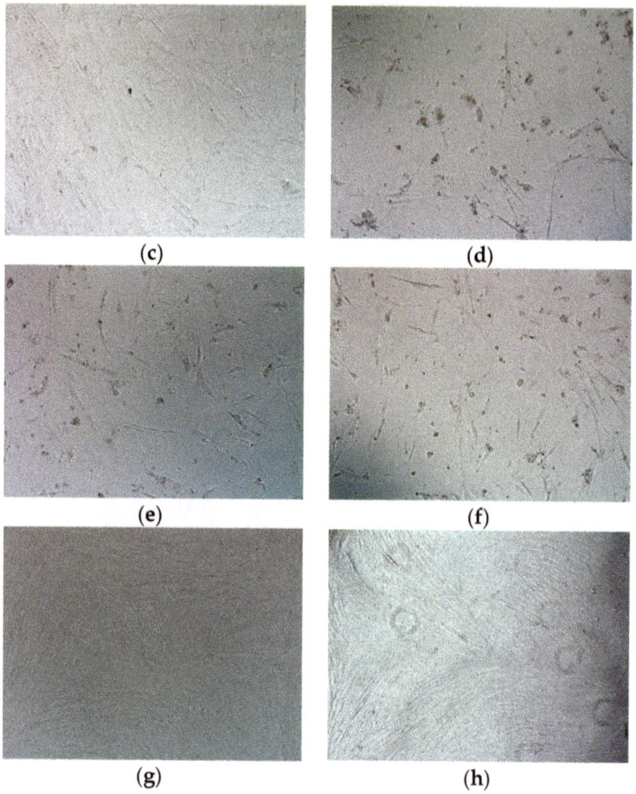

Figure 9. Representative microphotographs in bright field made 96 h after GCV treatment. The UF cells were transfected with R6p-cRGD/pCMV-lacZ polyplexes at N/Pratios of 8/1 (**a**,**b**) and 12/1 (c) with 0.7 µg (**a**) and 0.35 µg (**b**,**c**) of DNA per well; with R6p-cRGD/pPTK1 complexes at 8/1 (**d**,**e**) and 12/1 (**f**) charge ratios with 0.7 µg (**d**) and 0.35 µg (**e**,**f**) of DNA per well. Control wells contained GCV-treated intact cells (**g**) and untreated intact ones (**h**).

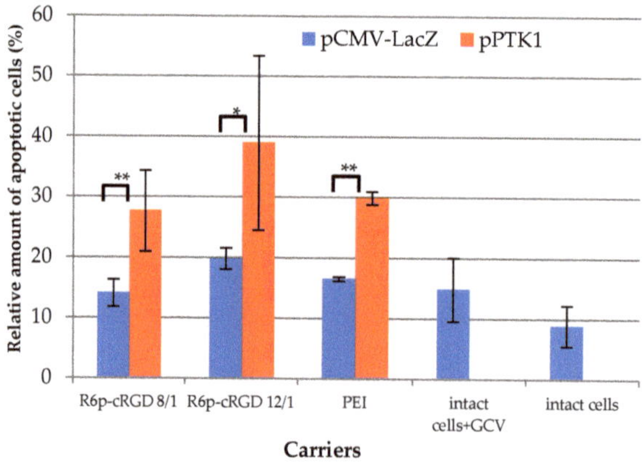

(a)

Figure 10. *Cont.*

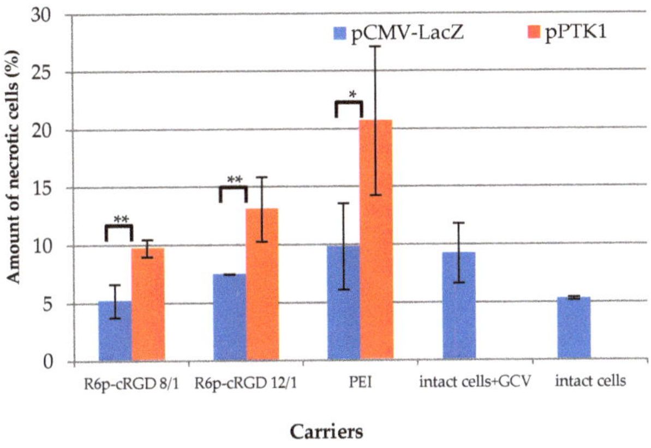

(**b**)

Figure 10. Relative amount of apoptotic (**a**) and necrotic (**b**) UF cells induced by GCV treatment after cell transfection with R6p-cRGD/DNA polyplexes formed with pPTK and pCMV-lacZ plasmids. Values are the mean ± SEM of the mean of four independent experiments. * $p < 0.05$, ** $p < 0.01$, compared to pCMV-lacZ-complexes.

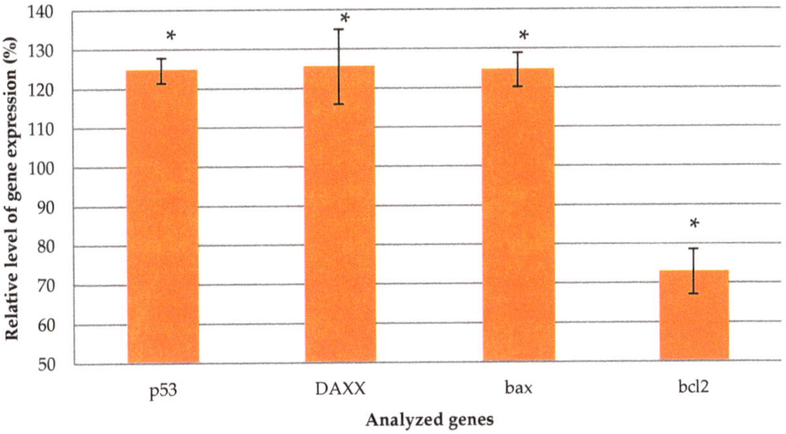

Figure 11. Relative gene expression levels of pro-apoptotic (p53, DAXX, bax) and anti-apoptotic factors in UF cells induced by GCV treatment after cell transfection with R6p-cRGD/DNA polyplexes formed with the pPTK plasmid. Values are the mean ± SEM of the mean of three independent experiments. * $p < 0.05$, compared to intact cells.

The Trypan Blue dye exclusion method showed a similar tendency in the assessment of the therapeutic effects (Figure 8b). No significant difference in the number of UF cells was observed upon cell transfection with pCMV-lacZ complexes and intact cells treated with GCV only. However, the number of viable UF cells transfected with R6p-cRGD/pPTK polyplexes decreased to 23–44% compared to R6p-cRGD/pCMV-lacZ complexes. Cell transfection with PEI/pPTK1 complexes resulted only in a 1.4-fold decrease in the population of viable UF cells.

Accordingly, in micrographs, we demonstrate a significant decrease in the number of UF cells after transfection with R6p-cRGD/pPTK polyplexes and GCV treatment compared to controls (Figure 9).

The success of suicide gene therapy is largely associated with the "bystander effect", when the phosphorylated GCV migrates to non-transfected cells through gap junctions or by endocytosis of apoptotic vesicles [60,61]. Activation of apoptosis, which was triggered by HSV-TK gene expression after UF cells transfection and subsequent treatment with GCV, was estimated with anApoDETECT annexin V-FITC kit (Figure 10a). Apoptosis is one of the best-studied forms of programmed cell death. Important signs of the apoptotic pathway are the exposition of phospholipid phosphatidylserine on the cell surface, which occurs at the initial or early stages of apoptosis, followed by membrane blebbing, nuclear fragmentation, a decrease in cell volume, and the formation of apoptotic bodies [62]. Annexin V is a protein that binds to phosphatidylserine, making it possible to detect the early apoptotic cells. The flow cytometry analysis after cell labeling with theApoDETECT annexin V-FITC kit showed that 28–39% of UF cells transfected by R6p-cRGD/pPTK polyplexes with 0.35 µg of DNA were annexin V-positive. This percentage was significantly different from that of R6p-cRGD/pCMV-lacZ complexes (14–20%), but detection of some annexin V-positive cells after the control polyplex transfection could be associated with adherent cell detachment via RGD-$\alpha v\beta 3$ integrin interaction [63]. Staurosporine treatment showed the presence of approximately 50% of annexin V-positive UF cells (data not shown). UF cells transfection by control PEI/pPTK complexes also resulted in the appearance of cells at an early stage of apoptosis (29%), although these polyplexes did not induce a largedecrease in UF cells proliferative activity (Figure 8).

After apoptosis, cells undergo programmed necrosis, which is accompanied by outer membrane disruption. Due to the membrane damage, propidium iodide can penetrate into the cells and be detected [64]. We found that 5% of intact cells were PI-positive (Figure 10b). UF cells transfection with R6p-cRGD/pCMV-lacZ polyplexes formed at 8/1 and 12/1 charge ratios resulted in 5.2% and 7.5% of PI-positive cells, respectively. After the cells' treatment with R6p-cRGD/pPTK complexes, the percentage of necrotic cells was significantly higher and reached 9.7–13%. However, the percentage of necrotic UF cells was demonstrated to be lower than that of the apoptotic cells. The highest amount of necrotic cells (21%) was registered for PEI/pPTK complexes, whichis caused not only by suicide effects but mainly by the cytotoxicity of PEIpolyplexes [65].

The key point of suicide gene therapy is the termination of DNA replication which leads to apoptosis. Thus, we decided to evaluate whether there is a difference in the expression level of the apoptotic factor transcripts (p53, Bax, DAXX, Bcl-2). Herein, we demonstrated the increased level of transcription of pro-apoptotic factors (DAXX, Bax, p53) and a decrease in anti-apoptotic Bcl-2 transcripts, which is further evidence of successful TK gene delivery by the carrier and suicide gene therapy (Figure 11). p53 is one of the key molecules regulating apoptosis and is involved in both the extrinsic (deathreceptor-activating) and intrinsic (mitochondrial) pathways [66]. The regulating function of p53 is presented by transcription activation of a variety of apoptotic factors, such as pro-apoptotic Bcl-2 family proteins (such as Bax) and transcription suppression of anti-apoptotic Bcl-2 family proteins (Bcl-2) [67]. Death domain-associated protein (DAXX) is involved in the extrinsic pathway and its expression was increased after UF treatment in Eker rats by Ad-DNER vector [12]. The same work presented the difference in the protein level of Bcl-2 and Bax between Ad-DNER-treated and control UFs [12].

Taken together, our results indicate that UF cell incubation with R6p-cRGD/pPTK complexes stimulated cell death in a suicide-specific manner and that the cells were registered mostly at the early apoptosis stage rather than the necrosis stage. Moreover, the polyplex treatment resulted in a significant decrease in the proliferation activity of UF cells and the number of viable cells. So, the developed complexes are capable of targeted therapy and represent a promising delivery system for the successful application of suicide gene therapy of uterine leiomyoma.

3. Materials and Methods

3.1. Cell Lines

Human pancreatic (PANC-1) and human kidney (293T) cell lines were obtained from Cell Collection of the Institute of Cytology RAS (Saint Petersburg, Russia). The cells were maintained according to the standard method, "Fundamental Techniques in Cell Culture", Sigma-Aldrich (Sailsbury, Wiltshire, UK). Primary UF cell lines were obtained previously after myomectomy in the D.O. Ott Research Institute of Obstetrics, Gynecology and Reproductology (Saint Petersburg, Russia) and maintained as described earlier [31,68].

3.2. Peptide Synthesis and Characterization

R6 (CHRRRRRRHC), cRGD(cyclic RGDyC(*Npys*)), and cyclo(RGDfK) peptides were synthesized with solid phase Boc-chemistry in NPF Verta, LLC (Saint Petersburg, Russia) and were stored as a dry powder at −20 °C (Table 1). The purity of the peptides was determined by high-performance liquid chromatography as 90–95%. The R6p carrier was obtained as described previously [30]. Briefly, the peptide was dissolved at a 30 mM concentration with 30% of DMSO, followed by oxidative polycondensation reaction for the next 96 h. In the case of R6p-cRGD, the cRGD moiety dissolved at 30 mM with 30% of DMSO added to R6p before polycondensation at the ratio of 1 cRGD molecule to 2 R6 molecules. The resultant R6p and R6p-cRGD were stored dissolved in water, 2 mg/mL at −70 °C. The relative number of free thiol groups was estimated by Ellman's assay and subsequently calculated as (P/Pf) × 100%, where Pf is the absorbance of unpolymerized peptide R6 [30]. The carrier molecular weight was analyzed on an AB Sciex 5800 TOF/TOF tandem time-of-flight mass spectrometer (AB Sciex, Foster City, CA, USA) in a linear mode; sinapic acid served as a matrix substance (Tables S1 and S2).

3.3. Reporter Plasmids

The pCMV-lacZ plasmid with the β-galactosidase gene was gifted from Prof. B. Sholte, Erasmus University Rotterdam, Netherlands. The pEXPR-IBA5-eGFP plasmid with green fluorescence protein (GFP) gene was obtained from IBA GmbH, Göttingen, Germany. The pPTK1 plasmid containing the HSV1 herpes virus thymidine kinase gene was provided by Dr. S.V. Orlov from the Institute of Experimental Medicine, Saint Petersburg, Russia. The plasmids were isolated according to the standard alkaline lysis technique [69].

3.4. Complex Preparation, DNA-Binding and Protection Assays, DS and DTT Treatment

DNA/peptide complexes were prepared at various N/P ratios (peptide nitrogen/DNA phosphorus ratio), as described previously [30,31]. The plasmid DNA was diluted to 20 g/mL in Hepes-buffered mannitol (HBM) (5% (*w*/*v*) mannitol, 5 mM Hepes, pH 7.5) and then anequal volume of peptides at the required charge ratio in HBM was added to the DNA solution, followed by vortexing. Complexes were incubated at room temperature for 30 min. Polyethyleneimine (branched PEI 25 kDa; Sigma-Aldrich, St. Louis, MO, USA) was used as 0.9 mg/mL (pH 7.5) aqueous stock solution, stored at +4 °C. The ratio of PEI to DNA was 8/1.

Peptide binding to DNA was studied using the ethidium bromide (EtBr) fluorescence quenching method at different nitrogen-to-phosphate ratios (0.1–5) [70]. Fluorescence measurements were performed in a Wallac 1420D scanning multilabel counter (PerkinElmer Wallac Oy, Turku, Finland) in emission fluorescence at 590 nm (544 nm excitation). EtBr displacement was calculated as (F − Ff)/(Fb − Ff), where Ff and Fb are the fluorescence intensities of EtBr in the absence and presence of DNA, respectively.

DNAse I protection assay was carried out for complexes formed at a 0.1–5 N/P ratio when adding to them 0.5 units of DNase I (Ambion, Austin, TX, USA) for 30 min at 37 °C with 2 min subsequently of DNAse I activation. Then, 0.1% trypsin was added for DNA overnight release at 37 °C and the DNA integrity was analyzed in 1% agarose gel [71].

Dextransulfate (DS)(Sigma–Aldrich, St. Louis, MO, USA) was added to the complexes at three-fold charge excess relative to the carrier for the next 24 h of incubation and analyzed by EtBrexclusion assay.

The dithiotreitol(DTT) at 200 mM was incubated with the complexes for 1 h at 37 °C and analyzed by SYBR-Green exclusion assay in a Wallac 1420D at an emission fluorescence of590 nm (585 nm excitation). SybrGreen displacement was calculated similarly to the EtBr assay [34].

3.5. *Measurement of Size and ζ-Potential of Peptide/DNA Complexes*

The peptide/DNA complexes were prepared as described above in quantities of 5 µg of DNA per sample. The size of the complexes was determined using dynamic light scattering, and the zeta potential was determined by microelectrophoresis. Three independent measurements were performed on a zetasizer NANO ZS (Malvern Instruments, Malvern, UK).

3.6. *Gene Transfer*

PANC-1 cells were seeded at a density of 5.0×10^4 cells per well in 48-well plates and incubated overnight. Before transfection, the cell culture medium was replaced with serum-free medium. DNA complexes, prepared as above (2 µg of DNA in each well), were added and incubated with cells for 4 h. Then, the transfection medium was replaced with FBS-containing medium and the cells were incubated for the next 48 h.

The β-galactosidase activity in cell extracts was measured with methyl-umbelliferyl-β-D-galactopyranoside (MUG) and normalized by the total protein concentration, measured with Bradford reagent (Helicon, Moscow, Russia), as described previously [30].

For the competition transfection study a 10-fold excess of cyclo(RGDfK) peptide was added to the cells 15 min before complex treatment, followed by the procedures described above.

GFP expression was determined by flow cytometry using a BD FACS-Canto II cytofluorimeter. Transfection efficacy was evaluated as a percentage of GFP-positive cells.

3.7. *Cellular Uptake of Peptide/DNA Complexes*

PANC-1 or 293T cells were seeded at a density of 5×10^4 cells/well in 48-well plates a day before the experiments. Peptide/DNA complexes were formed with added YOYO-1 iodide based on 1 molecule of the dye per 50 base pairs. Transfection was performed according to the protocol described above. After 2 h of incubation with complexes, the cells were washed three times in 1× PBS (pH 7.2) and once with 1M NaCl (in 1× PBS). The cells were detached and incubated with propidium iodide solution, as described previously [34]. Then, the living cells, at a rate of 10,000 per sample, were analyzed by flow cytometry with a BD FACS-Canto II cytofluorimeter. The results were presented as RFU/cell.

3.8. *Cytotoxicity Assay*

The cytotoxicity of DNA/peptide complexes was evaluated in PANC-1 cells using AlamarBlue assay (BioSources International, San Diego, CA, USA), as described previously [71]. Carrier/DNA complexes were prepared at the rate of 0.7 µg of DNA per well of a 96-well plate. Fluorescence measurements were performed in a Wallac 1420D scanning multilabel counter in emission fluorescence at 590 nm (544 nm excitation). The relative fluorescence intensity was counted as $(F - Ff)/(Fb - Ff) \times 100\%$, where Fb and Ff are the fluorescence intensities in untreated controls and without cells, respectively.

3.9. *Suicide Gene Therapy*

The primary UF cells were seeded on 96-well plates at 1.5×10^4 cells per well for overnight incubation. Transfections were performed in serum-free medium with 0.7 µg and 0.35 µg of DNA (pPTK1 or pCMV-lacZ plasmid) per well. After 2 h of cell incubation with complexes, the medium was replaced with FBS-containing medium and the cells were incubated for the next 24 h. Further, the medium was replaced withfresh mediumbut with ganciclovir at a concentration of 50 μg/mL and the cells were allowed to grow for 96 or 24 h [34].

After 96 h of incubation, the medium was replaced withthe same one but with 10% Alamar Blue solution added, and the cells were incubated for the next 2 h. Fluorescence measurements were performed in a Wallac 1420D scanning multilabel counter in emission fluorescence at 590 nm (544 nm excitation). The cell proliferation activity was estimated by the number of living cells, calculated as (F − Ff)/(Fb − Ff), where Fb and Ff are the fluorescence intensities in untreated controls and without cells, respectively. Micrographs of the cells were taken at 100× magnification with a AxioObserver Z1 microscope (Carl Zeiss, Oberkochen, Germany) equipped with the AxioVision program.

Trypan Blue dye was used to count the number of living cells after 96 h of incubation. UF cells were harvested with 0.25% Trypsin-EDTA (Thermo Fisher Scientific, Carlsbad, CA, USA), followed by addition of 0.4% Trypan Blue solution (Sigma-Aldrich, Munich, Germany) at a 1:1 volume ratio for 15 min. The Trypan Blue-negative cells were counted with a hemocytometer (MiniMedProm, Dyatkovo, Russia).

The relative number of apoptotic and necrotic cells was determined after 24 h of incubation with the ApoDETECT annexin V-FITC kit (Invitrogen, Darmstadt, Germany), according to the manufacturer's recommendations. The cells were detached, treated with the kit and analyzed with aBD FACS-Canto II cytofluorimeter.

For the gene expression analysis, total RNA extraction and quantitative real-time PCR analysis were performed as previously described [72]. The following primers were used: Bax forward primer 5′-TTC TGA CGG CAA CTT CAA CTG G-3′, reverse primer 5′-AGG AAG TCC AAT GTC CAG CC-3′ [73]; p53 forward primer 5′-TAA CAG TTC CTG CAT GGG CGG C-3′, reverse primer 5′-AGG ACA GGC ACA AAC ACG CAC C-3′ [74];DAXX forward primer 5′-CTG AAA TCC CCA CCA CTT CC-3′, reverse primer 5′-CTG AGCAG CTG CTT CAT CTT C-3′ [75]; Bcl-2 forward primer 5′-GAG GAT TGT GGC CTT CTT TG-3′, reverse primer 5′-GCC GGT TCA GGT ACT CAG TC-3′ [76]; and the endogenous reference gene GAPDH was detected using forward 5′-CGC CAG CCG AGC CAC ATC-3′, reverse primer 5′-CGC CCA ATA CGA CCA AAT CCG-3′. The samples were measured three times and a final result was inferred by averaging the data. The values are presented as mean ± SEM of the means obtained from two independent experiments.

3.10. Statistical Analysis

Statistically significant differences were obtained with the Mann–Whitney Utest and the Student's *t*-test using Instat 3.0 (GraphPad Software Inc., San Diego, CA, USA). $p < 0.05$ was considered statistically significant.

4. Conclusions

In the current study, we developed a promising non-viral gene delivery system based on cysteine-flanked arginine-rich peptides. The developed carrier, R6p-cRGD, was modified with a cyclic RGD ligand during a polycondensation reaction, as was proved by massspectrometry. Physicochemical experiments reveal that R6p-cRGD can form small-sized stable complexes with DNA thatprotect it from nuclease degradation. Cell transfection experiments confirmed the important role of the ligand modification for the specificity of DNA delivery to αvβ3 integrin-expressing cells. Complexes of the developed carrier and HSV-1 thymidine kinase-encoding plasmid were extensively studied in model experiments on suicide gene therapy of uterine leiomyoma in vitro. We showed that R6p-cRGD-mediated HSV-1 thymidine kinase gene expression in uterine leiomyoma cells reduced their proliferative activity and increased the number of apoptotic and necrotic cells. These findings were confirmed by the increased expression of pro-apoptotic factors and a decreased expression of anti-apoptotic factor Bcl-2 in uterine leiomyoma cells after R6p-cRGD-mediated transfection. Thus, we can conclude that the developed R6p-cRGD carrier can be used for further efforts in the development of uterine leiomyoma suicide gene therapy.

Supplementary Materials: The following are available online at https://www.mdpi.com/article/10.3390/ijms23031164/s1.

Author Contributions: Conceptualization, A.E., A.K. and V.B.; methodology, A.E. and A.S.; formal analysis, A.K. and A.E.; investigation, A.E., S.S. (Sofia Shtykalova), M.M. and D.D.; resources, S.S. (Sergey Selkov) and N.S.; writing—original draft preparation, A.E.; writing—review and editing, A.K.; supervision, V.B.; project administration, A.K.; funding acquisition, A.E. All authors have read and agreed to the published version of the manuscript.

Funding: This research was funded by the Russian Science Foundation, grant number 21-15-00111.

Institutional Review Board Statement: The study was conducted according to the guidelines of the Declaration of Helsinki and approved by the Ethics Committee of the D.O. Ott Research Institute of Obstetrics, Gynecology and Reproductology (protocol 89 was approved 22 December 2017).

Informed Consent Statement: Informed consent was obtained from all subjects involved in the study.

Data Availability Statement: The data presented in this study are available on request from the corresponding author. The data are not publicly available due to restrictions of the subjects' agreement.

Acknowledgments: The authors are grateful to A.G. Mittenberg (Institute of Cytology, Russian Academy of Sciences) for the assistance with massspectrometry analysis.

Conflicts of Interest: The authors declare no conflict of interest.

References

1. Wirth, T.; Parker, N.; Ylä-Herttuala, S. History of gene therapy. *Gene* **2013**, *525*, 162–169. [CrossRef] [PubMed]
2. Walker, W.J.; Barton-Smith, P. Long-term follow up of uterine artery embolisation-an effective alternative in the treatment of fibroids. *BJOG Int. J. Obstet. Gynaecol.* **2006**, *113*, 464–468. [CrossRef] [PubMed]
3. Stewart, E.A. Uterine fibroids. *Lancet* **2001**, *357*, 293–298. [CrossRef]
4. Dubuisson, J. The current place of mini-invasive surgery in uterine leiomyoma management. *J. Gynecol. Obstet. Hum. Reprod.* **2019**, *48*, 77–81. [CrossRef] [PubMed]
5. Donnez, J.; Dolmans, M.-M. Hormone therapy for intramural myoma-related infertility from ulipristal acetate to GnRH antagonist: A review. *Reprod. Biomed. Online* **2020**, *41*, 431–442. [CrossRef]
6. Cook, H.; Ezzati, M.; Segars, J.H.; McCarthy-Keith, D. The impact of uterine leiomyomas on reproductive outcomes. *Minerva Ginecol.* **2010**, *62*, 225–236.
7. Shtykalova, S.V.; Egorova, A.A.; Maretina, M.A.; Freund, S.A.; Baranov, V.S.; Kiselev, A.V. Molecular Genetic Basis and Prospects of Gene Therapy of Uterine Leiomyoma. *Russ. J. Genet.* **2021**, *57*, 1002–1016. [CrossRef]
8. Moolten, F.L.; Wells, J.M. Curability of Tumors Bearing Herpes Thymidine Kinase Genes Transfered by Retroviral Vectors. *JNCI J. Natl. Cancer Inst.* **1990**, *82*, 297–300. [CrossRef]
9. Andersen, J.; Grine, E.; Eng, C.L.Y.; Zhao, K.; Barbieri, R.L.; Chumas, J.C.; Brink, P.R. Expression of connexin-43 in human myometrium and leiomyoma. *Am. J. Obstet. Gynecol.* **1993**, *169*, 1266–1276. [CrossRef]
10. Salama, S.A.; Kamel, M.; Christman, G.; Wang, H.Q.; Fouad, H.M.; Al-Hendy, A. Gene therapy of uterine leiomyoma: Adenovirus-mediated herpes simplex virus thymidine kinase/ganciclovir treatment inhibits growth of human and rat leiomyoma cells in vitro and in a nude mouse model. *Gynecol. Obstet. Investig.* **2007**, *63*, 61–70. [CrossRef]
11. Bakhtiar, A.; Chowdhury, E.H. PH-responsive strontium nanoparticles for targeted gene therapy against mammary carcinoma cells. *Asian J. Pharm. Sci.* **2021**, *16*, 236–252. [CrossRef] [PubMed]
12. Hassan, M.H.; Khatoon, N.; Curiel, D.T.; Hamada, F.M.; Arafa, H.M.; Al-Hendy, A. Toward gene therapy of uterine fibroids: Targeting modified adenovirus to human leiomyoma cells. *Hum. Reprod.* **2008**, *23*, 514–524. [CrossRef] [PubMed]
13. Nair, S.; Curiel, D.T.; Rajaratnam, V.; Thota, C.; Al-Hendy, A. Targeting adenoviral vectors for enhanced gene therapy of uterine leiomyomas. *Hum. Reprod.* **2013**, *28*, 2398–2406. [CrossRef] [PubMed]
14. Abdelaziz, M.; Sherif, L.; Elkhiary, M.; Nair, S.; Shalaby, S.; Mohamed, S.; Eziba, N.; El-Lakany, M.; Curiel, D.; Ismail, N.; et al. Targeted Adenoviral Vector Demonstrates Enhanced Efficacy for in Vivo Gene Therapy of Uterine Leiomyoma. *Reprod. Sci.* **2016**, *23*, 464–474. [CrossRef]
15. Shalaby, S.M.; Khater, M.K.; Perucho, A.M.; Mohamed, S.A.; Helwa, I.; Laknaur, A.; Lebedyeva, I.; Liu, Y.; Diamond, M.P.; Al-Hendy, A.A. Magnetic nanoparticles as a new approach to improve the efficacy of gene therapy against differentiated human uterine fibroid cells and tumor-initiating stem cells. *Fertil. Steril.* **2016**, *105*, 1638–1648.e8. [CrossRef]
16. Lukashev, A.N.; Zamyatnin, A.A. Viral vectors for gene therapy: Current state and clinical perspectives. *Biochemistry* **2016**, *81*, 700–708. [CrossRef]
17. Niu, H. Nonviral Vector-Mediated Thymidine Kinase Gene Transfer and Ganciclovir Treatment in Leiomyoma Cells. *Obstet. Gynecol.* **1998**, *91*, 735–740. [CrossRef]

18. Kato, T.; Yamashita, H.; Misawa, T.; Nishida, K.; Kurihara, M.; Tanaka, M.; Demizu, Y.; Oba, M. Plasmid DNA delivery by arginine-rich cell-penetrating peptides containing unnatural amino acids. *Bioorg. Med. Chem.* **2016**, *24*, 2681–2687. [CrossRef] [PubMed]
19. Zhang, D.; Wang, J.; Xu, D. Cell-penetrating peptides as noninvasive transmembrane vectors for the development of novel multifunctional drug-delivery systems. *J. Control. Release* **2016**, *229*, 130–139. [CrossRef] [PubMed]
20. Thomas, E.; Dragojevic, S.; Price, A.; Raucher, D. Thermally Targeted p50 Peptide Inhibits Proliferation and Induces Apoptosis of Breast Cancer Cell Lines. *Macromol. Biosci.* **2020**, *20*, 2000170. [CrossRef]
21. Zhou, S.; Watanabe, K.; Koide, S.; Kitamatsu, M.; Ohtsuki, T. Minimization of apoptosis-inducing CPP-Bim peptide. *Bioorg. Med. Chem. Lett.* **2021**, *36*, 127811. [CrossRef] [PubMed]
22. Kosuge, M.; Takeuchi, T.; Nakase, I.; Jones, A.T.; Futaki, S. Cellular Internalization and Distribution of Arginine-Rich Peptides as a Function of Extracellular Peptide Concentration, Serum, and Plasma Membrane Associated Proteoglycans. *Bioconjug. Chem.* **2008**, *19*, 656–664. [CrossRef] [PubMed]
23. van Asbeck, A.H.; Beyerle, A.; McNeill, H.; Bovee-Geurts, P.H.M.; Lindberg, S.; Verdurmen, W.P.R.; Hällbrink, M.; Langel, Ü.; Heidenreich, O.; Brock, R. Molecular Parameters of siRNA–Cell Penetrating Peptide Nanocomplexes for Efficient Cellular Delivery. *ACS Nano* **2013**, *7*, 3797–3807. [CrossRef] [PubMed]
24. Wen, Y.; Guo, Z.; Du, Z.; Fang, R.; Wu, H.; Zeng, X.; Wang, C.; Feng, M.; Pan, S. Serum tolerance and endosomal escape capacity of histidine-modified pDNA-loaded complexes based on polyamidoamine dendrimer derivatives. *Biomaterials* **2012**, *33*, 8111–8121. [CrossRef]
25. Chou, S.-T.; Hom, K.; Zhang, D.; Leng, Q.; Tricoli, L.J.; Hustedt, J.M.; Lee, A.; Shapiro, M.J.; Seog, J.; Kahn, J.D.; et al. Enhanced silencing and stabilization of siRNA polyplexes by histidine-mediated hydrogen bonds. *Biomaterials* **2014**, *35*, 846–855. [CrossRef] [PubMed]
26. Tai, Z.; Wang, X.; Tian, J.; Gao, Y.; Zhang, L.; Yao, C.; Wu, X.; Zhang, W.; Zhu, Q.; Gao, S. Biodegradable Stearylated Peptide with Internal Disulfide Bonds for Efficient Delivery of siRNA In Vitro and In Vivo. *Biomacromolecules* **2015**, *16*, 1119–1130. [CrossRef]
27. McKenzie, D.L.; Kwok, K.Y.; Rice, K.G. A potent new class of reductively activated peptide gene delivery agents. *J. Biol. Chem.* **2000**, *275*, 9970–9977. [CrossRef] [PubMed]
28. Oupický, D.; Parker, A.L.; Seymour, L.W. Laterally stabilized complexes of DNA with linear reducible polycations: Strategy for triggered intracellular activation of DNA delivery vectors. *J. Am. Chem. Soc.* **2002**, *124*, 8–9. [CrossRef]
29. Soundara Manickam, D.; Bisht, H.S.; Wan, L.; Mao, G.; Oupicky, D. Influence of TAT-peptide polymerization on properties and transfection activity of TAT/DNA polyplexes. *J. Control. Release* **2005**, *102*, 293–306. [CrossRef] [PubMed]
30. Kiselev, A.; Egorova, A.; Laukkanen, A.; Baranov, V.; Urtti, A. Characterization of reducible peptide oligomers as carriers for gene delivery. *Int. J. Pharm.* **2013**, *441*, 736–747. [CrossRef]
31. Egorova, A.A.; Shtykalova, S.V.; Maretina, M.A.; Selyutin, A.V.; Shved, N.Y.; Krylova, N.V.; Ilina, A.V.; Pyankov, I.A.; Freund, S.A.; Selkov, S.A.; et al. Cys-Flanked Cationic Peptides For Cell Delivery of the Herpes Simplex Virus Thymidine Kinase Gene for Suicide Gene Therapy of Uterine Leiomyoma. *Mol. Biol.* **2020**, *54*, 436–448. [CrossRef]
32. Liu, J.; Cheng, X.; Tian, X.; Guan, D.; Ao, J.; Wu, Z.; Huang, W.; Le, Z. Design and synthesis of novel dual-cyclic RGD peptides for $\alpha v\beta 3$ integrin targeting. *Bioorg. Med. Chem. Lett.* **2019**, *29*, 896–900. [CrossRef]
33. Malik, M.; Segars, J.; Catherino, W.H. Integrin $\beta 1$ regulates leiomyoma cytoskeletal integrity and growth. *Matrix Biol.* **2012**, *31*, 389–397. [CrossRef] [PubMed]
34. Egorova, A.; Shtykalova, S.; Selutin, A.; Shved, N.; Maretina, M.; Selkov, S.; Baranov, V.; Kiselev, A. Development of irgd-modified peptide carriers for suicide gene therapy of uterine leiomyoma. *Pharmaceutics* **2021**, *13*, 202. [CrossRef] [PubMed]
35. Kanazawa, T.; Endo, T.; Arima, N.; Ibaraki, H.; Takashima, Y.; Seta, Y. Systemic delivery of small interfering RNA targeting nuclear factor κB in mice with collagen-induced arthritis using arginine-histidine-cysteine based oligopeptide-modified polymer nanomicelles. *Int. J. Pharm.* **2016**, *515*, 315–323. [CrossRef] [PubMed]
36. Ibaraki, H.; Kanazawa, T.; Takashima, Y.; Okada, H.; Seta, Y. Transdermal anti-nuclear kappaB siRNA therapy for atopic dermatitis using a combination of two kinds of functional oligopeptide. *Int. J. Pharm.* **2018**, *542*, 213–220. [CrossRef]
37. Egorova, A.; Selutin, A.; Maretina, M.; Selkov, S.; Baranov, V.; Kiselev, A. Characterization of iRGD-Ligand Modified Arginine-Histidine-Rich Peptides for Nucleic Acid Therapeutics Delivery to $\alpha v\beta 3$ Integrin-Expressing Cancer Cells. *Pharmaceuticals* **2020**, *13*, 300. [CrossRef] [PubMed]
38. Spears, R.J.; McMahon, C.; Chudasama, V. Cysteine protecting groups: Applications in peptide and protein science. *Chem. Soc. Rev.* **2021**, *50*, 11098–11155. [CrossRef]
39. Lechardeur, D.; Sohn, K.-J.; Haardt, M.; Joshi, P.B.; Monck, M.; Graham, R.W.; Beatty, B.; Squire, J.; O'Brodovich, H.; Lukacs, G.L. Metabolic instability of plasmid DNA in the cytosol: A potential barrier to gene transfer. *Gene Ther.* **1999**, *6*, 482–497. [CrossRef] [PubMed]
40. Xiang, S.; Tong, H.; Shi, Q.; Fernandes, J.C.; Jin, T.; Dai, K.; Zhang, X. Uptake mechanisms of non-viral gene delivery. *J. Control. Release* **2012**, *158*, 371–378. [CrossRef]
41. Khalil, I.A.; Kogure, K.; Akita, H.; Harashima, H. Uptake pathways and subsequent intracellular trafficking in nonviral gene delivery. *Pharmacol. Rev.* **2006**, *58*, 32–45. [CrossRef] [PubMed]
42. Grosse, S.; Aron, Y.; Thévenot, G.; François, D.; Monsigny, M.; Fajac, I. Potocytosis and cellular exit of complexes as cellular pathways for gene delivery by polycations. *J. Gene Med.* **2005**, *7*, 1275–1286. [CrossRef] [PubMed]

43. Litzinger, D.C.; Buiting, A.M.J.; van Rooijen, N.; Huang, L. Effect of liposome size on the circulation time and intraorgan distribution of amphipathic poly(ethylene glycol)-containing liposomes. *Biochim. Biophys. Acta Biomembr.* **1994**, *1190*, 99–107. [CrossRef]
44. Ganbold, T.; Han, S.; Hasi, A.; Baigude, H. Receptor-mediated delivery of therapeutic RNA by peptide functionalized curdlan nanoparticles. *Int. J. Biol. Macromol.* **2019**, *126*, 633–640. [CrossRef] [PubMed]
45. Egorova, A.; Kiselev, A. Peptide modules for overcoming barriers of nucleic acids transport to cells. *Curr. Top. Med. Chem.* **2016**, *16*, 330–342. [CrossRef] [PubMed]
46. Guo, X.D.; Tandiono, F.; Wiradharma, N.; Khor, D.; Tan, C.G.; Khan, M.; Qian, Y.; Yang, Y.-Y. Cationic micelles self-assembled from cholesterol-conjugated oligopeptides as an efficient gene delivery vector. *Biomaterials* **2008**, *29*, 4838–4846. [CrossRef]
47. Yue, Y.; Wu, C. Progress and perspectives in developing polymeric vectors for in vitro gene delivery. *Biomater. Sci.* **2013**, *1*, 152–170. [CrossRef] [PubMed]
48. Kim, S.; Kim, J.-H.; Jeon, O.; Kwon, I.C.; Park, K. Engineered polymers for advanced drug delivery. *Eur. J. Pharm. Biopharm.* **2009**, *71*, 420–430. [CrossRef]
49. Belting, M.; Petersson, P. Protective role for proteoglycans against cationic lipid cytotoxicity allowing optimal transfection efficiency in vitro. *Biochem. J.* **1999**, *342*, 281. [CrossRef]
50. Ruponen, M.; Ylä-Herttuala, S.; Urtti, A. Interactions of polymeric and liposomal gene delivery systems with extracellular glycosaminoglycans: Physicochemical and transfection studies. *Biochim. Biophys. Acta Biomembr.* **1999**, *1415*, 331–341. [CrossRef]
51. Ruponen, M.; Honkakoski, P.; Tammi, M.; Urtti, A. Cell-surface glycosaminoglycans inhibit cation-mediated gene transfer. *J. Gene Med.* **2004**, *6*, 405–414. [CrossRef]
52. Schaffer, D.V.; Fidelman, N.A.; Dan, N.; Lauffenburger, D.A. Vector unpacking as a potential barrier for receptor-mediated polyplex gene delivery. *Biotechnol. Bioeng.* **2000**, *67*, 598–606. [CrossRef]
53. Saar, K.; Lindgren, M.; Hansen, M.; Eiríksdóttir, E.; Jiang, Y.; Rosenthal-Aizman, K.; Sassian, M.; Langel, Ü. Cell-penetrating peptides: A comparative membrane toxicity study. *Anal. Biochem.* **2005**, *345*, 55–65. [CrossRef] [PubMed]
54. El-Andaloussi, S.; Järver, P.; Johansson, H.J.; Langel, Ü. Cargo-dependent cytotoxicity and delivery efficacy of cell-penetrating peptides: A comparative study. *Biochem. J.* **2007**, *407*, 285–292. [CrossRef] [PubMed]
55. Wiradharma, N.; Khan, M.; Tong, Y.W.; Wang, S.; Yang, Y.-Y. Self-assembled Cationic Peptide Nanoparticles Capable of Inducing Efficient Gene Expression In Vitro. *Adv. Funct. Mater.* **2008**, *18*, 943–951. [CrossRef]
56. Deng, K.; Tian, H.; Zhang, P.; Zhong, H.; Ren, X.; Wang, H. PH-temperature responsive poly(HPA-Co-AMHS) hydrogel as a potential drug-release carrier. *J. Appl. Polym. Sci.* **2009**, *114*, 176–184. [CrossRef]
57. Kargaard, A.; Sluijter, J.P.G.; Klumperman, B. Polymeric siRNA gene delivery—Transfection efficiency versus cytotoxicity. *J. Control. Release* **2019**, *316*, 263–291. [CrossRef]
58. Schiffelers, R.M. Cancer siRNA therapy by tumor selective delivery with ligand-targeted sterically stabilized nanoparticle. *Nucleic Acids Res.* **2004**, *32*, e149. [CrossRef]
59. Al-Hendy, A.; Salama, S. Gene therapy and uterine leiomyoma: A review. *Hum. Reprod. Update* **2006**, *12*, 385–400. [CrossRef]
60. Freeman, S.M.; Abboud, C.N.; Whartenby, K.A.; Packman, C.H.; Koeplin, D.S.; Moolten, F.L.; Abraham, G.N. The "Bystander Effect": Tumor Regression When a Fraction of the Tumor Mass Is Genetically Modified. *Cancer Res.* **1993**, *53*, 5274–5283.
61. Andrade-Rozental, A.F.; Rozental, R.; Hopperstad, M.G.; Wu, J.K.; Vrionis, F.D.; Spray, D.C. Gap junctions: The "kiss of death" and the "kiss of life". *Brain Res. Rev.* **2000**, *32*, 308–315. [CrossRef]
62. Krammer, P.H.; Kamiński, M.; Kießling, M.; Gülow, K. No Life Without Death. In *Advances in Cancer Research*; Academic Press Inc. Elsevier Science: San Diego, CA, USA, 2007; Volume 97, pp. 111–138.
63. Cheng, K.; Kothapalli, S.-R.; Liu, H.; Koh, A.L.; Jokerst, J.V.; Jiang, H.; Yang, M.; Li, J.; Levi, J.; Wu, J.C.; et al. Construction and Validation of Nano Gold Tripods for Molecular Imaging of Living Subjects. *J. Am. Chem. Soc.* **2014**, *136*, 3560–3571. [CrossRef] [PubMed]
64. Crowley, L.C.; Scott, A.P.; Marfell, B.J.; Boughaba, J.A.; Chojnowski, G.; Waterhouse, N.J. Measuring Cell Death by Propidium Iodide Uptake and Flow Cytometry. *Cold Spring Harb. Protoc.* **2016**, *2016*, pdb.prot087163. [CrossRef] [PubMed]
65. Fischer, D. *In Vivo Fate of Polymeric Gene Carriers*; CRC Press: Boca Raton, FL, USA, 2004; ISBN 9780203492321.
66. Haupt, S.; Berger, M.; Goldberg, Z.; Haupt, Y. Apoptosis—The p53 network. *J. Cell Sci.* **2003**, *116*, 4077–4085. [CrossRef] [PubMed]
67. Jan, R.; Chaudhry, G.-E.-S. Understanding apoptosis and apoptotic pathways targeted cancer therapeutics. *Adv. Pharm. Bull.* **2019**, *9*, 205–218. [CrossRef] [PubMed]
68. Shved, N.; Egorova, A.; Osinovskaya, N.; Kiselev, A. Development of primary monolayer cell model m odel and organotypic model of uterine leiomyoma. *Methods Protoc.* **2021**, *41*, 1–12.
69. Sambrook, J.; Fritsch, E.F.; Maniatis, T. *Molecular Cloning: A Laboratory Manual*, 2nd ed.; Cold Spring Harbor Laboratory Press: Cold Spring Harbor, NY, USA, 1989. [CrossRef]
70. Kiselev, A.V.; Il'ina, P.L.; Egorova, A.A.; Baranov, A.N.; Guryanov, I.A.; Bayanova, N.V.; Tarasenko, I.I.; Lesina, E.A.; Vlasov, G.P.; Baranov, V.S. Lysine dendrimers as vectors for delivery of genetic constructs to eukaryotic cells. *Russ. J. Genet.* **2007**, *43*, 593–600. [CrossRef]
71. Egorova, A.; Bogacheva, M.; Shubina, A.; Baranov, V.; Kiselev, A. Development of a receptor-targeted gene delivery system using CXCR4 ligand-conjugated cross-linking peptides. *J. Gene Med.* **2014**, *16*, 336–351. [CrossRef]

72. Egorova, A.; Petrosyan, M.; Maretina, M.; Balashova, N.; Polyanskih, L.; Baranov, V.; Kiselev, A. Anti-angiogenic treatment of endometriosis via anti-VEGFA siRNA delivery by means of peptide-based carrier in a rat subcutaneous model. *Gene Ther.* **2018**, *25*, 548–555. [CrossRef]
73. Saed, G.M.; Jiang, Z.; Fletcher, N.M.; Diamond, M.P. Modulation of the BCL-2/BAX ratio by interferon-γ and hypoxia in human peritoneal and adhesion fibroblasts. *Fertil. Steril.* **2008**, *90*, 1925–1930. [CrossRef] [PubMed]
74. Weglarz, L.; Molin, I.; Orchel, A.; Parfiniewicz, B.; Dzierzewicz, Z. Quantitative analysis of the level of p53 and p21(WAF1) mRNA in human colon cancer HT-29 cells treated with inositol hexaphosphate. *Acta Biochim. Pol.* **2006**, *53*, 349–356. [CrossRef] [PubMed]
75. Huang, L.; Xu, G.; Zhang, J.; Tian, L.; Xue, J.; Chen, J.; Jia, W. Daxx interacts with HIV-1 integrase and inhibits lentiviral gene expression. *Biochem. Biophys. Res. Commun.* **2008**, *373*, 241–245. [CrossRef] [PubMed]
76. Li, J.; Ma, Y.; Mu, L.; Chen, X.; Zheng, W. The expression of Bcl-2 in adenomyosis and its effect on proliferation, migration, and apoptosis of endometrial stromal cells. *Pathol. Res. Pract.* **2019**, *215*, 152477. [CrossRef] [PubMed]

MDPI
St. Alban-Anlage 66
4052 Basel
Switzerland
www.mdpi.com

International Journal of Molecular Sciences Editorial Office
E-mail: ijms@mdpi.com
www.mdpi.com/journal/ijms

MDPI

www.ingramcontent.com/pod-product-compliance
Lightning Source LLC
LaVergne TN
LVHW070710100526
838202LV00013B/1065
* 9 7 8 3 0 3 6 5 8 7 1 1 0 *